Seismicity Associated with Mines, Reservoirs and Fluid Injections

Edited by
Shahriar Talebi

1998

Springer Basel AG

Reprint from Pageoph
(PAGEOPH), Volume 150 (1997), No. 3/4

The Editor:

Prof. Dr.-Ing.
Shahriar Talebi
CANMET
1079 Kelly Lake Rd.
Sudbury, Ontario
Canada P3E 5P5

A CIP catalogue record for this book is available from the Library of Congress,
Washington D.C., USA

Deutsche Bibliothek Cataloging-in-Publication Data

Seismicity associated with mines, reservoirs and fluid injections / ed. by Shahriar Talebi. –
Basel ; Boston ; Berlin : Birkhäuser, 1998
  (Pageoph topical volumes)
  Aus: Pure and applied geophysics ; Vol. 150. 1997
  ISBN 978-3-7643-5878-5   ISBN 978-3-0348-8814-1 (eBook)
  DOI 10.1007/978-3-0348-8814-1

© 1998 Springer Basel AG
Originally published by Birkhäuser Verlag in 1998
Printed on acid-free paper produced from chlorine-free pulp

ISBN 978-3-7643-5878-5

9 8 7 6 5 4 3 2 1

Contents

Pure appl. geophys. 150 (1997) 379–380
0033–4553/97/040379–02 $ 1.50 + 0.20/0

Introduction

For purposes of motivating preparation for the 4th International Symposium on Rockbursts and Seismicity in Mines, held in Krakow, Poland in August 1997, a workshop was held on June 18, 1996, as part of the 2nd North American Rock Mechanics Symposium (NARMS'96), in Montreal, Québec, Canada. Twenty-eight reports on different cases of triggered and induced seismicity were presented at this workshop which was attended by over 50 professionals. A similar one-day workshop organized by Art McGarr in 1992 (materials published in Pure and Applied Geophysics, vol. 139, nos. 3/4, 1992) was quite effective in "setting the stage" for the 3rd International Symposium. The Montreal Workshop proved to be similarly stimulating and provocative with some lively interdisciplinary discussion between scientists and engineers studying different types of seismic events due to completely different causes. Some of the flavor of these debates can be gleaned from the transcript of the panel discussion, published here, that took place in the late afternoon of the workshop.

This special issue of Pure and Applied Geophysics contains 18 papers dealing with different aspects of induced and triggered seismicity and is intended to convey the essence of the material covered during the workshop. These papers divide naturally into four categories. The first group of six papers entails seismicity induced by mining and the next three groups of four papers focus on earthquakes triggered by reservoir impoundment, seismicity stimulated by fluid injections and seismic techniques for assessing the state of the rock mass. The balance of subject matter in these papers reflects my intent to broaden the interest to include induced and triggered earthquakes in different situations.

Many people and organizations contributed to the workshop and this topical issue. I would like to take this opportunity to thank the authors for their valuable contributions. I also thank Luc Vandamme and the organizing committee of NARMS'96 who provided an efficient infrastructure for holding the workshop, and the management of Canada Centre for Mineral and Energy Technology (CANMET) for providing support needed for the preparation of this volume. I am indebted to Renata Dmowska, editor-in-chief for topical issues at Pure and Applied Geophysics, for inviting me to serve as a guest editor and for her invaluable technical and editorial advice during the preparation of this volume.

I am grateful to many scientists for their time and efforts devoted to reviewing the manuscripts, particularly Art McGarr, Pradeep Talwani and Cezar Trifu who provided superb assistance during the review process. Other conscientious reviewers who contributed to the maintenance of high scientific standards are Behrouz Bazargan-Sabet, Tom Boone, Bruno Feignier, Jon B. Fletcher, Diane I. Doser, Don J. Gendzwill, Slawomir J. Gibowicz, Ewa Głowacka, Jean-Robert Grasso, Harsh K. Gupta, Ferri P. Hassani, Karel Holub, Tsuyoshi Ishida, Andrzej Kijko, A. A. Kozyrev, Charles A. Langston, Christophe Maisons, Mo Momayez, Hiroaki Niitsuma, Yang Qingyuan, B. K. Rastogi, Evelyn A. Roeloffs, Jim Rutledge, Kaveh Saleh, Christopher H. Scholz, Steve Spottiswoode, Ted I. Urbancic and Yizhang Zhong.

Shahriar Talebi
CANMET
1079 Kelly Lake Rd.
Sudbury, Ontario
Canada P3E 5P5

Pure appl. geophys. 150 (1997) 381–391
0033–4553/97/040391–11 $ 1.50 + 0.20/0

Pure and Applied Geophysics

A Mechanism for High Wall-rock Velocities in Rockbursts

A. McGarr[1]

Abstract—Considerable evidence has been reported for wall-rock velocities during rockbursts in deep gold mines that are substantially greater than ground velocities associated with the primary seismic events. Whereas varied evidence suggests that slip across a fault at the source of an event generates nearby particle velocities of, at most, several m/s, numerous observations, in nearby damaged tunnels, for instance, imply wall-rock velocities of the order of 10 m/s and greater. The common observation of slab buckling or breakouts in the sidewalls of damaged excavations suggests that slab flexure may be the mechanism for causing high rock ejection velocities. Following its formation, a sidewall slab buckles, causing the flexure to increase until the stress generated by flexure reaches the limit S that can be supported by the sidewall rock. I assume here that S is the uniaxial compressive strength. Once the flexural stress exceeds S, presumably due to the additional load imposed by a nearby seismic event, the slab fractures and unflexes violently. The peak wall-rock velocity $\mathbf{v}$ thereby generated is given by

$$\mathbf{v} = \left(3 + \frac{1 - v^2}{2} \right)^{1/2} \frac{S}{\sqrt{\rho E}}$$

for rock of density ρ, Young's modulus E, and Poisson's ratio v. Typical values of these rock properties for the deep gold mines of South Africa yield $\mathbf{v} = 26$ m/s and for especially strong quartzites encountered in these same mines, $\mathbf{v} > 50$ m/s. Even though this slab buckling process leads to remarkably high ejection velocities and violent damage in excavations, the energy released during this failure is only a tiny fraction of that released in the primary seismic event, typically of magnitude 2 or greater.

Key words: Wall-rock velocities, rockbursts, slab buckling.

Introduction

As reviewed by GIBOWICZ (1990), it is largely, although not universally, thought that mining-induced seismic events entail slip across faults. McGARR (1991, 1993) argued that particle velocities adjacent to these faults are unlikely to exceed 4 m/s, at least in the deep gold mines of South Africa.

In apparent contradiction to this, ORTLEPP (1993) presented evidence indicating wall-rock velocities associated with mine tremors, of the order of 10 m/s and greater (see also ORTLEPP and STACEY, 1994). Ortlepp went on to suggest that the relatively-high wall-rock velocities may be due to a rock-failure phenomenon quite distinct from that of the primary seismic event. This suggestion by Ortlepp is pursued in this report.

[1] U.S. Geological Survey, MS 977, Menlo Park, CA 94025, U.S.A.

Among the six mechanisms for mining-induced seismic events outlined by HASEGAWA *et al.* (1989; see also GIBOWICZ, 1990), a mechanism corresponding to the commonly-observed phenomenon of slab buckling of excavation sidewalls is not to be found. Surficial slab buckling has been proposed as a mechanism for rock bursting, however, by HOLZHAUSEN (1978), NEMAT-NESSER and HORII (1982), BARDET (1990) and MÜHLHAUS (1990). Field and laboratory observations, as well as continuum mechanics analyses, provide abundant confirmation that slab buckling is an expected outcome in highly-stressed rock adjacent to a free surface (e.g., TALEBI and YOUNG, 1992).

Accordingly, for purposes of the analysis to follow, excavation sidewalls are assumed to react to high levels of compressive stress, parallel to the free surface, by forming slabs and buckling as reported by the authors just mentioned. I will show here that highly-stressed sidewalls are capable of generating material velocities comparable to those suggested by ORTLEPP (1993), that is, 10 m/s or greater. As I will also demonstrate, though, this phenomenon, although commonly observed, is, at least in most cases, a secondary failure mechanism in deep mines in that its associated energy change is several orders of magnitude smaller than that of the attendant mining-induced earthquake whose effects must be withstood by the stope support (WAGNER, 1984).

Maximum Wall-rock Velocities

Energy Considerations

Before considering a specific mechanism to produce relatively high wall-rock velocities, it is worthwhile exploring the possibilities in terms of the available energy for this process. Specifically, here we ask how much energy is available at the free surface of an excavation to be converted into the kinetic energy of an ejected fragment of sidewall rock?

Of the principal stresses acting on the wall rock, the maximum σ_1 and intermediate σ_2 act parallel to the free surface and the minimum principal stress σ_3, normal to the surface, is zero. Presume that, due to the applied stresses, the sidewall somehow fails and ejects a rock fragment of mass m. Upper and lower bounds on the speed of the ejected rock can be estimated from the end-member situations of infinitely soft and infinitely stiff loading conditions.

Considering first the infinitely soft loading, the corresponding upper bound on the ejection speed can be calculated from

$$\sigma_1 \Delta V = \frac{1}{2} m v^2, \tag{1}$$

where ΔV is the volume of the fragment and v is its ejection speed. The term on the left represents the energy released due to the complete closure of the void as a consequence of the soft loading. For rock of density ρ, then, (1) can be solved for v to obtain, with $m = \rho \Delta V$,

$$v = \sqrt{\frac{2\sigma_1}{\rho}}. \tag{2}$$

For infinitely stiff loading the only energy available to propel the rock fragment is its elastic strain energy, which is released when it exits the sidewall; the void left behind in the sidewall retains the volume of the ejected fragment. Thus, equating the elastic energy of the fragment while still embedded in the sidewall to its kinetic energy after ejection gives

$$\frac{\Delta V \sigma_1^2}{2E} \le \frac{1}{2} m v^2, \tag{3}$$

where E is Young's modulus. Note that the contribution to the elastic strain energy from the lower principal sidewall stress has been neglected here, which is why (3) is an inequality. Solving for v yields

$$v \ge \sigma_1 / \sqrt{\rho E}. \tag{4}$$

Thus, from the energetics of the two end-member loading situations, infinitely soft or stiff, one concludes that fragments ejected from stressed wall rock will have initial speeds v in the range

$$\sigma_1 / \sqrt{\rho E} \le v \le \sqrt{\frac{2\sigma_1}{\rho}}. \tag{5}$$

To evaluate (5), I note that for sidewall failure to occur, σ_1 is likely to be close to the uniaxial compressive strength S of the rock. As an example, consider the quartzites of the deep gold mines in South Africa for which values of S, E and ρ are typically 200 MPa, 80 GPa, and 2.8×10^3 kg/m^3, respectively (McGARR et al., 1975; GAY et al., 1984). Then, replacing σ_1 with S in (5) gives

$$13 \text{ m/s} \le v \le 378 \text{ m/s}.$$

Accordingly, we see from these simple energy arguments that fragments ejected during the process of wall-rock failure are likely to have speeds exceeding 10 m/s. The next step is to consider a mechanical process for imparting such impressive fragment speeds.

Slab Buckling

Having shown that in sidewall rock, stressed to the point of failure, there is plenty of energy available to generate high wall-rock speeds, I now consider the specific mechanism of slab buckling to accomplish this. This mechanism seems a likely candidate because observations of slab buckling are ubiquitous in excavations at depth (Fig. 1).

As shown by BRACE and BOMBALAKIS (1963), FAIRHURST and COOK (1966), and NEMAT-NASSER and HORII (1982), among others, within rock samples subjected to uniaxial compression, as the axial stress approaches the uniaxial compressive strength S, axial cracks form, extend and finally link up. This results in failure by splitting. Based on these experimental results, I analyze slab buckling in excavation sidewalls, assuming that the maximum stress in the excavation sidewall σ is sufficiently close in magnitude to the uniaxial compressive strength to generate numerous cracks parallel to the free surface. These cracks then can link up to form slabs (Fig. 1). This process is illustrated quite nicely by DYSKIN and GERMANOVICH (1993, Fig. 1).

The model analyzed here entails an excavation sidewall of dimension L subject to a maximum sidewall stress σ, which is high enough to generate sidewall failure; that is, σ is assumed to be close to the uniaxial compressive strength S. As a consequence, a slab of thickness h forms and buckles (Fig. 2). The solution to this plate-buckling problem is (TURCOTTE and SCHUBERT, 1982, pp. 118–119)

$$w = c_1 \sin \left(\frac{\sigma h}{D} \right)^{1/2} x, \tag{6}$$

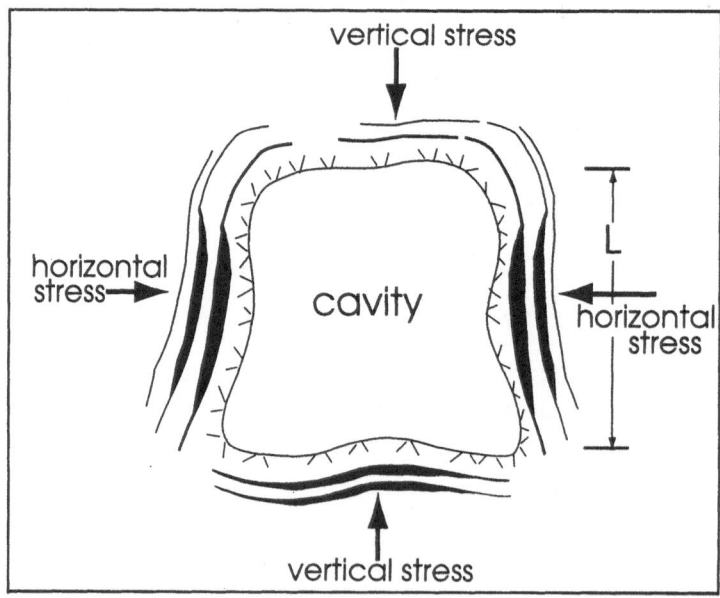

Figure 1
Slab formation around underground excavations (from MÜHLHAUS, 1993).

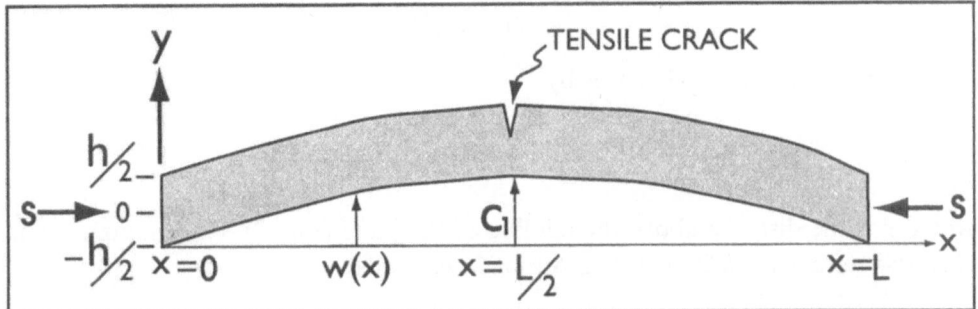

Figure 2
Configuration of slab thickness h and dimension L loaded by stress S.

where w is the plate deflection as a function of x, the distance from one edge of the plate, and D is the flexural rigidity, given by

$$D = \frac{Eh^3}{12(1 - v^2)},$$

where E is Young's modulus and v is Poisson's ratio. Because the deflection of the plate is zero at its ends $(x = 0, L)$,

$$\left(\frac{\sigma h}{D}\right)^{1/2} L = n\pi, \quad n = 1, 2, 3, \ldots. \tag{7}$$

Clearly, $n = 1$ is the only realistic solution here (Figs. 1 and 2), and consequently

$$\frac{h}{L} = \frac{2\sqrt{3}\sqrt{1 - v^2}}{\pi} \sqrt{\frac{\sigma}{E}}. \tag{8}$$

Thus, h can be calculated if we know the sidewall dimension, or plate length (Fig. 2), L and we set σ at the uniaxial compressive strength S of the sidewall rock.

From (7), with $n = 1$, (6) can be written

$$w = c_1 \sin \frac{\pi x}{L} \tag{9}$$

and it remains to specify c_1. To do this, I assume that the strength of the slab in tension is negligible compared to its strength in uniaxial compression, an assumption well supported by laboratory evidence (e.g., JAEGER and COOK, 1976). Thus, I assume that when the amplitude of flexure (c_1 in (9)) causes an extensional strain that exceeds the compressive strain associated with the applied stress σ, presumed near the uniaxial compressive strength, then a tensile crack will propagate through the plate, breaking it in two, at which time all of the stored elastic energy is released.

The applied stress σ is related to the corresponding compressive strain e_c according to (TURCOTTE and SCHUBERT, 1982, p. 114)

$$\sigma = \frac{E}{(1 - v^2)} e_c \qquad (10)$$

and the flexural strain e_f is given by

$$e_f = -y \frac{d^2 w}{dx^2}, \qquad (11)$$

where y is the distance above the midline of the plate (Fig. 2). Combining (9) and (11) and setting $y = h/2$, at the upper surface, yields

$$e_f(x) = (h/2)c_1 \left(\frac{\pi}{L}\right)^2 \sin \frac{\pi x}{L} \qquad (12)$$

which has a maximum value at $x = L/2$. Thus,

$$e_f(\text{max}) = c_1(h/2) \left(\frac{\pi}{L}\right)^2. \qquad (13)$$

Setting $\sigma = S$, the uniaxial compressive strength in (10) and then equating e_c to e_f (max) yields for c_1

$$c_1 = (2/h) \left(\frac{L}{\pi}\right)^2 \frac{(1 - v^2)S}{E}. \qquad (14)$$

That is, (9) combined with (14) gives the configuration w of the buckled slab at the point of failure when a tensile crack propagates through the flexed plate. At this time the elastic energy stored in the flexed plate is suddenly converted into the kinetic energy of the fractured plate. I estimate peak slab, or wall-rock, velocities by equating the elastic to kinetic energy; I show later that propagating the tensile crack through the flexed slab consumes a negligible amount of the available energy.

There are two contributions to the elastic energy. First, the energy associated with the applied stress $\sigma = S$ acting on the slab in the absence of buckling is

$$W_c = \frac{1}{2} \frac{S^2}{E} hL. \qquad (15)$$

The second contribution arises because the flexure of the plate, given by (9) and (14), results in shortening Δu along x direction (Fig. 2)

$$\Delta u = \frac{(1 - v^2)^2}{\pi^2} \left(\frac{S}{E}\right)^2 \frac{L^3}{h^2}. \qquad (16)$$

The elastic energy absorbed into the plate due to flexure is

$$W_f = Sh\Delta u = \frac{(1 - v^2)^2}{\pi^2} \frac{S^3}{E^2} \frac{L^3}{h} = \frac{1 - v^2}{12} \frac{S^2}{E} hL. \qquad (17)$$

The total stored elastic energy W_e is obtained by adding (15) and (17) to obtain

$$W_e = W_c + W_f = \frac{S^2}{E} hL \left[\frac{1}{2} + \frac{1-v^2}{12} \right]. \tag{18}$$

As soon as the tensile crack propagates through the flexed plate, at $x = L/2$, the two halves of the plate are free to unflex, thereby converting the stored elastic energy W_e into kinetic energy W_{KE}. For each half of the plate, the kinetic energy is maximal when it is completely unflexed, extended, and rotating about the fixed endpoint ($x = 0$ or L in Fig. 2). The velocity of each half must vary linearly from zero at the fixed point, say $x = 0$, to a maximum value at the fractured end, $x = L/2$. Integrating the kinetic energy density over the two half-plates gives

$$W_{KE} = (1/6)\rho hL v^2, \tag{19}$$

where v represents the peak velocity at the free end of either fractured plate.

Equating (18) to (19) and solving for v yields

$$v = \frac{S}{\sqrt{\rho E}} \left[3 + \frac{1-v^2}{2} \right]^{1/2}. \tag{20}$$

From GAY et al. (1984, Table 1), median (or typical) rock properties for the Witwatersrand rocks are as follows:

$$S = 206 \text{ MPa}$$

$$E = 77 \text{ GPa}$$

$$v = 0.19 \tag{21}$$

$$S/E = 0.0027$$

$$h/L = 0.057, \quad \text{from (8)}.$$

Table 1

Energy changes

Event	ΔW (event)[1], J	M	$2r_0$, m	ΔW (slab failure), J	$\dfrac{\Delta W(\text{slab failure})}{\Delta W(\text{event})}$
3102038	2.91×10^8	1.7	192	1.40×10^7	0.05
3121332	7.86×10^8	1.9	90	6.56×10^6	0.008
3153552	3.20×10^9	2.2	224	1.63×10^7	0.005
3151554	4.67×10^9	2.5	206	1.50×10^7	0.003
3241523	2.71×10^9	2.2	108	7.87×10^6	0.003
3241624	5.57×10^9	2.5	168	1.22×10^7	0.002
3251605	4.07×10^9	2.3	178	1.30×10^7	0.003
0301411	1.20×10^{11}	3.3	446	3.25×10^7	0.0003
0301411a	2.49×10^{10}	3.0	336	2.45×10^7	0.001
0341528	5.16×10^9	2.4	206	1.50×10^7	0.003
0271046	8.19×10^9	2.5	206	1.50×10^7	0.002

[1] Total energy changes for these events were calculated from ΔW (event) $= \sigma_v \Delta V$, where σ_v is the stress due to the weight of overburden and ΔV is the volumetric reduction determined from the measured moment tensor (McGARR, 1993).

Inserting these values into (20) yields a peak velocity of

$$\mathbf{v} = 26 \text{ m/s.} \tag{22}$$

Interestingly, $\mathbf{v}$ depends only on a few rock properties and nothing else. In fact, the uniaxial compressive strength is by far the most essential factor. In any event, (22) is compatible, at least to some extent, with the observations of high wall-rock velocities reviewed by ORTLEPP (1993) and is approximately an order of magnitude in excess of particle-velocity estimates associated with slip across faults as reviewed by McGARR (1991, 1993). Note, moreover, that (22) falls within the velocity range given by (5) from simple energy considerations.

As mentioned earlier, I neglected the energy consumed in propagating the tensile crack through the slab because it is tiny compared to the stored elastic energy W_e. From experimental results presented by BRACE and WALSH (1962), the surface energy of quartzite, for example, is unlikely to exceed $1 \ J/m^2$. For a tunnel dimension $L = 2$ m and with the rock properties of (21) it turns out that the energy required for the tensile crack is $0.23 \ J/m$ and that stored elastically (from (18)) is $7.3 \times 10^4 \ J/m$, more than five orders of magnitude greater.

Energetics of Slab Buckling and Mining-induced Earthquakes

Guided by observations (e.g., ORTLEPP, 1984, 1993), I assume here that the sidewall slab of a tunnel buckles and fails simultaneously with a nearby tremor due to the substantial increment in loading imposed within the source region of the tremor. Additionally, I assume that the extent of sidewall buckling is comparable to the source dimension of the associated tremor. How do the energy changes of these two phenomena compare?

To make this comparison I assume a tunnel cross-section dimension L (Fig. 1) of 2 m and the rock properties given by (21) and also that the length of sidewall buckling along the tunnel is the same as the source dimension of the associated tremor. With these assumptions, (18) gives the energy change per unit length of tunnel as $W_e = 7.29 \times 10^4 \ J/m$ and the total energy change, ΔW (slab failure), is given by multiplying W_e by the source dimension, $2r_0$, of the associated tremor.

A suite of mine tremors studied by McGARR (1993) is well suited for comparing the energetics of slab buckling and the associated tremors because the total energy changes associated with 11 of these tremors were calculated directly. With magnitudes ranging from 1.7 to 3.3 (Table 1), at least the larger tremors of this suite seem capable of being associated with slab-buckling sidewall failure in nearby tunnels (Fig. 1), in view of evidence reviewed by KAISER (1993), among others.

Based on information presented in Table 1 of McGARR (1993), Table 1 here lists the total energy release ΔW (event) and the source dimension, $2r_0$ of an assumed circular fault plane of radius r_0 (BRUNE, 1970, 1971; SPOTTISWOODE and McGARR, 1975). Also listed is the energy release due to slab buckling along the assumed tunnel of dimension 2 m within the tremor source zone, of dimension $2r_0$.

As seen in Table 1, the ratio of slab-failure energy release to the energy release of the associated tremor is quite small ranging from 0.05 down to 0.003, with a clear tendency for this ratio to decrease with increasing event energy or magnitude. These ratios would be quite small even if the hypothetical tunnel sidewall failure involved multiple slabs (Fig. 1). From this energy comparison, then, it is clear that tunnel sidewall failure is a secondary effect of the causative tremor and is of little consequence to the overall energy budget.

Conclusions

The buckling of slabs in the sidewalls of tunnels is a common manifestation of rockburst damage. The analysis of such slab flexure and failure implies peak wall-rock velocities v that depend on just a few rock properties, and in particular, v is proportional to the uniaxial compressive strength. For typical rocks found in the deep gold mines of South Africa the slab-buckling model predicts that v is about 26 m/s, as estimated from (20). For especially strong siliceous quartzites found in these mines (e.g., McGARR et al., 1975, Table 2) v could be somewhat in excess of 50 m/s.

Such high wall-rock velocities, associated with the unflexing of buckled slabs, would yield evidence of violent damage in tunnels and other excavations within the source zone of a sizable mine tremor (e.g., magnitude 2 or greater) (ORT-LEPP, 1984). The results of the slab-buckling analysis are compatible with both the observations of high wall-rock velocities reviewed by ORTLEPP (1993) as well as his suggestion that these phenomena are distinct from the causative mine tremor, for which near-fault ground velocities are unlikely to exceed several m/s.

Finally, to the extent that tunnel sidewall failure, and attendant high wall-rock velocities, are associated with nearby tremors, it is clear that in terms of released energy, the slab failure in tunnels is only a tiny fraction of that released in the primary tremor, because a much smaller volume of rock is involved in the slab failure process. It is worth noting, incidently, that the question of why sidewall failure tends to coincide with nearby tremors has not been addressed here.

8192

Acknowledgments

I thank S. Talebi for his effort in motivating this analysis by organizing the 1996 Montreal Workshop on Induced Seismicity. I am grateful to C. Sullivan for editorial assistance and to H.-P. Liu, J. B. Fletcher, W. D. Ortlepp, and S. M. Spottiswoode for careful reviews of this paper.

REFERENCES

BARDET, J. P., *Numerical modeling of a rockburst as surface buckling*. In *Rockbursts and Seismicity in Mines* (ed. Fairhurst, C.) (Balkema, Rotterdam 1990) pp. 81–85.

BRACE, W. F., and WALSH, J. B. (1962), *Some Direct Measurements of the Surface Energy of Quartz and Orthoclase*, The American Mineralogist 47, 1111–1122.

BRACE, W. F., and BOMBOLAKIS, E. G. (1963), *A Note on Brittle Crack Growth in Compression*, J. Geophys. Res. 68, 3709–3713.

BRUNE, J. N. (1970), *Tectonic Stress and the Spectra of Seismic Shear Waves from Earthquakes*, J. Geophys. Res. 75, 4997–5009. (Correction (1971), J. Geophys. Res. 76, 5002.

DYSKIN, A. V., and GERMANOVICH, L. N., *Model of rockburst caused by cracks growing near free surface*. In *Rockbursts and Seismicity in Mines* (ed. Young, R. P.) (Balkema, Rotterdam 1993) pp. 169–174.

FAIRHURST, C., and COOK, N. G. W. (1966), *The Phenomenon of Rock Splitting Parallel to the Direction of Maximum Compression in the Neighbourhood of a Surface*, Proc. First. Congr. Int. Soc. Rock Mech. 1, 687–692.

GAY, N. C., SPENCER, D., VAN WYK, J. J., and VAN DER HEEVER, P. K., *The control of geological and mining parameters in the Klerksdorp gold mining district*. In *Rockbursts and Seismicity in Mines* (ed. Gay, N. C. and Wainwright, E. H.) (SAIMM, Johannesburg 1984) pp. 107–120.

GIBOWICZ, S. J., *Keynote lecture: The mechanism of seismic events induced by mining*. In *Rockbursts and Seismicity in Mines* (ed. Fairhurst, C.) (Balkema, Rotterdam 1990) pp. 3–27.

HASEGAWA, H. S., WETMILLER, R. J., and GENDZWILL, D. J. (1989), *Induced Seismicity in Mines in Canada—An Overview*, Pure Appl. Geophys. 129, 423–453.

HOLZHAUSEN, G. R. (1978), *Sheet Structure in Rock and Some Related Problems in Rock Mechanics*, Ph.D. Thesis, Stanford Univ., Stanford, Calif.

JAEGER, J. C., and COOK, N. G. W., *Fundamentals of Rock Mechanics* (Halsted, New York 1976).

KAISER, P. K., *Keynote address: Support of tunnels in burst-prone ground—Toward a rational design methodology*. In *Rockbursts and Seismicity in Mines* (ed. Young, R. P.) (Balkema, Rotterdam 1993) pp. 13–27.

MCGARR, A. (1991), *Observations Constrainng Near-source Ground Motion Estimated from Locally Recorded Seismograms*, J. Geophys. Res. 96, 16,495–16,508.

MCGARR, A., *Factors influencing the strong ground motion from mining-induced tremors*. In *Rockbursts and Seismicity in Mines* (ed. Young, R. P.) (Balkema, Rotterdam 1993) pp. 3–12.

MCGARR, A., SPOTTISWOODE, S. M., and GAY, N. C. (1975), *Relationship of Mine Tremors to Induced Stresses and to Rock Properties in the Focal Region*, Bull. Seismol. Soc. Am. 65, 981–993.

MÜHLHAUS, H.-B., *Exfoliation phenomena in pre-stressed rock*. In *Rockbursts and Seismicity in Mines* (ed. Fairhurst, C.) (Balkema, Rotterdam 1990) pp. 101–107.

NEMAT-NASSER, S., and HORII, H. (1982), *Compression-induced Nonplanar Crack Extension with Application to Splitting, Exfoliation, and Rockburst*, J. Geophys. Res. 87, 6805–6821.

ORTLEPP, W. D., *Rockbursts in South African gold mines: a phenomenological view*. In *Rockbursts and Seismicity in Mines* (ed. Gay, N. C and Wainwright, E. H.) (SAIMM, Johannesburg 1984) pp. 165–178.

ORTLEPP, W. D., *High ground displacement velocities associated with rockburst damage.* In *Rockbursts and Seismicity in Mines* (ed. Young, R. P.) (Balkema, Rotterdam 1993) pp. 101–106.

ORTLEPP, W. D., and STACEY, T. R., *The need for yielding support in rockburst conditions, and realistic testing of rockbolts.* In *International Workshop on Applied Rockburst Research* (ed. Cereceda, J. C. and Van Sint Jan, M.) (Santiago, Chile, May 12, 1994) pp. 249–259.

SPOTTISWOODE, S. M., and McGARR, A. (1975), *Source Parameters of Tremors in a Deep-level Gold Mine*, Bull. Seismol. Soc. Am. *65*, 93–112.

TALEBI, S., and YOUNG, R. P. (1992), *Microseismic Monitoring in Highly Stressed Granite: Relation Between Shaft-wall Cracking and in situ Stress*, Int. J. Rock Mech. Min. Sci. and Geomech. Abstr. *29*, 25–34.

TURCOTTE, D. L., and SCHUBERT, G., *Geodynamics* (John Wiley and Sons, New York 1982).

WAGNER, H., *Support requirements for rockburst conditions.* In *Rockbursts and Seismicity in Mines* (ed. Gay, N. C. and Wainwright, E. H.) (SAIMM, Johannesburg 1984) pp. 209–218.

(Received August 29, 1996, accepted January 17, 1997)

Pure appl. geophys. 150 (1997) 393–414
0033–4553/97/040393–22 $ 1.50 + 0.20/0

An Anatomy of a Seismic Sequence in a Deep Gold Mine

S. J. GIBOWICZ[1]

Abstract—An unusual swarm-like seismic sequence occurred in April 1993 at the Western Deep Levels gold mine, South Africa. Altogether 199 events with moment magnitude from −0.5 to 3.1 were recorded and located by the mine seismic network. The sequence lasted 12 days and was composed in fact of four main shock-aftershocks sequences, closely following each other in space and time. The events were confined to a volume of rock extending to 670 m in the N-S, 630 m in the E-W, and 390 m in the vertical directions. The first sequence lasted 179 hours and the second only 13 hours, being interrupted by the third sequence which lasted 31 hours, being in turn interrupted by the fourth sequence. The parameter p, describing the rate of occurrence of aftershocks, ranged from 0.7 to 1. The first sequence is characterized by the lowest value of the fractal correlation dimension $D = 1.75$ and the second by the highest value of $D = 2.4$, whereas the third and fourth sequences are characterized by the middle value of $D = 1.9$.

The corner frequencies of P and S waves are in close proximity and range from 14 to 220 Hz. A display of source parameters as a function of time shows that the four main shocks are most distinctly marked by their source radius. For 46 events a moment tensor inversion was performed. In most cases the double-couple component is dominant, ranging from 60 to 90 percent of the solution. The double-couple solutions correspond to the same number of normal and reverse faults and oblique-slip focal mechanisms. An analysis of space distribution of P, T and B axes reveals that the distribution of B axes is the most regular.

Key words: Induced seismicity, seismic sequence, fractal correlation dimension, source parameters, seismic moment tensor, focal mechanism.

Introduction

Little is known about seismic sequences in mines as only a few studies on this subject are available from South Africa and Poland. MCGARR and GREEN (1978) have studied foreshock-aftershocks sequences of two mine tremors of magnitudes 1.5 and 1.2, which occurred in May 1973 at the East Rand Propriety Mines, South Africa. In the first hour after each event some 140 microaftershocks were recorded with a rate of occurrence decreasing with time in a regular manner. Although the two events occurred within about 50 m of one another and within 12 days, the seismicity before the tremor of magnitude 1.5 was at its normal ambient level, whereas the seismicity before the event of magnitude 1.2 was unusually high. They

[1] Institute of Geophysics, Polish Academy of Sciences, ul. Ks. Janusza 64, 01-452 Warsaw, Poland.
E-mail: gibowicz@igf.edu.pl

have also shown that a main shock-aftershocks sequence and a sequence similar to a swarm can occur in the same volume of rock under similar strain conditions.

In contrast to South African gold mines, seismic sequences are seldom observed at Polish hard-rock mines. The 1977 tremor of magnitude 4.5, which occurred at the Lubin copper mine, was an exception. It was followed by a regular aftershock sequence (GIBOWICZ et al., 1979). The frequency-time distribution of aftershocks was typical with a rate of occurrence decaying hyperbolically in time. But the 1987 event of magnitude 4.3, which occurred at the same mine, was followed by a few small aftershocks, differing considerably from the level of seismicity observed during 3 months preceding the main shock (GIBOWICZ et al., 1989). The daily release of seismic moment from aftershocks during the first month after the main event was found to be twice smaller than that from the tremors preceding this event. In general, aftershocks to seismic events in mines are not as ubiquitous as those to natural earthquakes.

Western Deep Levels gold mine, the deepest mine in the world selected for our study, is situated in the Carletonville gold mining district, some 80 km south of Johannesburg. The mine area contains two major conglomerate formations, the Ventersdorp Contact Reef at an average depth of 2 km and the Carbon Leader Reef at a depth of 3 km. The Western Deep Levels lease area extends 10.8 km from east to west and 4 km from north to south. Since its inception, Western Deep Levels has developed into three mines, West, East and South, which operate as separate entities.

On April 7, 1993, an unusual swarm-like sequence of seismic events occurred in the Upper Carbon Leader Back Area at the East mine at a depth of 3 km. These events were associated with several pillars intersected by the Lesser and Greater Green Dykes and the Speckled Dyke. The pillars left along these dykes were no part of the regional stabilizing program in use at Western Deep Levels, but were left because of the large amount of waste mining that would have been required had they been removed (A. G. Butler, pers. comm., 1993). Very few seismic events have been recorded in the area prior to a moment magnitude 2.7 event which occurred on April 7 as the first event in the sequence. The sequence continued until April 19 and 199 seismic events have been recorded with moment magnitude down to −0.5.

Spectral analysis of seismic waves and moment tensor inversion techniques are used to study the source parameters and source mechanism of seismic events forming this swarm-like sequence. The aim of this work is to investigate space and time distribution of these events and possible time variations of their source parameters and source mechanism. It should be noted that only a few works have been published to date that are related to the use of moment tensor inversion in studies of seismicity induced by mining (SATO and FUJII, 1989; FUJII and SATO, 1990; FEIGNIER and YOUNG, 1992; McGARR, 1992a,b; WIEJACZ, 1992, 1995; GIBOWICZ and WIEJACZ, 1994).

Seismic Network and Data

The Integrated Seismic System (ISS) has been in operation at Western Deep Levels from the beginning of 1990. In 1993 the underground seismic network was composed of 22 three-component stations; several of them in close vicinity to the area where the sequence occurred. Their horizontal distribution in relation to the location of seismic events forming the sequence is shown in Figure 1, and their vertical distribution along the N-S direction is presented in Figure 2.

The ISS system has been described in some detail by MENDECKI (1993). It is comprised of transducers, remote stations, a communication system, and a central computer. With conversion to a digital format as close as possible to the sensors, maximum dynamic range is ensured. The seismometers allow for a dynamic range greater than 120 dB with a resolution of 12 bits. The standard sampling frequency is 2000 Hz. The system performs on-line the quality controlled seismological processing of three-component waveforms, developed to operate in an automatic mode.

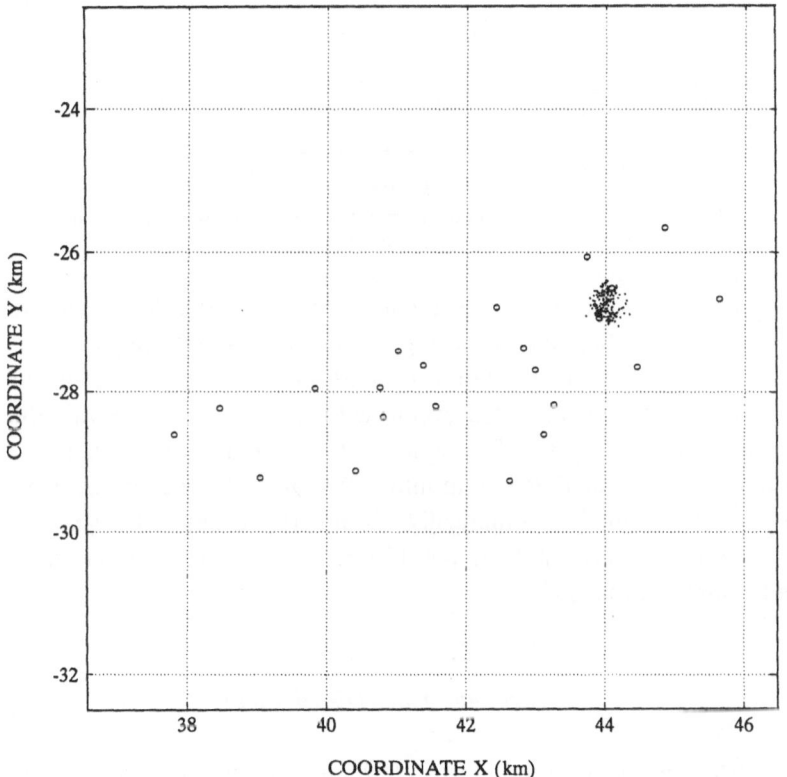

Figure 1
Horizontal distribution of seismic stations (open circles) and seismic events forming the sequences (dots) at Western Deep Levels.

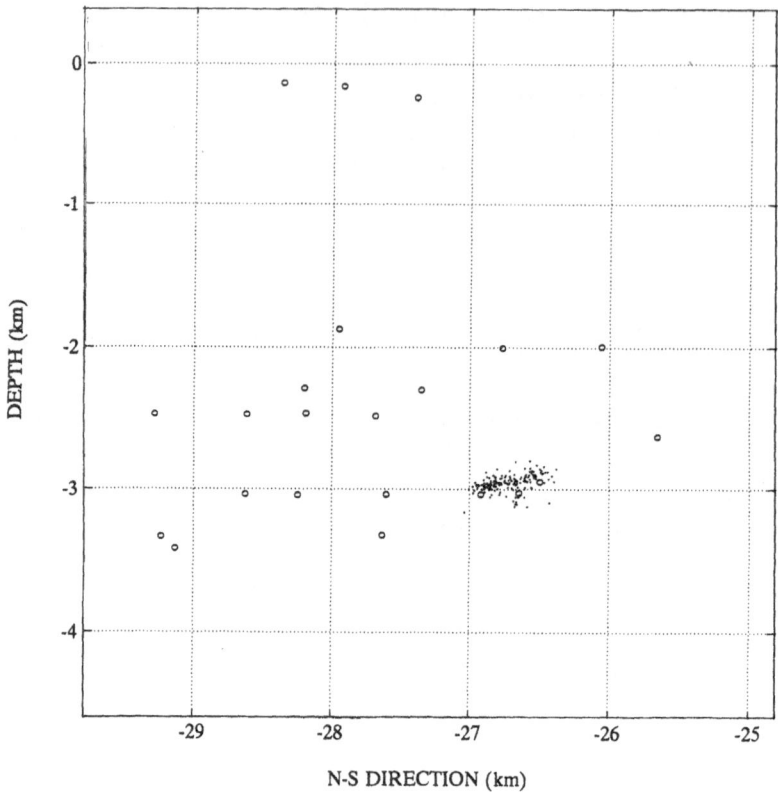

Figure 2
Vertical distribution along the N-S direction of seismic stations (open circles) and seismic events forming
the sequences (dots).

The sequence of April 1993 was recorded, located and processed by the ISS system. The smallest aftershock resulting from 199 recorded events was of moment magnitude -0.5, but the magnitude threshold, above which the set is complete, is -0.3 and the number of recorded events with magnitude above this threshold is 196. The source parameters of seismic events forming the sequence and their moment tensors were calculated in an interactive mode by the author, whereas their location was determined automatically. From 199 events, the general moment tensor inversion was successful for the 35 largest events and for 11 smaller events restrained double-couple solutions were found.

Space and Time Distributions

The sequence has started on April 7, 1993 at 09 h 09 m local time and ended on April 19 at 05 h 43 m. Although the sequence as a whole appears to form a swarm-like series, in the sense that no single event is dominant, it was in fact

Table 1

General characteristics of the four sequences

Sequence	Main shock magnitude	Number of aftershocks	Duration (hours)	Rate of occurrence	Fractal dimension
1	2.7	38	179	0.95	1.75
2	2.7	63	13	0.74	2.41
3	3.1	67	31	0.80	1.92
4	(2.6)	27	61	0.70	1.89
Total		195	284		

composed of four main shock-aftershocks sequences, closely following each other in space and time. The first main shock of moment magnitude 2.7 was followed by 38 recorded aftershocks, the second event of the same magnitude 2.7 occurred on April 14 at 20 h 17 m and was followed by 63 aftershocks, and the third main shock of magnitude 3.1 occurred on April 15 at 09 h 30 m and was followed by 67 aftershocks. The fourth sequence was composed of 25 aftershocks and three main shocks of magnitude 1.5, 2.5 and 2.0, which occurred close to each other on April 16 at 18 h 24 m, 18 h 25 m and 18 h 26 m, respectively. These events would correspond to a single main shock of magnitude 2.6 (Table 1).

The accuracy of hypocenter determination ranged between less than 10 m and 40 m; in most cases it was between 10 and 20 m. The events were confined to a volume of rock extending to 630 m in the E-W, 670 m in the N-S, and 390 m in the vertical directions. The horizontal distribution of seismic events forming the four sequences is shown in Figure 3, and the vertical cross section along the N-S direction is presented in Figure 4. The first sequence occurred at the southern edge of the rock volume containing the four sequences, the second and third sequences occurred in the middle of the volume, and the main shocks of the fourth sequence occurred at the northern edge of the volume, while their aftershocks moved back to the south. The vertical cross section delineates that most events occurred along a plane gently dipping from north to south.

A cumulative number of seismic events against the time after the occurrence of the first main shock is shown for the four sequences in Figure 5. Altogether, the sequences lasted 284 hours. The first sequence lasted 179 hours while the second sequence lasted only 13 hours and had no time for full development as it was followed immediately by the third sequence, which lasted 31 hours and was in turn interrupted by the fourth sequence persisting 61 hours (Table 1). From Figure 5 it appears that the last three sequences form one sequence only. Figure 6 illustrates that this is not the case. In this figure the rate of occurrence of seismic events after each main shock against the consecutive numbers of the time intervals used to calculate the rate of occurrence is presented. The numbers of time intervals instead of real time are used for the sake of clarity. The rate of occurrence was calculated

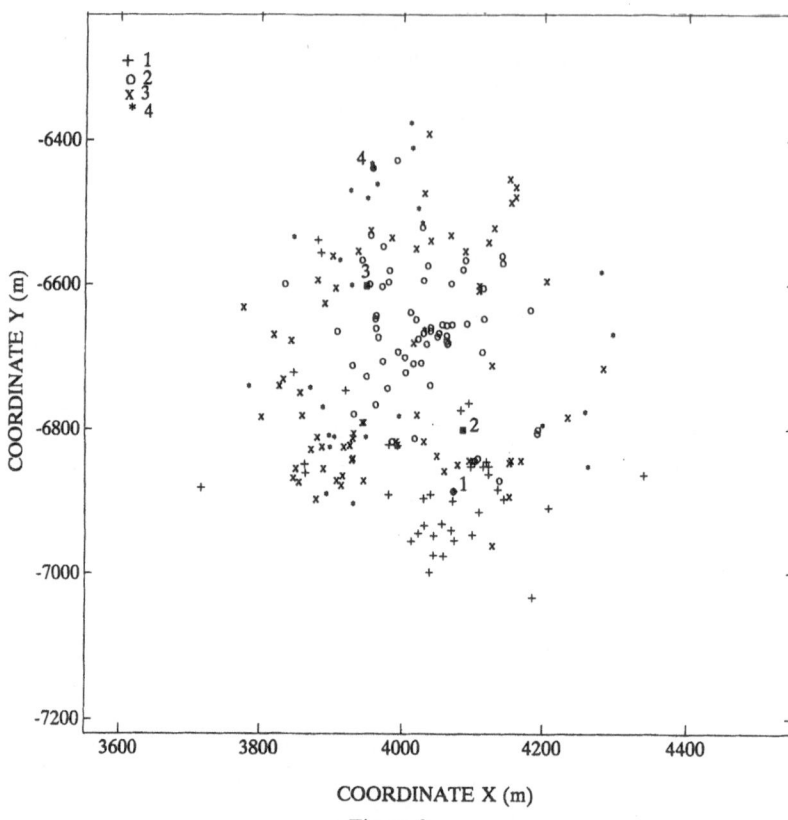

Figure 3
Horizontal distribution of seismic events forming four sequences, marked by different symbols. The main shocks of each sequence are marked by encircled points and are numbered from 1 to 4.

for a fixed number of seismic events and variable time windows. From Figure 6 it follows that the second sequence was suddenly interrupted by the third sequence, which possibly was preceded by a few foreshocks.

The parameter p from the Omori formula $n(t) = c/t^p$, where c is constant, describing the rate of occurrence $n(t)$ of aftershocks with time t from the occurrence of the main shock, was determined by the maximum likelihood method (OGATA, 1983). It should be noted that the original Omori formula fitted our data better than the modified Omori formula (UTSU, 1961), universally used to describe the rate of occurrence of tectonic aftershocks. This difference could be explained either by the different behavior of rock mass in underground mines from that in natural environment or by the fact that our data set is more complete than it is the case in most studies of natural earthquakes. The values of parameter p range from 1 for the first sequence, which had time for full development, to 0.7 for the shortest second sequence (Table 1).

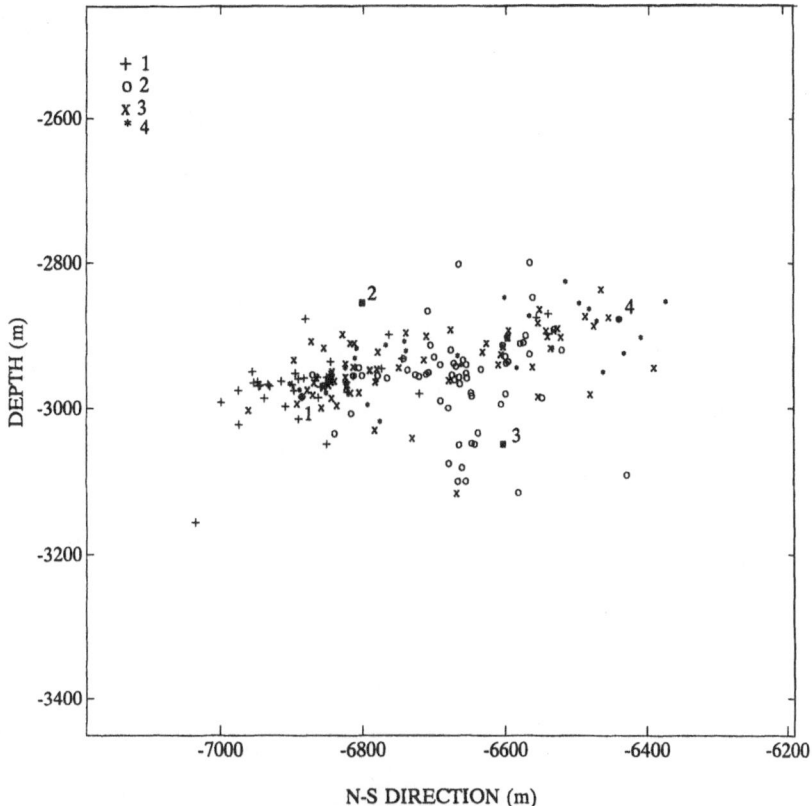

Figure 4
Vertical distribution along the N-S direction of seismic events forming four sequences, marked by different symbols. The main shocks of each sequence are marked by encircled points and are numbered from 1 to 4.

Fractal Dimension

Moment magnitude of all seismic events was calculated from seismic moment. Our data set is complete down to magnitude -0.3 and the number of complete observations is 196. The coefficient b in the Gutenberg-Richter relation was estimated by the maximum likelihood method (AKI, 1965; UTSU, 1965). The calculated value of b parameter, describing the ratio of small to large events within a given set, is 0.63 which is rather low in comparison with b values found for tectonic aftershocks (e.g., GUO and OGATA, 1995). The b values for the four sequences are also low, ranging from 0.52 to 0.64, though their accuracy is limited by the number of recorded events.

Following the fractal phenomena described by MANDELBROT (1982), the application of the concept in seismology has been reported by many authors and a number of models have been proposed to simulate the seismic process characterized

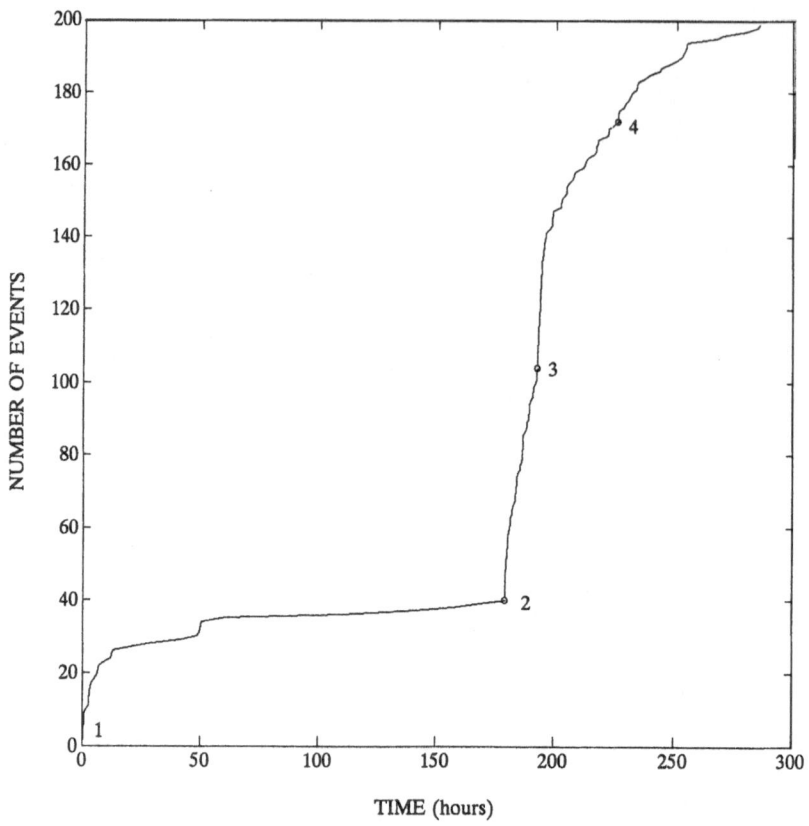

Figure 5
Cumulative number of seismic events versus time after the occurrence of the first main shock. The main
shocks are marked by open circles and numbered.

by the Gutenberg-Richter relation and the Omori formula, showing that they correspond to the self-organized critical behavior (e.g., BÅK and TANG, 1989; ITO and MATSUZAKI, 1990). Recently, GUO and OGATA (1995) have shown that there are positive correlations between p and b values and the fractal dimension of hypocenter distribution of aftershocks.

There are various definitions of the fractal dimension (e.g., HIRABAYASHI et al., 1992). The correlation fractal dimension was determined for our aftershock sequences, similarly as it was done by GUO and OGATA (1995). The correlation integral $C(r) = 2N(r)/N(N-1)$, where $N(r)$ is the number of pairs of hypocenters whose distances are smaller than r and N is the total number of aftershocks, is considered to determine the fractal dimension of a given set of hypocenters. If the hypocenter distribution has a fractal structure, then the correlation integral $C(r)$ should be proportional to r^D, where D is the fractal dimension.

The distances between the pairs of hypocenters were calculated for each sequence. The first sequence is the most extensive in space. The largest distance

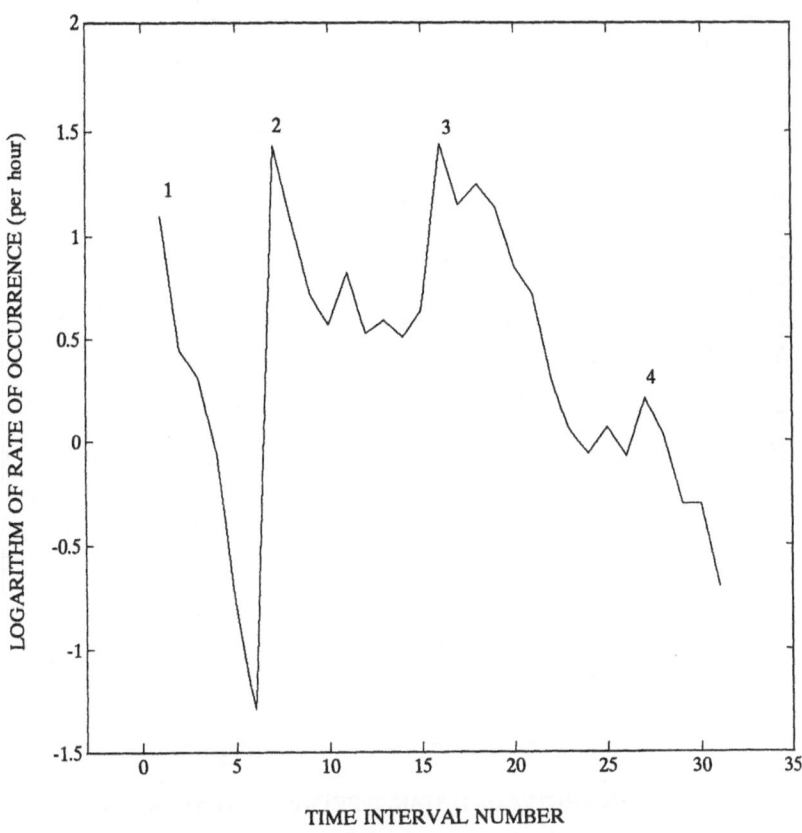

Figure 6

Logarithm of the rate of occurrence versus the consecutive number of time intervals used to calculate the
rate of occurrence for the four sequences numbered from 1 to 4.

between two events is 650 m and the smallest distance is 13 m, while the second
sequence is the most compact with the largest distance of 480 m and the smallest of
12 m. The fractal dimension D was calculated by the linear regression of the
correlation integral $C(r)$ versus the distance r displayed on a logarithmic scale for
each sequence in Figure 7. The encircled points are used to estimate the fractal
dimension. The deviation from the linearity at large distances is caused by the
limited size of the aftershock volume (e.g., GUO and OGATA, 1995). No deviation
from the linearity at short distances is observed, which confirms the high precision
of the location of hypocenters. The values of fractal dimension for the four
sequences are significantly different. The first sequence is characterized by the
lowest value of $D = 1.75$ and the second by the highest value of $D = 2.4$, with the
third and fourth sequences characterized by the middle value of $D = 1.9$ (Table 1).

From the laboratory experiments on rock fracturing, HIRATA (1987) found that
bursts of acoustic emissions decayed exponentially during the early stage of
fracturing and changed to decay hyperbolically at the last stage. The decay of

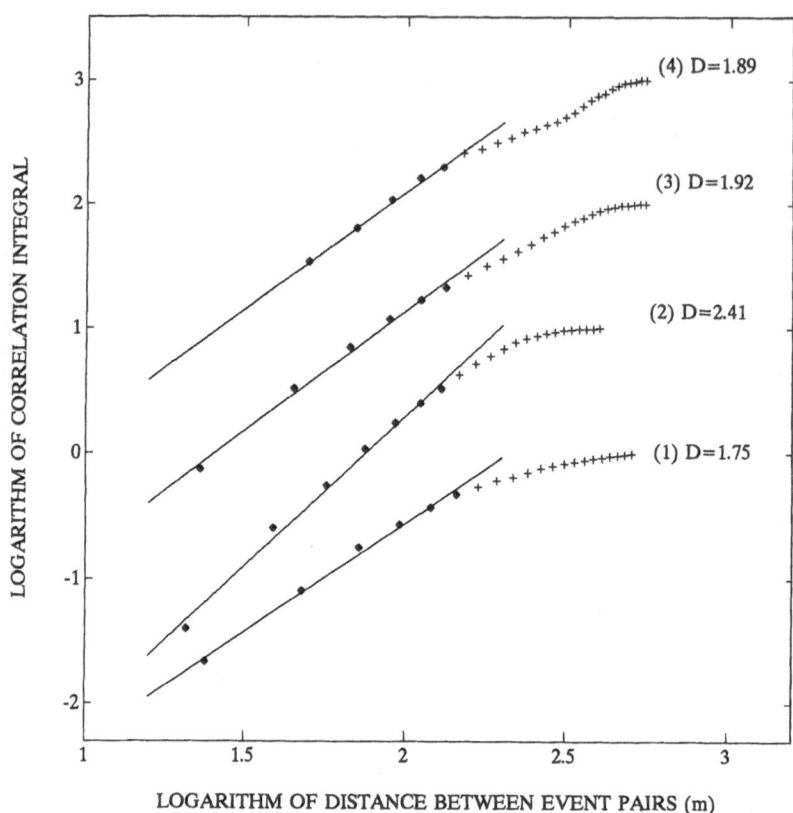

Figure 7
Logarithm of the correlation integral versus logarithm of the distance between the pairs of events from sequences 1, 2, 3, and 4. Encircled points are used to estimate the fractal correlation dimension D by the linear regression.

bursts in the rock may correspond to the decay of aftershocks (ITO, 1992). The transmission from the exponential decay to the hyperbolic decay indicates that fracture events at the early stage were rather independent but they became correlated with progressing fracturing. HIRATA *et al.* (1987) also found that the space distribution of acoustic emissions had the fractal dimension of 2.75 and was almost homogeneous at the primary stage of fracturing after which it gradually organized to become a fractal with decreased dimension of 2.25 at the final stage of fracturing. The decrease of fractal dimension indicates that fracture events became more strongly correlated with advancing fracturing (ITO, 1992).

Thus, the seismic events from our first, the best developed sequence, are the most strongly correlated and are characterized by the lowest value of fractal dimension, whereas the events from the second sequence had no time for stronger correlation as a result of their interruption by the third sequence, and are characterized by the highest value of fractal dimension. The third sequence was in turn interrupted by the fourth sequence although less suddenly as the second sequence (Figure 5) and is characterized by a middle value of fractal dimension, similarly as the fourth sequence.

Source Parameters

The traditional three-component seismograms N, E, Z were rotated into the local ray coordinate system with one longitudinal component in the P-wave direction and the transverse components in the SH and SV directions to reduce the number of seismic pulses and spectra to be processed. The seismograms were filtered between the frequency of a given sensor, which is in our case 3 Hz, and the frequency corresponding to the sampling frequency divided by five, which is 400 Hz. The spectra were calculated by a FFT routine and corrected for instrumental effect. Finally, the original ground velocity spectra were transformed into the displacement spectra. Since only global features of the spectra, described by a few spectral parameters, are of practical interest, the spectra were smoothed before further processing. In the ISS system, the spectra are smoothed by spectral filtering, which was found to be a highly efficient technique.

The observed spectra must be corrected for attenuation and scattering effects. In the ISS system, the average quality factor Q along the ray path is one of a few parameters set free in the spectrum inversion procedure. The simplest model of displacement spectra, described by the low-frequency level Ω_0 and the corner frequency f_0 above which the spectrum is assumed to fall off as a second power of frequency (BRUNE, 1970, 1971), is accepted. The spectra are multiplied by the exponential term correcting attenuation and scattering effects. The three free parameters Ω_0, f_0 and Q are estimated by an iterative linear inversion of spectral data, expanding the model spectrum into a Taylor series around some trial estimate (e.g., SCHERBAUM, 1990; FEHLER and PHILLIPS, 1991).

From the inversion of seismic spectra, two independent spectral parameters Ω_0 and f_0 are obtained from the three spectra of P, SH and SV waves. The third parameter, the energy flux, is also calculated from each spectrum at each station, and is used for the calculation of seismic energy of P and S waves. The energy flux of a plane wave is the product of the medium density ρ, the wave velocity V, and the integral I of the square of the ground velocity (e.g., BULLEN and BOLT, 1985). The integral I was calculated using the method described by SNOKE (1987), and the values of $\rho = 2750$ km/m^3, $V_p = 5700$ m/s and $V_s = 3300$ m/s were accepted for the computation of the energy flux and other source parameters.

From the low-frequency level of the spectrum either of P or S waves seismic moment is computed, taking into account the radiation pattern, the free-surface amplification and the site correction for either P or S waves. When the focal mechanism was not determined, the *rms* averages of radiation coefficients over the entire focal sphere (BOORE and BOATWRIGHT, 1984) were used. A free-surface correction can be neglected for sensors located in underground mines, and site corrections are not as important as those for sensors located at the earth's surface. The average values from the estimates of P and S waves were accepted as the final values of seismic moment. They range from $1.8 \cdot 10^8$ to $5.2 \cdot 10^{13}$ N · m. The energy radiated in the P and S waves was calculated from the relation between the energy and the energy flux, derived by BOATWRIGHT and FLETCHER (1984). The values of total seismic energy, the sum of P- and S-wave energy, range from $6.6 \cdot 10^1$ to $9.4 \cdot 10^9$ J.

The corner frequencies of P and S waves are in remarkably close proximity. They range for P waves from 14.4 to 220 Hz and for S waves from 14.5 to 197 Hz. The corner frequencies of S waves correspond to the source radius ranging from 6 to 85 m, based on the source model of BRUNE (1970, 1971). The stress drop, often called the Brune stress drop, is rather high, ranging from 0.07 to 34.4 MPa. If the P-wave contribution to the seismic energy and the azimuthal dependence of the energy flux are neglected, the Brune stress drop is a constant multiple of the apparent stress (SNOKE, 1987). For mine tremors the energy of P waves cannot be neglected and the apparent stress becomes an independent parameter. The apparent stress ranges from 0.0065 to 5.8 MPa.

The ratio of S- to P-wave energy appears as an important indicator of the type of mechanism responsible for the generation of seismic events in mines. The energy radiated in P waves from natural earthquakes is a small fraction of that in S waves, with the ratio of S- over P-wave energy ranging between 10 and 30 (e.g., BOATWRIGHT and FLETCHER, 1984). In contrast, it has been found that the energy ratio for the majority of mine tremors in the Ruhr Basin, Germany, was smaller than 10 (GIBOWICZ et al., 1990). A similar result has also been reported from the Underground Research Laboratory in Manitoba, Canada, where small seismic events induced by the excavation of a shaft in granite were observed (GIBOWICZ et al., 1991), and from two coal mines in Poland (GIBOWICZ and WIEJACZ, 1994). The observed energy depletion in S waves could possibly be explained by a nonshearing mechanism of mine tremors, enriching the energy radiated in P waves, and implying that shear failures with tensile components are often generated in mines. P-wave energy versus S-wave energy from our seismic events is shown in Figure 8. The ratio ranges from about 1 to 30 and for about three quarters of events it is less than 10.

An analysis of several source parameters as a function of time for the four sequences shows that the four main shocks are most distinctly marked by their source radius rather than by their seismic moment, energy or stress drop. The source radius versus the time after the occurrence of the first main shock is displayed in Figure 9. The four main shocks are clearly marked by the highest values of the source radius ranging from 50 to 85 m, while the source radius of aftershocks is smaller than 30 m. The duration of the four sequences is highly diverse and the distribution of their source parameters in real time is not distinct.

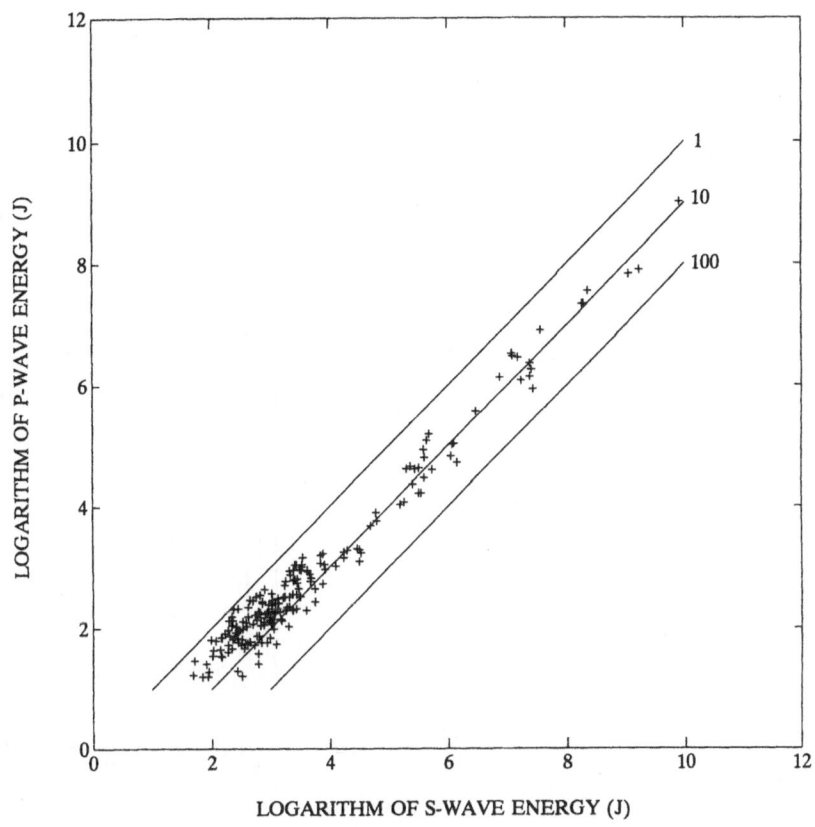

Figure 8

Logarithm of *P*-wave energy versus logarithm of *S*-wave energy. The ratio of *S*- over *P*-wave energy equal to 1, 10, and 100 is shown by straight lines.

A better presentation of time distribution of the source parameters is achieved when the source parameters are displayed against the number of consecutive seismic events forming the four sequences. Such a presentation for the seismic moment, energy and source radius is shown in Figure 10. The second sequence, which occurred in the middle of the volume containing the four sequences (Figures 3 and 4), is characterized by a number of large events in terms of their seismic moment and energy, not much smaller than the main shock. Thus the second sequence is more similar to a seismic swarm than to a main shock-aftershocks series. There are still a few large events in the third sequence, whereas the fourth sequence displays a regular main shock-aftershocks pattern similar to that also characterizing the first sequence.

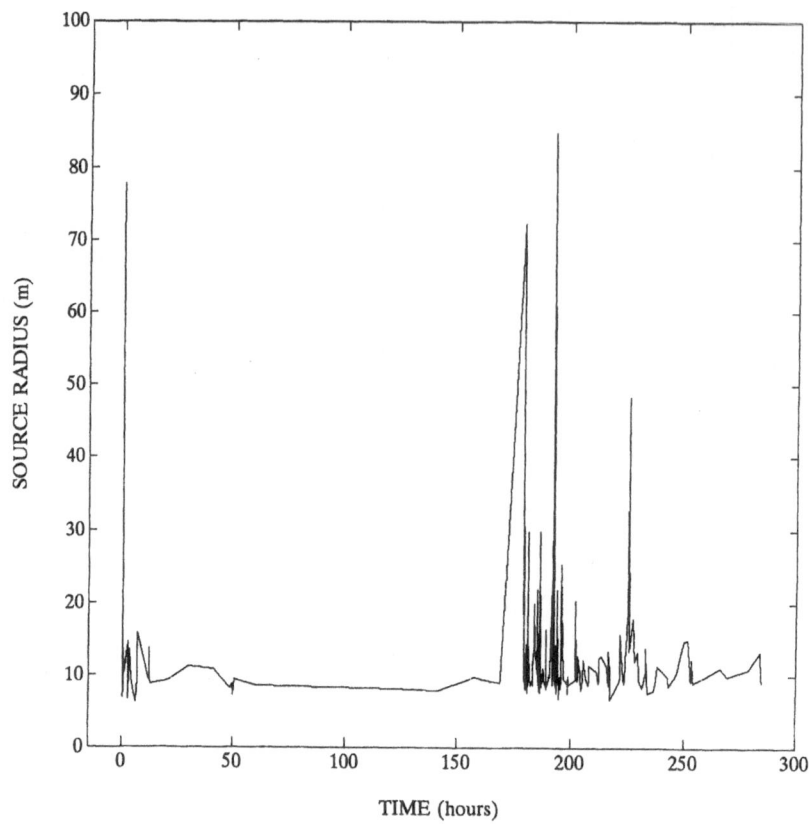

Figure 9
Source radius versus time after the occurrence of the first main shock for the four sequences.

Source Mechanism

The inversion of moment tensor in the ISS system is accomplished in the time domain. The source time function is assumed to be a step function. The simplest approach for the calculation of Green's function is used, which is the source radiation formulation for P, SH and SV waves. The input data for the inversion contain the polarities of P waves and the values of the spectral low-frequency level from P, SH and SV waves.

From several tests on real data it was found that the deviatoric moment tensor provides the most stable solution in terms of variable configuration of seismic stations and variable depth. In the ISS system, therefore, the deviatoric moment tensor solution is accepted as a starting point for the full moment tensor inversion. The initial deviatoric moment tensor components are found by solving by the singular value decomposition method the system of linear equations for selected sets of observations. Then the initial solution becomes the starting point for the final deviatoric moment tensor inversion carried out by the simplex method. The

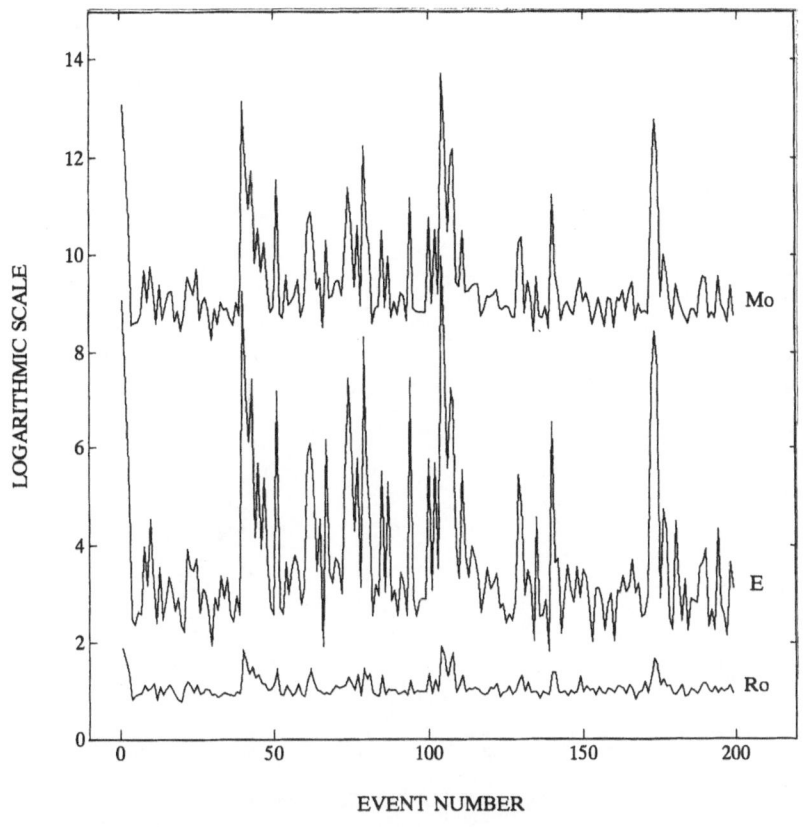

Figure 10

Logarithm of seismic moment *Mo* (in N · m), seismic energy *E* (in J) and source radius *Ro* (in m) versus the number of consecutive seismic events forming the four sequences.

final full moment tensor is decomposed into the isotropic component, the compensated linear vector dipole (CLVD) and the double-couple (DC) components.

The double-couple solution can be obtained separately and independently from the general moment tensor inversion. Several starting points (defined by the values of the strike, dip and rake angles of nodal planes) are used for an iterative inversion by the simplex method, and the best solution is sought in terms of norm L_1, i.e., the smallest sum of residuals.

From 199 seismic events, forming our four sequences, full moment tensor solutions were found for 35 events and constrained double-couple solutions were obtained for another 11 events. In most cases the shearing component of the source mechanism, represented by a double-couple, is dominant. It ranges from 60 to 90 percent of the solution for 80 percent of the events for which the general moment tensor solution was found. The isotropic component ranges from 0 to 15 percent of the solution for 83 percent of the events, and the CLVD component is the largest nonshearing component ranging from 0 to 40 percent of the solution. The CLVD

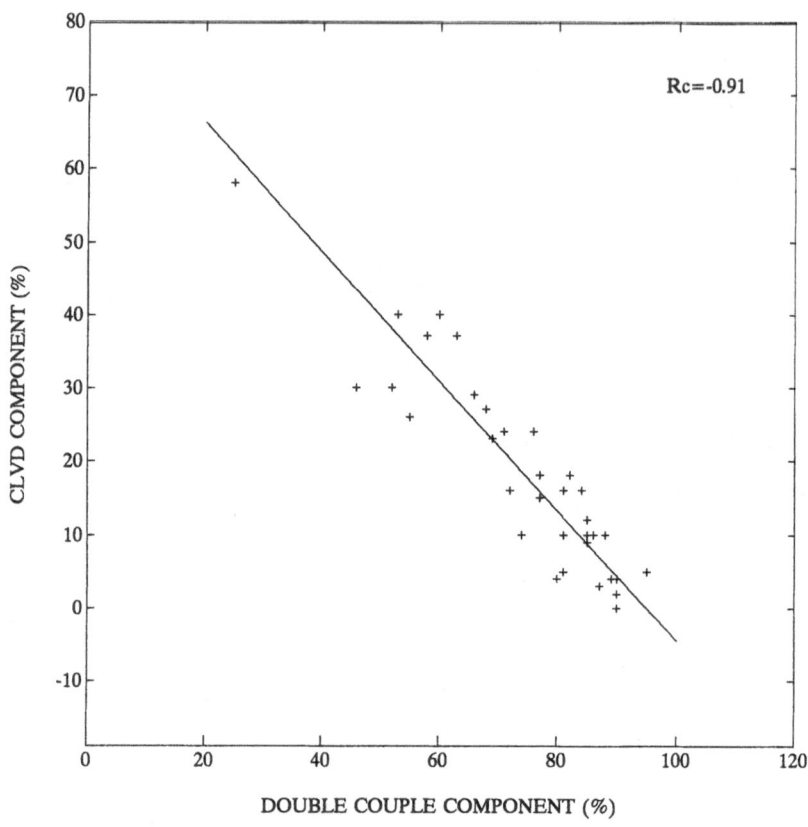

Figure 11

The CLVD component versus the double-couple component of the general moment tensor solution for 35 seismic events from the four sequences. Linear approximation and the correlation coefficient Rc are also indicated.

component versus the DC component of the general solution is shown in Figure 11, where a clear negative correlation is observed with the correlation coefficient as high as -0.9. A negative correlation is also observed between the isotropic and DC components with a much smaller correlation coefficient of about -0.5, but which is significant with more than 98 percent confidence by the Fisher's z-test.

The restrained double-couple moment tensor solutions found for 46 events represent normal faults for 15 events, including the main shocks from the second and third sequences, reverse faults for 17 events, including the main shock from the first sequence, and oblique-slip focal mechanism for 14 events, including the main shock from the fourth sequence (Table 2). The horizontal distribution of these events is shown in Figure 12. Linear correlations between the X and Y coordinates of the events with normal fault and those with reverse fault mechanisms seem to be evident. They are characterized by the correlation coefficients of about 0.5, which are significant with more than 95 percent confidence. The two regression lines are

Table 2

Number of fault plane solutions for the four sequences

Sequence	Number of solutions	Normal fault	Reverse fault	Mixed solution
1	4	0	1	3
2	25	8	11	6
3	9	4	2	3
4	8	3	3	2
Total	46	15	17	14

shown in Figure 12. The normal fault, defined in such a way, is close to the Lesser Green Dyke. Its strike coincides with that of the Lesser Green Dyke and the Green Dyke, which are about 200 m apart. This fault is almost perpendicular to the

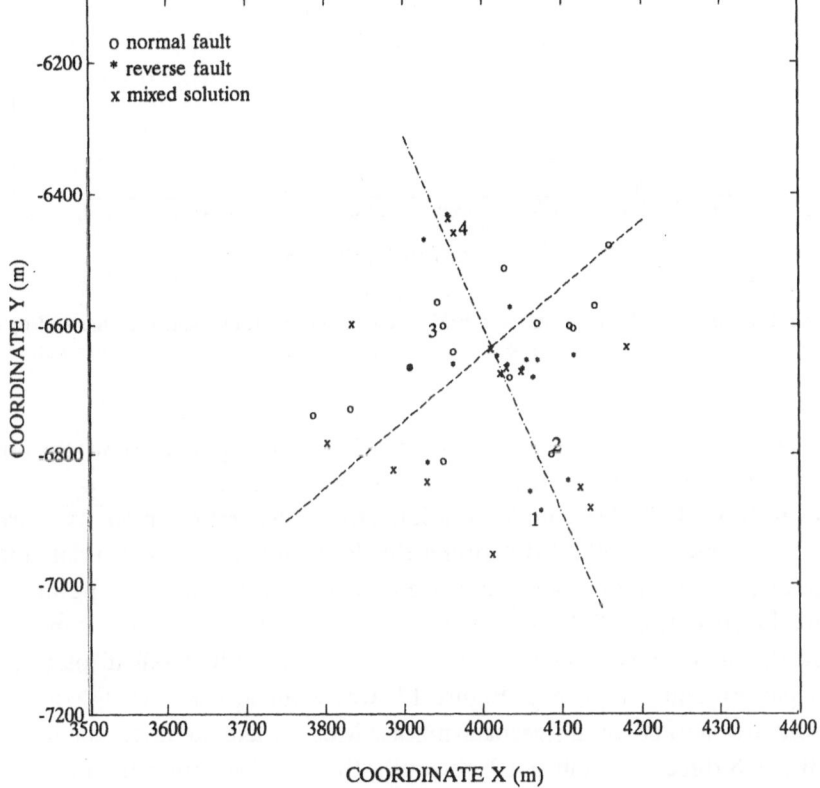

Figure 12

Horizontal distribution of 46 seismic events with known double-couple solutions. Linear regressions are shown by a dashed line for the events with focal mechanism corresponding to normal faults and by a dashed-dotted line for the events with reverse faulting. The four main shocks are numbered.

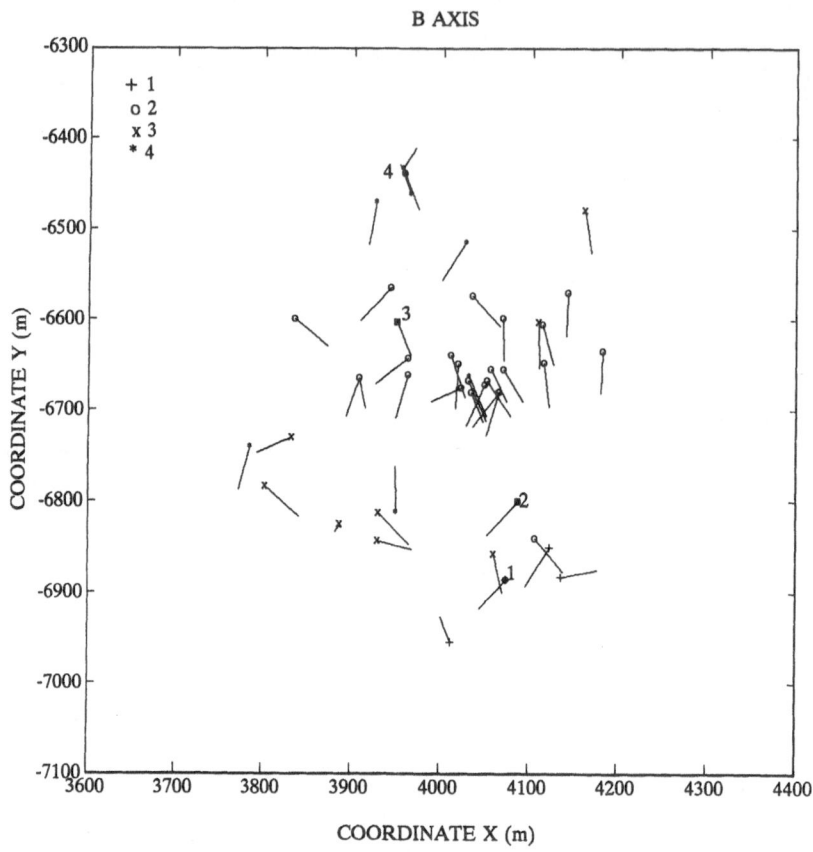

Figure 13
Horizontal distribution of *B* axes from 46 seismic events from the four sequences, marked by different symbols. The four main shocks are marked by encircled points and numbered.

reverse fault and they cross each other in a highly complex area where one more dyke and pillars left around the major dykes are present.

The restrained double-couple solutions were also represented by normalized eigenvectors (length of 50 m) describing the *P*, *T* and *B* axes, and their horizontal and vertical distributions were analyzed. There is no distinct pattern in the horizontal distribution of *P* axes, whereas the distribution of *T* axes is more regular with the dominant E-W direction. The distribution of null *B* axis displays the most regular pattern and is shown in Figure 13, where the seismic events from the four sequences are marked by different symbols. Most *B* axes tend to act horizontally along the N-S direction from north to south, that is in the opposite direction to the movement of the four sequences from south to north (Figure 3). The vertical distributions along the N-S direction of all three axes show that the *P* axes, with a few exceptions, tend to act vertically down, and the *T* axes tend to act vertically up. The distribution of *B* axes, shown in Figure 14, displays again the most regular

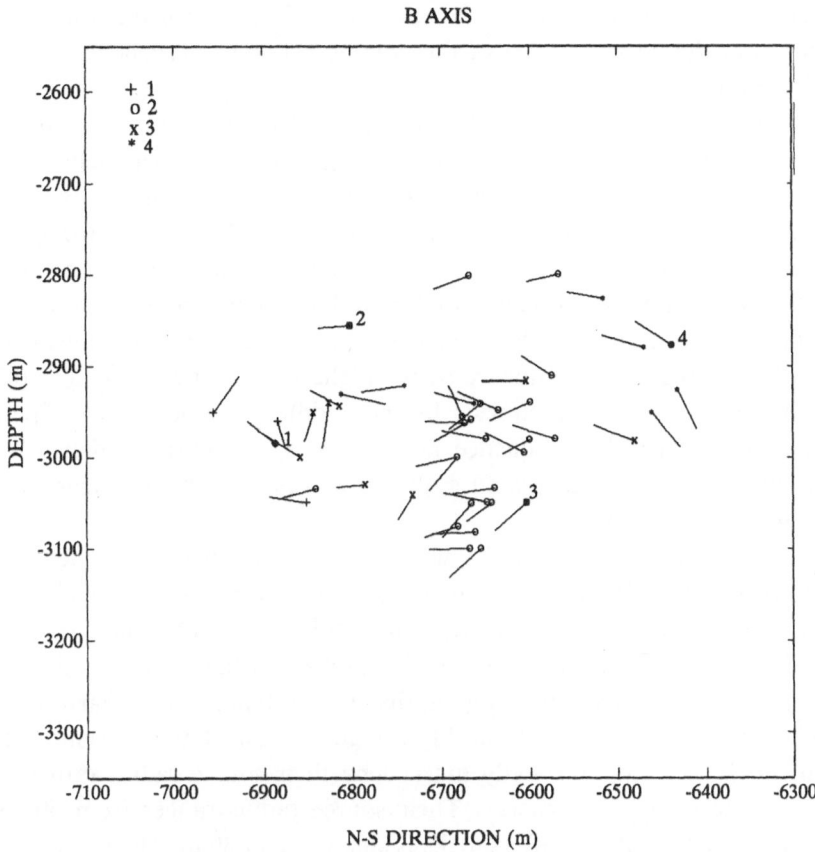

Figure 14
Vertical distribution along the N-S direction of *B* axes from 46 seismic events from the four sequences, marked by different symbols. The four main shocks are marked by encircled points and numbered.

pattern. Most *B* axes are distributed horizontally and are directed from north to south. No regular patterns are apparent in the distributions of azimuth and plunge of the three axes with time.

Conclusions

1. The 1993 swarm-like seismic sequence occurred suddenly in the area where very little seismicity has been recorded prior to the sequence. The events were associated with several pillars intersected by dykes. The sequence lasted 12 days and was composed in fact of four main shock-aftershocks sequences, closely following each other in space and time. The events were confined to a volume of rock extending to 670 m in the N-S, 630 m in the E-W, and 390 m in the vertical directions. The first sequence occurred at the southern edge of the rock volume,

the second and third in the middle, and the main shock of the fourth sequence occurred at the northern edge of the volume, while its aftershocks moved back to the south.

2. The first sequence continued 179 hours and had time for full development. The second sequence lasted only 13 hours and was followed immediately by the third sequence which lasted 31 hours and was in turn interrupted by the fourth sequence lasting 61 hours. The parameter p ranged from 0.7 for the second sequence to 1 for the first sequence. The parameter b maintained the value of 0.6 for the entire series. The fractal correlation dimension D is significantly different for the four sequences. The first sequence is characterized by the lowest value of $D = 1.75$ and the second sequence by the highest value of $D = 2.4$. The third and fourth sequences are characterized by the middle value of $D = 1.9$. The seismic events forming the first sequence are therefore the most strongly correlated, while the events from the second sequence had no time for stronger correlation with advancing fracturing.

3. The corner frequencies of P and S waves are remarkably close to each other and range from 14 to 220 Hz. The corresponding source radius ranges from 6 to 85 m. The seismic energy ranges from $6.6 \cdot 10^1$ to $9.4 \cdot 10^9$ J and the ratio of S-over P-wave energy ranges from 1 to 30. For about three quarters of events the energy ratio is less than 10, implying that tensile failures, or shear failures with tensile components, were responsible for generation of these events. The four main shocks are most distinctly marked by their source radius rather than by their seismic moment or energy. Their source radius ranges from 50 to 85 m, while the source radius of aftershocks is smaller than 30 m. The second sequence is characterized by a number of large events in terms of their magnitude and energy, only slightly smaller than those of the main shock.

4. Full moment tensor solutions were found for 36 events and constrained double-couple solutions were obtained for another 11 events. In most cases the shearing component of the source mechanism is dominant, ranging from 60 to 90 percent of the solution for 80 percent of the events. The isotropic component ranges from 0 to 15 percent of the solution and the CLVD component ranges from 0 to 40 percent of the solution. Correlations between the isotropic, CLVD and double-couple components are evident.

5. The double-couple solutions correspond to normal, reverse and oblique-slip focal mechanism. Linear correlations between the horizontal coordinates of the events with normal fault and those with reverse fault mechanisms were found. The normal fault is close to a dyke present in the area. Its strike coincides with that of two dykes, which are about 200 m apart. This fault is almost perpendicular to the reverse fault and they cross each other in a highly complex area where one more dyke and several pillars are present. An analysis of space distribution of P, T and B axes reveals that the distribution of B axes is the most regular. Most B axes tend to act horizontally from north to south, in opposite

direction to the movement of the four sequences. The observed pattern is most probably associated with several pillars intersected by dykes in the area, however, details are not available since not all pillars left in the area have been digitized at the time of the sequence occurrence.

Acknowledgements

The data from Western Deep Levels gold mine were collected during my visit at the ISS International Limited in Welkom in 1993. I am grateful to Dr. A. J. Mendecki, Managing Director, for his assistance and permission to use these data in this study and to his staff for their aid with data processing. The work on interpretation of these data was financially supported by the European Office of Aerospace Research and Development in London, U.S. Department of the Air Force, under special contract SPC-94-4073.

References

AKI, K. (1965), *Maximum Likelihood Estimate of b in the Formula log N = a − bM and its Confidence Limits*, Bull. Earthq. Res. Inst. Tokyo Univ. *43*, 237–239.

BÅK, P., and TANG, C. (1989), *Earthquakes as a Self-organized Critical Phenomenon*, J. Geophys. Res. *94*, 15,635–15,637.

BOATWRIGHT, J., and FLETCHER, J. B. (1984), *The Partition of Radiated Energy between P and S Waves*, Bull. Seismol. Soc. Am. *74*, 361–376.

BOORE, D. M., and BOATWRIGHT, J. (1984), *Average Body-wave Radiation Coefficients*, Bull. Seismol. Soc. Am. *74*, 1615–1621.

BRUNE, J. N. (1970), *Tectonic Stress and the Spectra of Seismic Shear Waves from Earthquakes*, J. Geophys. Res. *75*, 4997–5009.

BRUNE, J. N. (1971), *Correction*, J. Geophys. Res. *76*, 5002.

BULLEN, K. E., and BOLT, B. A., *An Introduction to the Theory of Seismology* (Cambridge University Press, Cambridge 1985).

FEHLER, M., and PHILLIPS, W. S. (1991), *Simultaneous Inversion for Q and Source Parameters of Microearthquakes Accompanying Hydraulic Fracturing in Granitic Rock*, Bull. Seismol. Soc. Am. *81*, 553–575.

FEIGNIER, B., and YOUNG, R. P. (1992), *Moment Tensor Inversion of Induced Microseismic Events: Evidence of Non-shear Failures in the −4 < M < −2 Moment Magnitude Range*, Geophys. Res. Lett. *19*, 1503–1506.

FUJII, Y., and SATO, K., *Difference in seismic moment tensors between microseismic events associated with a gas outburst and those induced by longwall mining activity*. In *Rockbursts and Seismicity in Mines* (ed. Fairhurst, C.) (Balkema, Rotterdam 1990) pp. 71–75.

GIBOWICZ, S. J., and WIEJACZ, P. (1994), *A Search for the Source Non-shearing Components of Seismic Events Induced in Polish Coal Mines*, Acta Geophys. Pol. *42*, 81–110.

GIBOWICZ, S. J., BOBER, A., CICHOWICZ, A., DROSTE, Z., DYCHTOWICZ, Z., HORDEJUK, J., KAZIMIER-CZYK, M., and KIJKO, A. (1979), *Source Study of the Lubin, Poland, Tremor of 24 March 1977*, Acta Geophys. Pol. *27*, 3–38.

GIBOWICZ, S. J., NIEWIADOMSKI, J., WIEJACZ, P., and DOMANSKI, B. (1989), *Source Study of the Lubin, Poland, Mine Tremor of 20 June 1987*, Acta Geophys. Pol. *37*, 111–132.

GIBOWICZ, S. J., HARJES, H.-P., and SCHAFER, M. (1990), *Source Parameters of Seismic Events at Heinrich Robert Mine, Ruhr Basin, Federal Republic of Germany: Evidence for Nondouble-couple Events*, Bull. Seismol. Soc. Am. *80*, 88–109.

GIBOWICZ, S. J., YOUNG, R. P., TALEBI, S., and RAWLENCE, D. J. (1991), *Source Parameters of Seismic Events at the Underground Research Laboratory in Manitoba, Canada: Scaling Relations for the Events with Moment Magnitude Smaller than −2*, Bull. Seismol. Soc. Am. *81*, 1157–1182.

GUO, Z., and OGATA, Y. (1995), *Correlation between Characteristic Parameters of Aftershock Distributions in Time, Space and Magnitude*, Geophys. Res. Lett. *22*, 993–996.

HIRABAYASHI, T., ITO, K., and YOSHII, T. (1992), *Multifractal Analysis of Earthquakes*, Pure Appl. Geophys. *138*, 591–610.

HIRATA, T. (1987), *Omori's Power Law Aftershock Sequences of Microfracturing in Rock Fracturing Experiments*, J. Geophys. Res. *92*, 6215–6221.

HIRATA, T., SATOH, T., and ITO, K. (1987), *Fractal Structure of Spatial Distribution of Microfracturing in Rock*, Geophys. J. Roy. Astr. Soc. *67*, 697–717.

ITO, K. (1992), *Towards a New View of Earthquake Phenomena*, Pure Appl. Geophys. *138*, 531–548.

ITO, K., and MATSUZAKI, M. (1990), *Earthquakes as Self-organized Critical Phenomena*, J. Geophys. Res. *95*, 6853–6860.

McGARR, A. (1992a), *An Implosive Component in the Seismic Moment Tensor of a Mining-induced Tremor*, Geophys. Res. Lett. *19*, 1579–1582.

McGARR, A. (1992b), *Moment Tensors of Ten Witwatersrand Mine Tremors*, Pure Appl. Geophys. *139*, 781–800.

McGARR, A., and GREEN, R. W. E. (1978), *Microtremor Sequences and Tilting in a Deep Mine*, Bull. Seismol. Soc. Am. *68*, 1679–1697.

MANDELBROT, B. B., *The Fractal Geometry of Nature* (Freeman, San Francisco 1982).

MENDECKI, A. J., *Real time quantitative seismology in mines*. In *Rockbursts and Seismicity in Mines* (ed. Young, R. P.) (Balkema, Rotterdam 1993) pp. 287–295.

OGATA, Y. (1983), *Estimation of the Parameters in the Modified Omori Formula for Aftershock Frequencies by Maximum Likelihood Procedure*, J. Phys. Earth *31*, 115–124.

SATO, K., and FUJII, Y. (1989), *Source Mechanism of a Large Scale Gas Outburst at Sunagawa Coal Mine in Japan*, Pure Appl. Geophys. *129*, 325–343.

SCHERBAUM, F. (1990), *Combined Inversion for the Three-dimensional Q Structure and Source Parameters Using Microearthquake Spectra*, J. Geophys. Res. *95*, 12,423–12,438.

SNOKE, J. A. (1987), *Stable Determination of (Brune) Stress Drops*, Bull. Seismol. Soc. Am. *77*, 530–538.

UTSU, T. (1961), *Statistical Study on the Occurrence of Aftershocks*, Geophys. Mag. *30*, 521–605.

UTSU, T. (1965), *A Method for Determining the Value of b in the Formula log n = a − bM Showing the Magnitude-frequency Relation for Earthquakes*, Bull. Hokkaido Univ. *13*, 99–103.

WIEJACZ, P. (1992), *Calculation of Seismic Moment Tensor for Mine Tremors from the Legnica-Głogów Copper Basin*, Acta Geophys. Pol. *40*, 103–122.

WIEJACZ, P., *Moment tensors for seismic events from Upper Silesian coal mines, Poland*. In *Mechanics of Jointed and Faulted Rock* (ed. Rossmanith, H.-P.) (Balkema, Rotterdam 1995) pp. 667–672.

(Received June 6, 1996, accepted November 12, 1996)

Pure appl. geophys. 150 (1997) 415–434
0033–4553/97/040415–20 $ 1.50 + 0.20/0

❙ Pure and Applied Geophysics

Scaling Laws for the Design of Rock Support

PETER K. KAISER[1] and SEAN M. MALONEY[1]

Abstract—The dynamic loads imposed by rockbursts initiated by remote seismic events are directly related to the resultant ground motions. Consequently, the first step in the design of systems for the support of openings subject to such loading is to assess the ground motion characteristics; i.e., to evaluate the peak particle velocity *ppv* in the design region. Design scaling laws, relating seismic source intensity (event magnitude, radiated energy or a combination of seismic moment and stress drop at the source) to the peak ground motion characteristics at a target (e.g., a development drift or a stope) some distance from the source, are developed, based upon theoretical considerations and supported by a world-wide database. These scaling laws reflect the most relevant, critical conditions for engineering design.

Key words: Scaling laws, rockburst, induced seismicity, ground motion, peak particle velocity.

Introduction

Safe, efficient and economical methods for the recovery of ore reserves at great depth need to be developed to maintain the viability of many mining operations throughout the world. At depth the rockmass is highly stressed and excavations will often become temporarily unstable. The purpose of ground control measures then is to contain the failing rockmass in a safe manner. One key component of such ground control methods is the mitigation of damage from mining-induced seismicity by appropriate support systems. The design of rock support requires consideration of the nature of the seismic hazard (i.e., rockburst), the additional demand placed on the support by dynamic forces and the capacity of the support system to meet that demand.

Rockbursts can be broadly classified according to their genesis as either self-initiated or remotely-triggered. In the former class, the locations of damage and causative event are coincident whereas in the latter, the rockburst is a result of energy transmission from a distant event. The seismic events, in both cases, are a consequence of stress redistribution processes associated with mining activities.

[1] Geomechanics Research Centre, Laurentian University, Sudbury, Ontario, P3E 2C6, Canada.

Rockburst Mechanisms

Self-initiated rockbursts, often termed strainbursts, occur when the stresses near the boundary of an excavation exceed the rockmass strength and failure proceeds in an unstable, violent manner (see Fig. 1a). This happens if the stored strain energy cannot be dissipated gradually via the formation of new fracture surfaces and frictional slip along existing joints or fractures. The principle underlying this process is presented in the stress strain curve of Figure 1b. When the stress exceeds the strength (A) in a particular element of the rockmass, failure occurs and the stress reduces locally as the rock continues to deform. How this stress reduction occurs is governed by the unloading stiffness of the surrounding material (slope of line segment AB). If the unloading stiffness of the failing rock (slope of line segment AC) is greater (i.e., steeper) than the system's unloading stiffness, then excess energy (shaded zone) will be released seismically leading to rock ejection. Hence, damage from this type of rockburst is related to the amount of fractured rock and the energy released during this brittle failure. This type of rockburst is not the focus of this article.

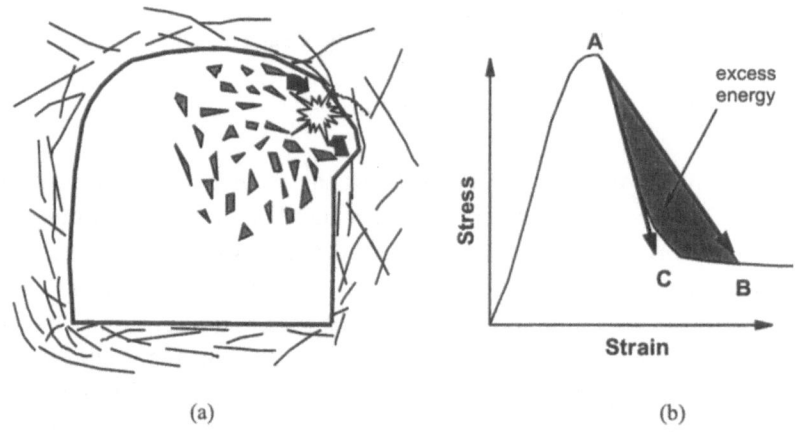

(a)   (b)

Figure 1
Damage mechanism corresponding to a self-initiated rockburst (strainburst).

Rockbursts triggered by remote, relatively large magnitude seismic events (e.g., fault slip) are a common occurrence in some hard rock mines, particularly in the later stages of the mine's life when regional changes to the stress field have been created. The seismic energy radiated from these events may cause damage to excavations at some distance from the source, i.e., at a remote target. Three damage mechanisms, each requiring different design considerations, have been identified by KAISER et al. (1995).

- *Rock bulking due to fracturing*—Seismic waves, radiating away from an event (i.e., source), impose dynamic stress changes on the already highly stressed rock which are magnified near underground openings. This can lead to local over-stressing and the development of a zone of fractured rock. When rock fractures, it increases in volume or bulks. If this process occurs in an unstable, violent manner, then rock may be ejected as in a strainburst. If the rock is well confined and fracturing consumes most of the liberated strain energy, bulking occurs in a controlled, stable manner. However, if sufficient energy is liberated from the failing rock, it may be transformed into kinetic energy and blocks may then be ejected into the opening. This latter situation is termed "bulking with ejection." It is important to note that, in this case, the cause of ejection is not the energy of the seismic source but the energy stored in the failing annulus. Hence, the ejection velocity cannot be directly related to the energy of the source. Another form of bulking occurs when the rock surrounding the excavation fails as a result of structural instability; buckling of slabs constitute one example of such a process. In these cases, geometric effects will tend to magnify the bulking, leading to high bulking factors.

- *Rock ejection by energy transfer from seismic sources*—A seismic stress wave reaching an underground opening may accelerate blocks of rock causing their dislodgement and ejection from the periphery of the excavation. In this case, the ejection velocity *is* related to the energy of the seismic source. In a mining environment with relatively low local mine stiffness, additional energy may be released during failure and generate very high ejection velocities. Momentum transfer from large to small blocks can also magnify ejection velocities.

- *Rockfalls*—A seismically induced rockfall occurs when the radiated seismic wave accelerates a volume of marginally stable rock. While the seismic shaking triggers the failure, it is gravity that drives it. This mechanism is facilitated where deep-seated fracturing has already loosened the rockmass or where weak geological structures exist that permit kinematic movement of blocks or wedges into the opening. Dynamic stress changes associated with the seismic wave may also temporarily reduce the clamping stress on geological structures, thereby reducing their shear strength and initiating a rockfall.

In these three cases, shown schematically in Figure 2, damage can be related to the ground motions generated at the target by the seismic wave of a remote event.

Use of Seismic Monitoring Data for Engineering Design

Sophisticated monitoring systems, now in operation in many mines, can accurately record the ground motions caused by the emissions from a seismic source: a brittle, failing weakness in the rockmass. Seismic signals emanating from earthquakes have been studied for decades by seismologists, to arrive at a better understanding of source mechanisms and for the prediction of occurrence. More recently, this accumulated know-how has been applied in the study of mining-

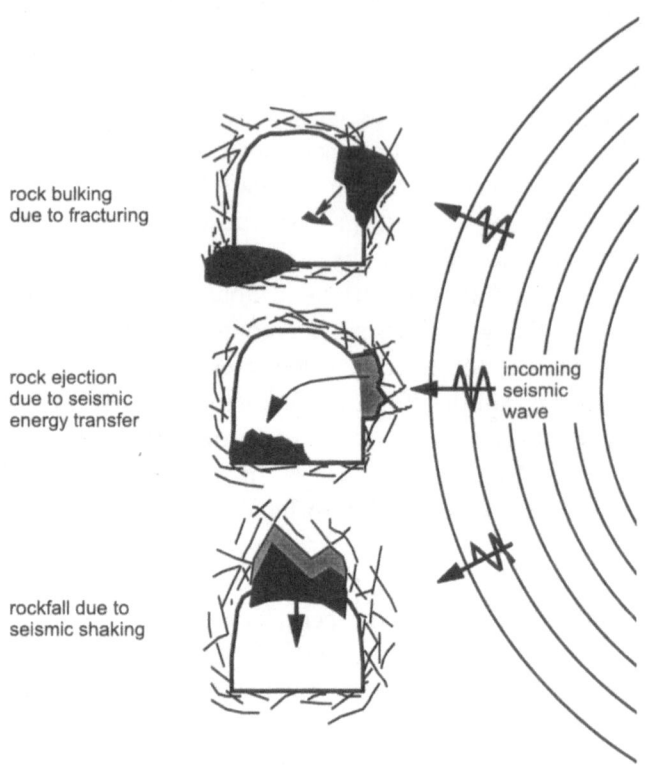

Figure 2
Damage mechanisms related to ground motions generated by a remote seismic event.

induced seismicity with respect to source locations, orientations and sizes, as well as to identify the characteristics of mining-induced source mechanisms. For these types of analyses, particularly seismic moment tensor analyses, high quality waveforms over the full frequency spectrum are required. The influence of attenuation and radiation patterns, as well as the effects of near-versus far-field approximations and source models, also need to be considered. The quality of the interpretation depends largely on the quality of the signal and the match with underlying assumptions. Furthermore, calculation of source parameters requires spectral analysis of the full waveform and the results are often only valid if very restrictive conditions are met.

From an engineering perspective, particularly when using waveform information for excavation damage prediction or for support design, only specific parts of the waveform are used. Conventional engineering design for dynamic loads is based on enhanced gravitational acceleration or directly on assumed ground motion thresholds, particularly for damage assessments. For example, slopes or earth dams have traditionally been designed against gravity plus a horizontal acceleration that is assumed to be some percentage of the gravitational acceleration g. Since the waveforms can be reasonably well represented by sinusoidal functions, at least in the relevant, low frequency range, the peak particle velocity ppv rather than the peak particle acceleration ppa is commonly used as the basis for engineering design, i.e., to define the dynamic design loads.

Because high frequency wave components act in opposing directions on representative, large volumes of rock (i.e., large relative to the wavelength), their effects may self-cancel and the resultant forces will be zero. Opposing, high frequency particle accelerations cannot accelerate a block and, consequently, no extra force is exerted on the support. Hence, only low frequency accelerations, with wavelengths sufficiently long to accelerate the entire volume of rock in one direction, are relevant for damage prediction and support design. For a sinusoidal shaped waveform, accelerations and velocities are theoretically related as:

$$ppa = 2 \cdot \pi \cdot f \cdot ppv \qquad (1)$$

where f is the dominant frequency of the incoming wave.

If the net acceleration is zero, the velocity is also zero and a block of rock is not ejected, even if locally accelerated at high ppa. As a consequence, for engineering design purposes, ground motion parameters in the high frequency range are of little value. For typical mine opening and block sizes that could, for example, suffer seismically induced falls of ground, dominant frequencies below 100 Hz are most critical. Furthermore, if damage is not initiated by the early part of the incoming wave, it will not likely be triggered by the latter portion of the waveform either as both accelerations and velocities will be lower due to attenuation. Thus, only the peak particle acceleration or velocity at a frequency that accelerates the entire volume is relevant for engineering design.

Because of these distinct engineering needs, which differ from those for seismological data processing, empirical rules and guidelines about waveform quality and near-field effects that are applicable for seismological data processing, are not directly transferrable. For example, what are considered to be near-field effects for seismological interpretation may not be at all relevant for support design purposes. Similarly, scaling laws that are justified for engineering purposes may not be valid for seismological purposes. This article only deals with ground motion determination and scaling for the engineering of rock support.

In summary, sound relationships must be established between the dominant characteristics of the remote seismic event and the resulting ground motions for use in support design. These relationships, commonly called scaling laws, are typically determined by regression analysis of field observations from sensors placed throughout a mine. By definition, these regression fits represent the average conditions encountered throughout the mine and not conditions at locations experiencing damage and, thus, are not suitable for design purposes. Engineering design must focus on conditions that cause damage. For this reason, it is necessary to establish 'design scaling laws', reflecting critical conditions for engineering decision making. These design scaling laws, relating seismic source intensity with the distance from the source to the target (e.g., development drift or stope) can then be used to obtain *ppv*s for design. Such design laws, justified on the basis of a detailed analysis of a world-wide database, together with the theoretical and statistical background for their recommendation, are presented here to bridge the gap between the available seismological information and the required support design parameters.

Seismic Activity Centre

The first step in engineering is to define the design loads. For support design, this means that a seismic activity centre must be defined in terms of its expected location and intensity. Large seismic events occur infrequently, and consequently can be expected to occur, on average, further from a given site. Conversely, as many more smaller seismic events occur in any given time period, there is a higher probability that a small seismic event occurs close to a particular site. Furthermore, multiple, small seismic events may cause incremental damage to the rockmass or the support. Hence, the design intensity will depend upon the chosen recurrence rate (i.e., the probable maximum intensity of an event within a specific return period) established from the mine's seismic record and modified on the basis of engineering judgement about the likely behaviour of the rockmass as the mining geometry is altered. The location can be either a single point or a volume containing many events with a given intensity distribution. In this latter case, the cumulative effect of the seismicity can be considered.

Once the location of a seismic activity centre and its intensity have been identified, the ground motion and, hence, the induced dynamic stress, can be determined. Ground motions are proportional to the energy radiated from the seismic source and inversely proportional to the distance from that source. The seismic energy does not necessarily emanate uniformly from the source; the radiation pattern depends on the source mechanism (AKI and RICHARDS, 1980). Seismic waves are also subject to reflection or refraction at geological boundaries or structural features and attenuation due to geometric spreading and inelastic straining of the rockmass. For design purposes, the actual distribution of ground motions is of little interest because only that in the direction of maximal energy transmission will produce conditions that are critical for design. Therefore, the event intensity must be defined in a manner that properly describes these extremes in ground motions, i.e., by a measure of the radiated energy.

Seismic Event Intensity for Support Design

With advanced seismic monitoring and signal processing, the seismic energy E_s of the source can be directly established. Since it is usually determined from the response of a limited number of sensors with the assumption of a spherical radiation pattern, it represents an average source energy. Some of the sensors will undoubtedly be located where little or no radiated energy will arrive due to non-spherical radiation or shielding. This average seismic energy may not be a good measure of the actual source intensity for support design because there will be locations in a mine, in the direction of maximum radiation, that will experience much higher than average ground motions. Additionally, attenuation effects caused by backfilled stopes, geological structures, and jointed or fractured rock vary in space and are not accounted for by conventional attenuation corrections. As a consequence, the energy flux at some locations in the mine may be considerably higher than determined from the average source energy obtained from the mine's seismic system. A further deterrent to the use of seismic energy as a measure of intensity is that the hardware and expertise required for its determination are not resident at most mines.

Currently, the most common parameter describing the intensity of a seismic event at mines is the event *magnitude*. The local Richter M_L or, in eastern North America, the Nuttli magnitude m_N, recorded at some distance from the source in a low frequency band, is one, albeit crude, measure of the radiated seismic energy (GUTENBERG and RICHTER, 1956). The term *local* seismic event magnitude, is often used to emphasize that the magnitude is a local estimate, derived from local instrumentation and using a local scale. It is not important which scale is used, as long as it is applied consistently. Comparisons between regions using different magnitude scales may, however, require empirical adjustments.

The event magnitude does have its limitations; it does not permit differentiation between disparate source and failure mechanisms, nor can it account for the radiation pattern. Differences in source mechanisms and radiation patterns can lead to ground motion levels at sensor locations that may easily vary by a factor of 5 to 10. Hence, if magnitudes alone are used as a measure of source intensity, a standard deviation in the ground motion parameter *ppv* of 25 to 35% would have to be anticipated. Despite these shortcomings, the event magnitude can be a meaningful and relevant measure of the seismic source intensity for support design. Indeed, for many mines it is the only information available and thus must be used for design decisions.

Seismologists often favour the seismic moment M_o to describe a seismic source. Since it is actually a measure of the deformation or slip at the source (AKI, 1966), the seismic moment alone does not uniquely describe the radiated energy; it must be combined with a measure of stress change at the source, e.g., the stress drop $\Delta\sigma$ or the apparent stress σ_a. This combination could potentially constitute a more robust measure of source intensity than magnitude. Unfortunately, it requires full-waveform analysis of high quality data and thus, advanced seismic monitoring and data interpretation skills.

For situations where only the seismic moment is given, expressions have been proposed to relate it to the magnitude. HANKS and KANAMORI (1979) proposed, for relatively large earthquakes, that the moment magnitude can be expressed by:

$$M = \tfrac{2}{3} \log M_o [GN \cdot m]. \tag{2}$$

Equation (2) was developed by averaging seismic moments from individual sensors; consequently, it tends to underestimate the source intensity and is not recommended for the purpose of determining ground motion levels for support design.

Consideration of published data and previously proposed relationships between magnitude and moment (e.g., HASEGAWA, 1983; HEDLEY, 1992), within the range relevant for the design of mine support (i.e., for magnitudes less than 4), led to the development of the following expressions (KAISER et al., 1995):

$$m_N = \log M_o [GN \cdot m] - (1.0 \pm 0.15)$$

$$M_L = \log M_o [GN \cdot m] - (1.5 \pm 0.15). \tag{3}$$

Equations (2) and (3) do not take into account the above-mentioned effect of the source strength and therefore do not provide a sufficiently accurate basis for support design. If information on the stress drop is available then the following, improved expressions may be applied (KAISER et al., 1995):

$$m_N = \log M_o [GN \cdot m] + \log \Delta\sigma [MPa] - (1.5 \pm 0.15)$$

$$M_L = \log M_o [GN \cdot m] + \log \Delta\sigma [MPa] - (2.0 \pm 0.15). \tag{4}$$

Similar expressions have been determined by KAISER et al. (1995) for the relationship between magnitude and radiated seismic energy as:

$$m_N = \log E_s [MJ] - \log \Delta\sigma [MPa] + (1.0 \pm 0.15)$$

$$M_L = \log E_s [MJ] - \log \Delta\sigma [MPa] + (0.5 \pm 0.15). \tag{5}$$

Equations (3) through (5) are recommended for use in Canadian mines but should be verified by on-site calibration.

In summary, seismic moments in combination with some measure of source strength (e.g., static stress drop or apparent stress) are recommended as a measure of source intensity for the determination of the ground motion necessary for support design. Where such sophisticated seismic data are unavailable, locally recorded magnitudes (M_L or m_N) or magnitudes derived from Equation (3), with meaningful confidence limits, can be used as approximations.

Ground Motion Determination

As previously discussed, for engineering purposes, the intensity of ground motion at a potential target caused by a stress wave emanating from a remote seismic source must be determined for both damage (potential) assessment and support design. The distribution of ground motion could be simulated in a deterministic manner by numerical modelling or be estimated from empirical scaling laws. As pointed out earlier, scaling laws to be used in design differ from conventional scaling laws determined by best-fit or regression analysis procedures. Design scaling laws must reflect the most relevant, critical engineering conditions.

Conventional Scaling Laws

A scaling law provides a relationship between the *ppv* or *v* and some measure of the seismic event's source intensity (seismic energy, seismic moment or event magnitude) as a function of the distance R from the source. In Canada, the seismic event intensity is commonly described by its magnitude on the Nuttli scale m_N and scaling laws are generally written in the form:

$$ppv = C \cdot SD^{-b} = C \cdot \left(\frac{R}{10^{aM}} \right)^{-b} \tag{6a}$$

where SD is the scaled distance $(R/10^{aM})$, a, b and C are empirical constants, and M is the magnitude of the seismic event. An alternate form of this expression is:

$$\log R^b v = a \cdot b \cdot M + \log C. \tag{6b}$$

In the latter case, the ground motion parameter v $(=ppv)$ is multiplied by the distance R from an assumed point source. The exponent b is often assumed to be unity to account for pure geometric spreading of the seismic wave (*ppv* is function of R^{-1}). Near a fault or a shear source of radius r_o, the ground motion is higher but dissipates more rapidly and v is proportional to $(R - r_o)^{-2}$. For cylindrical radiation patterns, v is theoretically proportional to $R^{-0.5}$.

The three scaling parameters a, b and C are often determined by regression analysis of data obtained from sensors positioned throughout a mine. Values found in the literature represent average conditions which, because they are determined by regression analysis, may lead to unconservative predictions on one extreme, far from the source, and possibly to conservative predictions at the other extreme, closer to the source.

At mines with full waveform systems, the seismic moment or the seismic energy of a source can be obtained and used instead of the source magnitude. MCGARR (1984) demonstrated, based on ANDREWS (1975), that the ground velocity should be scale-invariant if the stress regime is constant and the rockmass properties do not change. It is important to recognize that the condition of a 'constant stress regime' is rarely satisfied in mining situations close to excavations where the stresses are dominated by changes in opening geometry. McGarr also found that there was a robust dependence on M_o over more than ten orders of magnitude and wrote the scaling law in the form:

$$\log Rv = a^* \log M_o + \log C^* \tag{7a}$$

which can be rewritten as:

$$ppv = C^* \frac{M_o^{a^*}}{R}. \tag{7b}$$

By linear regression to a large set of observations, MCGARR (1984) determined the slope a^* to be 0.44 and $\log C^*$ to be -4.78 m^2/s when M_o has units of N · m.

These parameters provide average fits, corresponding to a 50% confidence, and are not recommended for support design. While the general form of Equation (7a) will be retained in the following, the constants a^* and C^* need to be determined differently for design purposes.

Design Scaling Laws

The seismic source was originally characterized by its magnitude M, a measure of the source's radiated seismic energy E_s, and later, as full waveform analysis became available, by source parameters such as seismic moment M_o, static stress drop $\Delta\sigma$, apparent stress σ_a, and source radius r_o. Sample data, plotted as E_s versus M_o from a world-wide database, including data from the Underground Research Laboratory (AECL, Pinawa, Manitoba, GIBOWICZ *et al.*, 1991), El Teniente (TRIFU, 1994), Brunswick and Strathcona Mines (URBANCIC and YOUNG, 1993), are shown in Figure 3.

Theoretical considerations suggest that both M_o and E_s increase with increasing slip D at the source. The seismic moment is proportional to the source area A and the slip D (and the rigidity G of the rock surrounding the fault; AKI, 1966):

$$M_o \propto A \cdot D. \tag{8}$$

The slip D depends on the excess shear stress at the source (RYDER, 1987) and is thus indirectly related to the stress drop.

The seismic energy is related to A and D but also depends directly on the static stress drop $\Delta\sigma$ at the source:

$$E_c \propto \Delta\sigma \cdot A \cdot D. \tag{9}$$

Hence,

$$\frac{E_s}{\Delta\sigma} \propto M_o \quad \text{or} \quad \frac{E_s}{\Delta\sigma} = K \cdot M_o. \tag{10a}$$

Alternatively,

$$E_s = K \cdot M_o \Delta\sigma \quad \text{or} \quad \log E_s = \log[M_o \Delta\sigma] + \log K. \tag{10b}$$

This is illustrated for the same data (shown in Fig. 3) in Figure 4. The relationship in Equation (10) for $K = 10$ is shown (solid line) for comparison. Deviation between data sets (e.g., El Teniente and Brunswick, both acquired with a similar monitoring system) can be attributed to changes in data processing software. Higher observations on the graph correspond to events with larger $A \cdot D$ values, or events with larger sources or more slip at the source. These events cause more co-seismic deformation in larger volumes of rock.

MADRIAGA (1976) established the relationship presented in Equation (10) and demonstrated that the constant K depends on the shear modulus G (in GPa) of the rockmass:

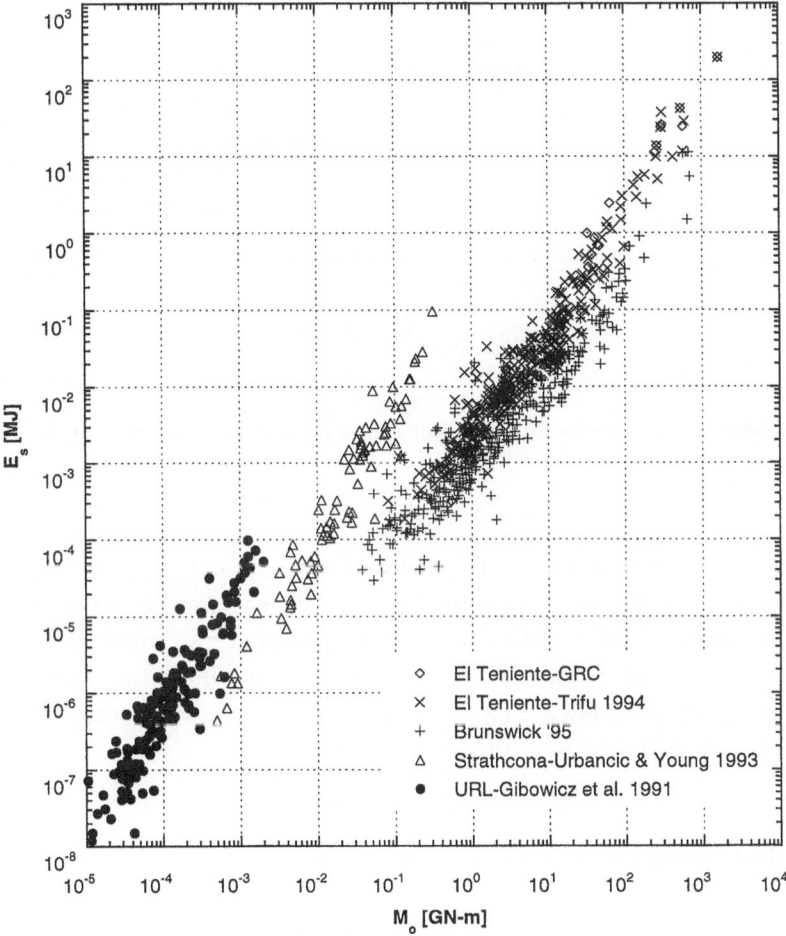

Figure 3
Radiated seismic energy E_s versus seismic moment M_o.

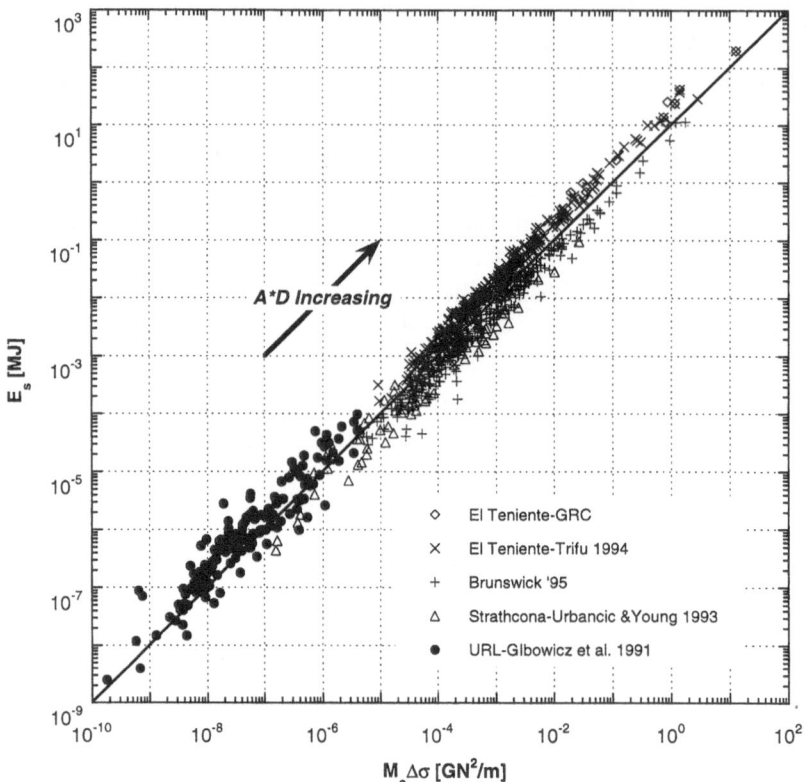

Figure 4

Radiated seismic energy E_s versus the product of seismic moment and stress drop $M_o \Delta\sigma$.

$$\log E_c = \log M_o + \log \Delta\sigma + \log \frac{0.75}{G}. \tag{11}$$

Figure 4 clearly illustrates that Equation (10) accurately describes the relationship between the radiated energy, the seismic moment and the stress drop at the source. The slope of the best-fit line is 1.0 as predicted by Equations (10) or (11).

The seismic energy radiated from a source and the time history of the ground velocity at a point are related. PERRET (1972) showed that, theoretically, the seismic energy is proportional to R^2v^2 or:

$$2 \log Rv \propto \log E_c \tag{12}$$

which, when coupled with Equation (10b), yields:

$$2 \log Rv \propto \log[M_o \Delta\sigma] \tag{13a}$$

or

$$\log Rv = 0.5 \log[M_o \Delta\sigma] + \log C \tag{13b}$$

where C is a constant. This has the same general form as the scaling law given in Equation (7a). From a comparison with Equation (13b), it follows that a^* should be 0.5 and log C^* should depend on the stress drop $\Delta\sigma$.

Ground motion data from Canada (TALEBI, 1993), Chile (TRIFU, 1994) and South Africa (MCGARR, 1984) are plotted in Figure 5 as a function of $M_o \Delta\sigma$. The same data, although filtered for $\Delta\sigma > 1$ MPa, are presented in Figure 6. The 50% and upper 95% confidence limits, assuming a slope of $a^* = 0.5$, are also plotted on these figures and demonstrate good agreement with the theoretical relationships presented above (Equation 13b).

From the ordinate axis on the right side of Figure 6, it can be seen that the data above log $Rv = -1$ are most relevant for design, i.e., the *ppv* at 100 m from source exceeds 1 mm/s. Hence, these data have been used to determine the design scaling law parameters. Since log E_s is proportional to log$[M_o \Delta\sigma]$, the same data are plotted in Figure 7, which again confirms the above given relationships (Equation 12).

Furthermore, since the seismic event magnitude is a function of the radiated seismic energy (Equation 5), any one of E_s, m_N or $M_o \Delta\sigma$ can be used to obtain R_v for support design. The relationship between all three of these event intensity parameters is illustrated in Figure 8. These relationships should be confirmed by site-specific calibration.

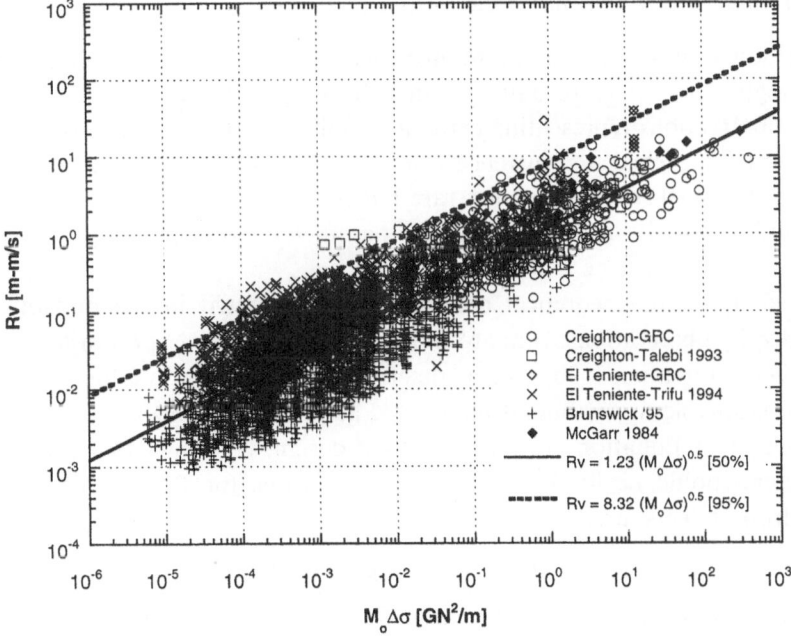

Figure 5
Ground motion parameter Rv versus the product of seismic moment and stress drop $M_o \Delta\sigma$ for all data.

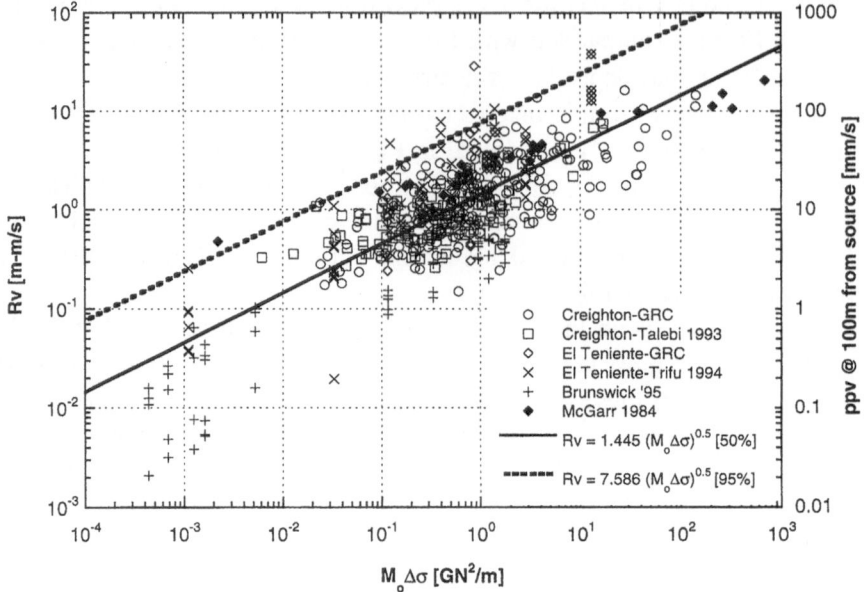

Figure 6
Ground motion parameter Rv versus the product of seismic moment and stress drop $M_o \Delta\sigma$ for data where the stress drop exceeds 1 MPa.

Recommended Design Scaling Law Parameters

For design, it is not meaningful, nor even acceptable, to prescribe conditions by fitting regression curves to data obtained from spatially distributed sensors. Only critical observations representing extreme conditions are of interest. By analogy, the flood level for a chosen recurrence rate is used to design a bridge instead of the average river level. In this manner, rare but relevant conditions for a safe structure are found. Similarly, design scaling law parameters corresponding to some upper confidence limit must be determined.

Based upon the theoretical considerations presented in the previous section, supported by observations from around the world, the most appropriate scaling law for design is Equation (13b) with C determined from $\log Rv$ vs. $\log(M_o \Delta\sigma)$ graphs with a reasonable upper bound limit (e.g., at 95% confidence).

However, if Equation (7) is adopted for design, it follows that the scaling law parameters should be fixed at $a^* = 0.5$ and a value for C^* that depends on the stress drop at the source:

$$\log C^* = 0.5 \log \Delta\sigma + \log C. \tag{14}$$

If the conventional scaling law (Equation 7) is adopted without consideration of the stress drop at the source and without mine specific data, the parameters listed in Table 1 are recommended for determining the anticipated ground motions.

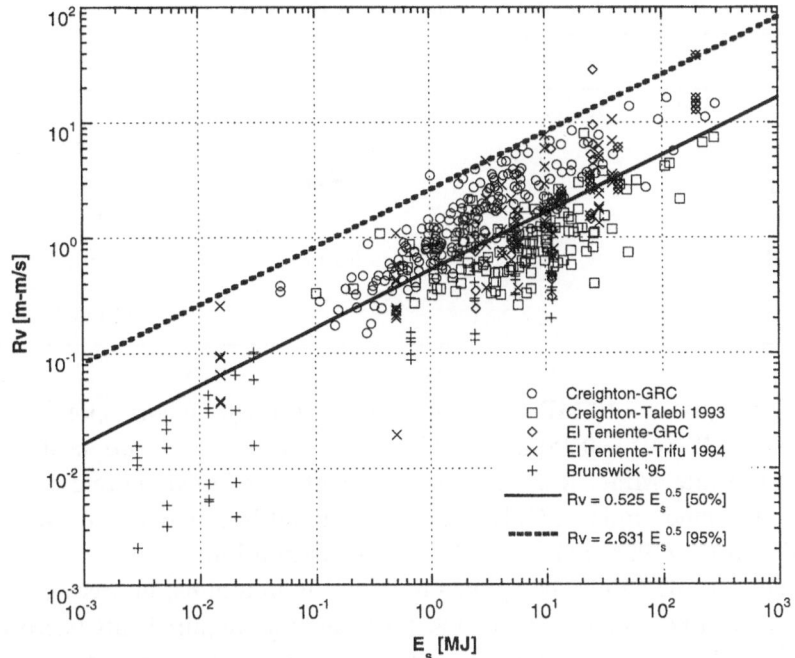

Figure 7

Ground motion parameter Rv versus the radiated seismic energy E_s for data where the stress drop exceeds 1 MPa.

Relationship I represents average conditions and should not be used for design because most of the damage causing ground motion would exceed that predicted by it. It can, however, be used for comparison with results from regression analyses of field observations. By definition, similar numbers of observations should plot below and above this relationship. This is illustrated, for example, by the data shown in Figure 5.

Relationship II is recommended for design if no unusual stress conditions are anticipated, e.g., as reflected by average stress drop observations of less than 2.5

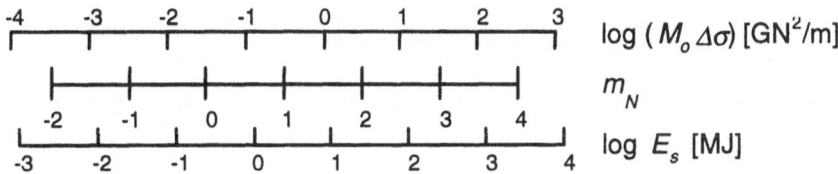

Figure 8

Relationships between the product of seismic moment and stress drop $M_o \Delta \sigma$, Nuttli magnitude m_N and radiated seismic energy E_s.

Table 1

Recommended parameters for scaling law (Equation 7)

	a^*	C^* [m²/s]	Confidence limit
I		0.1	50% at $\Delta\sigma \leq 5$ MPa
II	0.5	0.2 to 0.3	90 to 95% at $\Delta\sigma < 2.5$ MPa
III		0.5 to 1.0	90 to 95% at $\Delta\sigma =$ 10 to 25 MPa

MPa (maximum $\Delta\sigma = 10$ MPa). The dashed line in Figure 6 corresponds to a stress drop of 1.1 MPa ($C^* = 0.25$) and illustrates that, with the exception of some data from El Teniente Mine, Relationship II provides an upper bound for typically anticipated ground motions. Only a single data point from Creighton Mine and one from McGarr's (1984) data, fall above this relationship.

Relationship III should be applied only locally in a mine, in areas of extremely high stress with very high stress drops. Since support requirements based on these parameters will turn out to be heavy, a detailed investigation should be undertaken to justify their use. The fact that eight data points from El Teniente Mine fall above the dashed line (Figure 6; Relationship II for $\Delta\sigma = 1.1$ MPa) suggests that these events reflect seismicity from high stress areas and that Relationship III should be applied for support design in those areas affected by these high stress drop events. Examination of seismic records from El Teniente Mine indeed showed stress drops in excess of 10 MPa for those seismic events causing unusually high ground motions.

Alternatively, in terms of seismic energy:

$$\log Rv = 0.5 \log E_c + \log C' \tag{15}$$

is recommended for design with the constant C' determined from $\log Rv$ vs. $\log E_s$ graphs.

Similarly, for design based upon Nuttli magnitude:

$$\log Rv = 0.5m_N + \log C'' \tag{16}$$

is recommended, with C'' determined from plots of $\log Rv$ versus magnitude m_N.

Limitations of the Design Scaling Law

The design scaling law parameters presented in Table 1 were developed from a database of ground motions measured at some distance from the source, i.e., in the far-field. In the near-field, close to a seismic source, the radiation pattern is extremely non-uniform and higher ground motions must be anticipated.

KAISER *et al.* (1995) have shown that the far-field relationship dominates the peak ground motion for distances greater than about two times the source radius r_o from the centre of the seismic source. According to SCHOLZ (1990), the source radius for a circular fault depends on the seismic moment and the static stress drop as:

$$r_o^3 = \frac{7M_o}{16\Delta\sigma}. \tag{17}$$

Accordingly, the design scaling laws should only be applied for distances in excess of:

$$R_{\min} = 2 \cdot \left(\frac{7M_o}{16\Delta\sigma}\right)^{1/3} \approx 1.5 \cdot \left(\frac{M_o}{\Delta\sigma}\right)^{1/3}. \tag{18}$$

Thus, for typical mining-induced, fault-slip seismic events in the range of 1.5 to 4.0 on the Nuttli magnitude scale, the design scaling laws should not be applied to

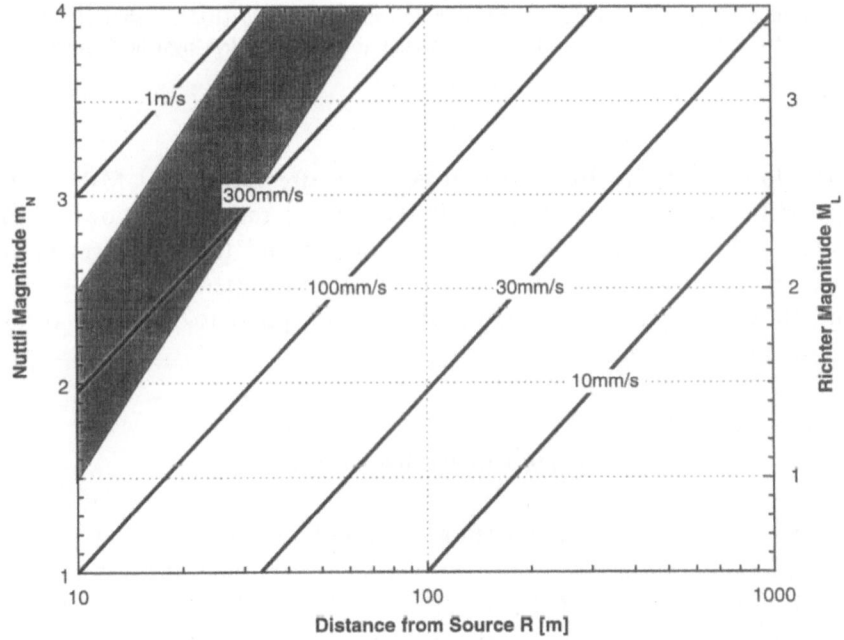

Figure 9
Average peak particle velocity distribution (Relationship I; $a^* = 0.5$, and $C^* = 0.1$ m^2/s). Higher average values are expected above the shaded 'near-field' zone.

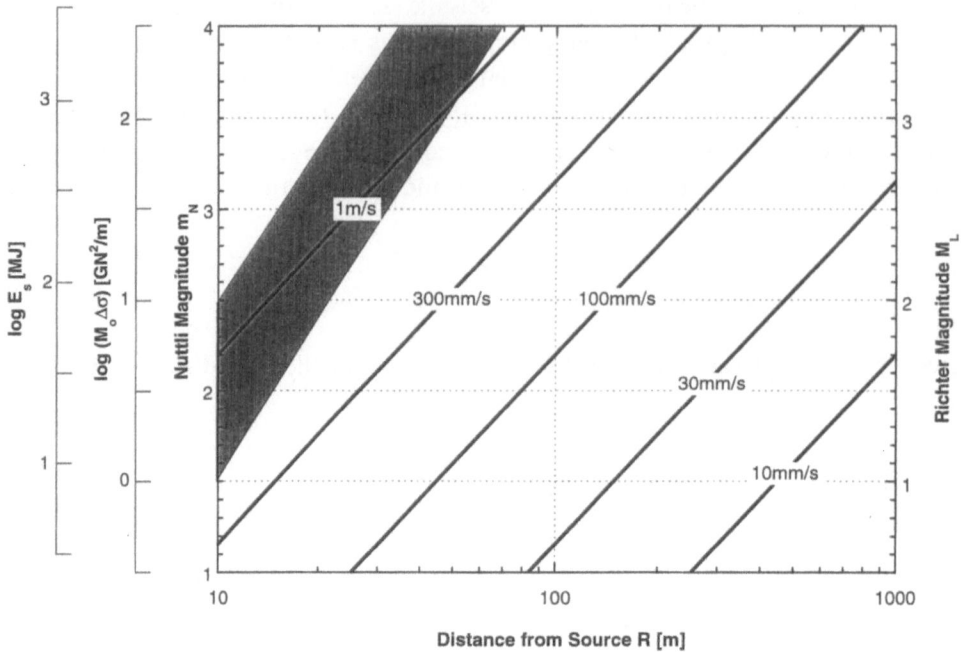

Figure 10
Recommended peak particle velocity distribution for support design (Relationship II; $a^* = 0.5$ and $C^* = 0.25$ m^2/s). Higher values are expected above the shaded 'near-field' zone.

locations closer than 5 to 10 m respectively for a stress drop of 1 MPa, and 10 to 70 m respectively for a stress drop of 10 MPa. These ranges are shown in Figures 9 and 10 as shaded zones. The frequency content and the waveform in general however, are influenced for considerably larger distances (ten times the wavelength) and the above limitations do not, therefore, correspond to near-field definitions from a seismological perspective.

Design Ground Motion Charts

Figure 9 presents average predicted ground motion levels, based upon Relationship I as a function of distance and seismic event magnitude. It provides *ppv*s that should, on average, be expected from seismic events with stress drops less than 5 MPa. For design, higher values such as those shown in Figure 10 for $C^* = 0.25$ m^2/s (Relationship II), must be assumed as explained earlier. In this case, the design *ppv*s at any given distance from the source are greater than the average by a factor of 2.5.

Unless unusual circumstances exist at a mine, the values presented in Figure 10 should only be exceeded in 5 to 10% of the total volume of rock affected by a seismic event. As mine openings likely constitute only a small fraction of this total volume, the probability of experiencing *ppv*s greater than the design values at underground excavations is much smaller.

While the design scaling law presented graphically in Figure 10 has been developed from a global database, it does not mean that it can be universally applied. Its application to a specific site requires validation by field data.

Conclusion

Supported by theoretical considerations and based on world-wide data, the scaling laws presented herein are recommended for the design of support in burst-prone ground. For support design, the ground motion parameters must be determined from source parameters that best represent the radiated seismic energy, e.g., the seismic event magnitude, the measured radiated seismic energy, or the product of seismic moment and stress drop. The seismic moment alone is not recommended for scaling purposes to determine ground motion parameters for support design unless the constant C^* is defined from data of a stress regime comparable to the design domain.

The recommended procedure is to plot $\log Rv$ against $\log(M_o \Delta\sigma)$ or $\log E_s$ or m_N or M_L and then to determine a reasonable upper bound limit (e.g., 95% confidence line) as shown in Figures 5 through 7 with the slope fixed at the theoretical value of $a^* = 0.5$.

Because design scaling laws are intended to define design conditions and not those actually encountered, observed peak particle velocities and thus, inflicted rockburst damage, should normally be less severe, less frequent and more random. This must be taken into account when support design recommendations are compared to *ppv* measurements and rockburst damage observations.

Acknowledgements

This work was funded through the Mining Research Directorate (now CAMIRO) by twelve mining companies sponsoring the Canadian Rockburst Research Program (1990–1995).

Special thanks are extended to Cezar Trifu of ESG, Shahriar Talebi of CAN-MET, Eduardo Rojas and Julio Cuevas of El Teniente Mine and Terry MacDonald of Brunswick Mine for their assistance in assembling the database.

REFERENCES

AKI, K. (1996), *Generation and Propagation of G Waves from the Niigata Earthquake of June 16, 1964, 2. Estimation of Earthquake Moment, Released Energy and Stress-strain Drop from the G Wave Spectrum*, Bull. Earthq. Res. Institute, Tokyo University *44*, 73–88.

AKI, K. and RICHARDS (1980), *The Seismic Source: Kinematic, Quantitative Seismology*, Theory and Methods *14*(2), 799–911.

ANDREWS, D. J. (1975), *From Antimoment to Moment: Plane Strain Models of Earthquakes that Stop*, Bull. Seismol. Soc. Am. *65*, 163–182.

GIBOWICZ, S. J., YOUNG, R. P., TALEBI, S., and RAWLENCE, D. J. (1991), *Source Parameters of Seismic Events at the Underground Research Laboratory in Manitoba, Canada: Scaling Relationships for Events with Moment Magnitude Smaller than −2*, Bull. Seismol. Soc. Am. *81*(4), 1157–1182.

GUTENBERG, B., and RICHTER, C. F. (1956), *Magnitude and Energy of Earthquakes*, Ann. Geofis. Rome *9*, 1–15.

HANKS, T. C., and KANAMORI, H. (1979), *A Moment Magnitude Scale*, J. Geophys. Res. *84*(B5), 2348–2350.

HASEGAWA, H. S. (1983), *Lg Spectra of Local Earthquakes Recorded by the Eastern Canada Telemetered Network and Spectral Scaling*, Bull. Seismol. Soc. Am. *73*(4), 1041–1061.

HEDLEY, D. G. F., *Rockburst Handbook for Ontario Hardrock Mines SP92-1E* (CANMET, Ottawa 1992).

KAISER, P. K., MCCREATH, D. R., and TANNANT, D. D., *Canadian Rockburst Support Handbook* (Mining Research Directorate, Sudbury 1995).

MADRIAGA, R. (1976), *Dynamics of an Expanding Circular Fault*, Bull. Seismol. Soc. Am. *66*, 639–666.

MCGARR, A. (1984), *Scaling of Ground Motion Parameters, State of Stress, and Focal Depth*, J. Geophys. Res. *89*(B8), 6969–6979.

PERRET, W. R. (1972), *Seismic Source Energies of Underground Nuclear Explosions*, Bull. Seismol. Soc. Am. *62*, 763–774.

RYDER, J. A., *Excess Shear Stress (ESS): An engineering criterion for assessing unstable slip and associated rockburst hazards*. In *Proc. ISRM Congress* (Montreal 1987) *3*, 1211–1215.

SCHOLZ, C. H., *The Mechanics of Earthquakes and Faulting* (Cambridge University Press 1990).

TALEBI, S., *Source studies of mine-induced seismic events over a broad magnitude range (−4 < M < 4)* (CANMET, Report MRL 93-046 (CL), 1993).

TRIFU, C.-I., *Characterization of the mining induced seismicity at the El Teniente Mine of Codelco (Chile), with special emphasis on the Sub-6 region* (ESG Report to Itasca 1994 plus data by pers. comm.).

URBANCIC, T. I., and YOUNG, R. P. (1993), *Space-time Variations in Source Parameters of Mining-induced Seismic Events with M < 0*, Bull. Seismol. Soc. Am. *83*, 378–397.

(Received December 23, 1996, accepted June 26, 1997)

Pure appl. geophys. 150 (1997) 435–450
0033–4553/97/040435–16 $ 1.50 + 0.20/0

⌐Pure and Applied Geophysics

Predisposition to Induced Seismicity in Some Czech Coal Mines

Karel Holub[1]

Abstract—Mining-induced seismicity associated with longwall face operations in the Ostrava-Karviná coal mines, Czech Republic, has been investigated in order to establish the conditions leading to a focal zone generation. The study, based on macroseismic and instrumental observations, proved that seismicity is influenced by natural as well as mining conditions. The first group includes the influence of faults, washouts and red beds, while the second one is represented by shaft and/or crosscut safety pillars and various types of remnant pillars. All the cases discussed show that many focal zones are generated in overstressed strata as a consequence of interaction of natural conditions and/or old workings with the active coal face.

Key words: Focal region origin, induced seismicity, geomechanics, Ostrava-Karviná Coal Basin.

1. Introduction

Rockbursts which represent a sudden failure of the rock mass underground are severe problems in coal and ore mining. These mining-induced seismic events occur in response to the accumulation of the stresses imposed, e.g., due to local different properties of rock mass and/or due to various consequences of the applied mining technologies. Most of the situations previously described (e.g., KANEKO et al., 1990; KNOLL and KUHNT, 1990; GAY, 1993; GILL et al., 1993 and PTÁČEK and TRÁVNÍČEK, 1994) proved that rockburst occurrence is related to geological structures as well as technological conditions of mining. The present paper reports results obtained in the course of investigations into natural and mining factors which affect the formation of focal zones. The investigations of mining-induced seismic activity are based on the use of long-term macroseismic assessments and instrumental data.

---

[1] Institute of Geonics, Academy of Sciences of the Czech Republic, Studentská 1768, 708 00 Ostrava-Poruba, Czech Republic.

2. Data

In the present analysis, the macroseismic data assembled by ZAMARSKÁ (1981) were used. These data were included in the catalogue of rockbursts and rockburst-like phenomena for the period 1912–1980. The catalogue contains all relevant data, i.e., date, approximate origin time, depth of the mined seam, its numbering and thickness, characteristics of the mining activity at the time of rockburst occurrence (drivage of gates, preparation phase of longwall panel operation, coal extraction), physical and mechanical properties of the coal seam and its overburden. A substantial part of this documentation consists of a detailed description of manifestations observed *in situ* and by a schematic map displayed mining situations and locations of the maximum rockburst damage.

Instrumental data have been obtained roughly since the 1980s from the local seismographic network which was gradually deployed at individual mines. At present, there is in operation a local seismographic system equipped with digital instrumentation in which an experimental network in the Lazy (former A. Zápotocký) Mine (see KNEJZLÍK *et al.*, 1992) is also included. This monitoring system comprised, at the end of 1996, one-component (vertical) stations situated underground (37) and on surface (6). Beside the local network, a regional monitoring system is in operation which consists of 10 three-component stations in the adjacent area of coal mines. The layouts of monitoring systems described are given in Figure 1.

All instrumental data of the local network, i.e., origin time, energy and source location of seismic events, have been gathered in a data base since 1989. New data are added continuously to this data base and applied for the purposes of geomechanical service. Current results of the monitoring and investigation of seismicity induced by coal mining in the Ostrava-Karviná district were described by HOLUB *et al.* (1995). For the present study, service location plots of observed and recorded seismic events have been used.

3. Natural Conditions

3.1. Tectonic Faults

The Ostrava-Karviná Coalfield represents a part of the Upper Silesian Basin which is situated on the northeastern territory of the Czech Republic between the inner parts of the Variscan mobile belt and the Carpathians. The basin's evolution was seriously affected during various structural stages, which resulted in the disruption of subhorizontally layered strata of sedimentary beds in the Karviná partial basin by numerous faults and faulted zones. The predominant orientation of these discontinuities is nearly east-west and north-south. Many other faults of lesser importance are parallel to the significant faults and faulted zones.

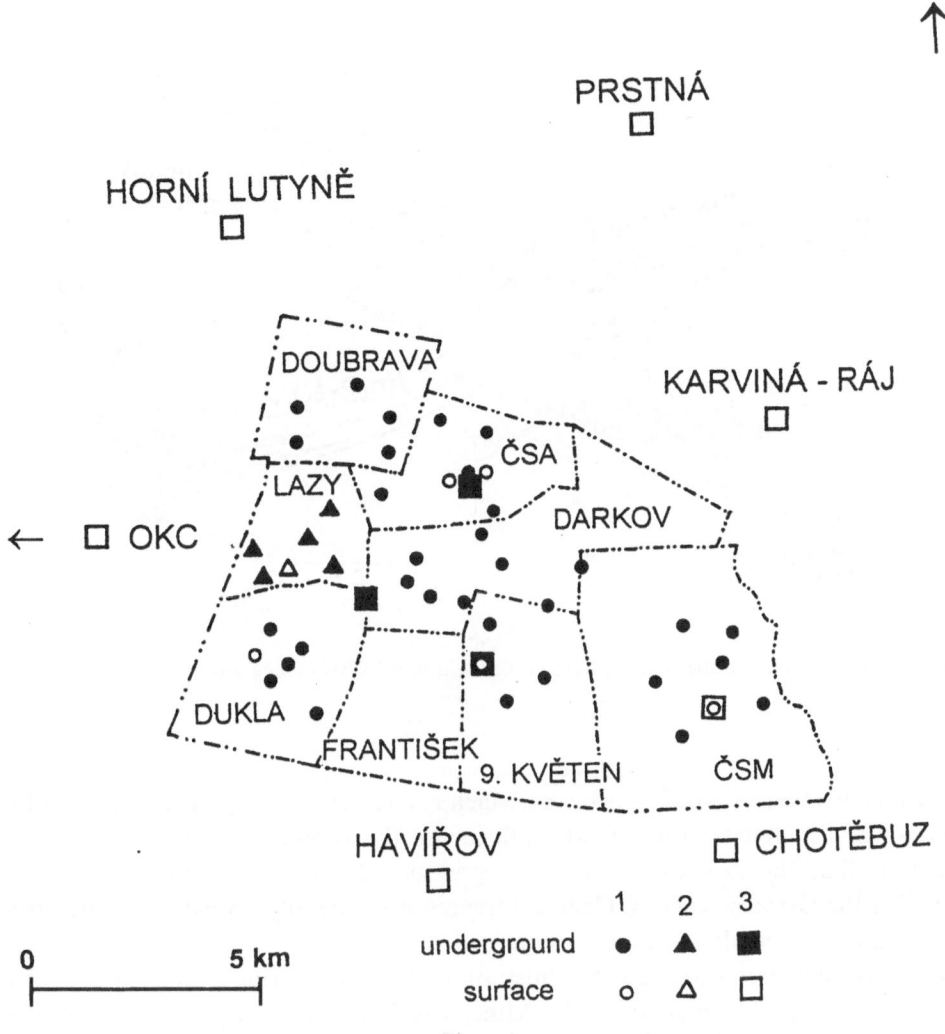

Figure 1

Seismic networks in the eastern part of the Ostrava-Karviná Coal Basin. 1—local seismographic network, 2—the PCM-3 experimental microarray, 3—regional network. Seismic station OKC is situated approximately 20 km westwards.

As seen in a block diagram (Fig. 2), related to the contact of the Doubrava and ČSA Mines, the system of faults oriented in the east-west direction could be constrained (e.g., Nepojmenovaná, Žofinská and Jindřišská faults) as well as that in the north-south orientation (e.g., Barbora, Hlubinská and Gabrielská faults). According to the foci plots, a focal region was also established within the area of the so-called Jindřich transtensile basin. This very complicated structural situation was described by GRYGAR (1987) and also reported by TAKLA and PTÁČEK (1990),

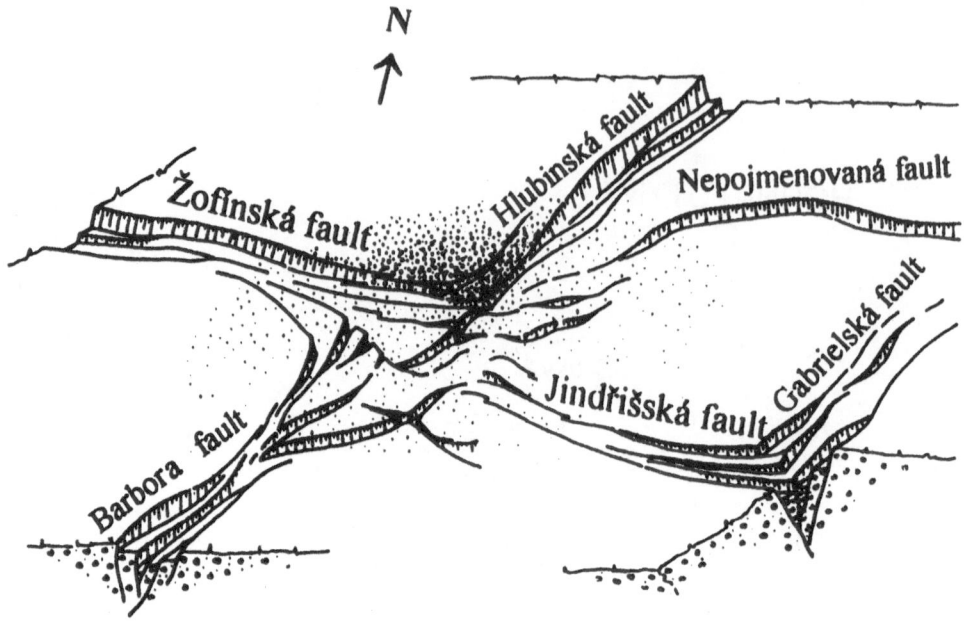

Figure 2
Genesis reconstruction of the Jindřich transtensile basin in the ČSA mining area after GRYGAR (1987).

and was denoted as a pull-apart zone which is characterized by the crossing and/or ending of different tectonic faults influencing the origin of seismic events. It is obvious that the existence of a dense grid of different joints and faults which transect the Ostrava-Karviná Coalfield, represents generally a weakness zone prone to rockbursts. In all the above-mentioned faulted zones, a clustering of seismic events foci has been observed. An illustration of foci clustering along some faults in the 3rd tectonic block in the ČSA Mine, which was discussed by HOLUB et al. (1988), is given in Figure 3.

The whole area of Ostrava-Karviná coal mines is divided by a series of faults into a number of small structural units denoted usually as tectonic and/or mining blocks which became the basis for a further subdivision of mine fields. In this manner, every mine field is divided in several blocks which are numbered consecutively within the respective mine. Though many of the foci were located in the strips following these faults, as reported (for example: PTÁČEK and TRÁVNÍČEK, 1994 and KONEČNÝ, 1989), no unambiguous evidence exists, to determine whether they originated from tectonic movements. In our opinion, the observed seismic activity is linked with changes of stresses induced by mining operations and their consequences occurring as a rheologic process of rock mass in the respective parts of this coal field, rather than with natural tectonic movements.

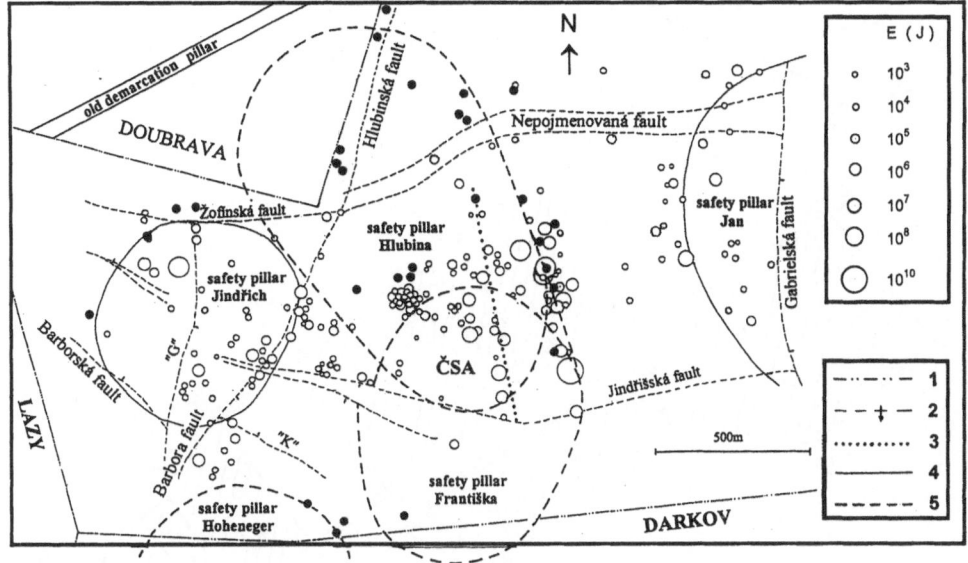

Figure 3

Plan view of the foci of seismic events in the 3rd tectonic block at the ČSA Mine and in its vicinity on the background of the structural geology. Solid circles show foci of rockbursts (after ZAMARSKÁ, 1981), open circles are foci determined from July 1979 through February 1987 (after HOLUB et al., 1988). 1—demarcation line of the mine fields, 2—tectonic fault, 3—man-made demarcation between ČSA collieries 1 and 3, boundary of the present 4—and abandoned 5—safety pillars, respectively.

3.2. Washouts

In the Carboniferous coal deposit in the Ostrava-Karviná Coal Basin, one can frequently find washouts which represent an inhomogeneity in the rock mass. A washout is usually characterized by a partial and/or complete reduction of the coal seam, the filling of which is created during further sedimentation cycle by a material which possesses physical and mechanical properties different from the coal seam.

A typical example of the creation of a rockburst focal zone in the area of a washout was described, e.g., by HOLUB (1997), and its *in situ* situation is given in Figure 4. The existence of unmined seams and remnant pillars in the overburden of longwall panel no. 63762 in the Doubrava Mine influenced to a great extent the mining conditions in this region.

Rockbursts nos. 1–3 occurred in July 1978 and no. 4 in September 1978 during development operations (drivage of roadways) and the subsequent ones (nos. 5–13) within the time interval of May–September 1979 during the mining of this longwall panel. However, all foci were situated only in the area of influence of the washout as reported by ZAMARSKÁ (1981) and TAKLA and PTÁČEK (1990). The maximum energy release of 2×10^7 J was observed for rockburst no. 12 which occurred on August 28, 1979.

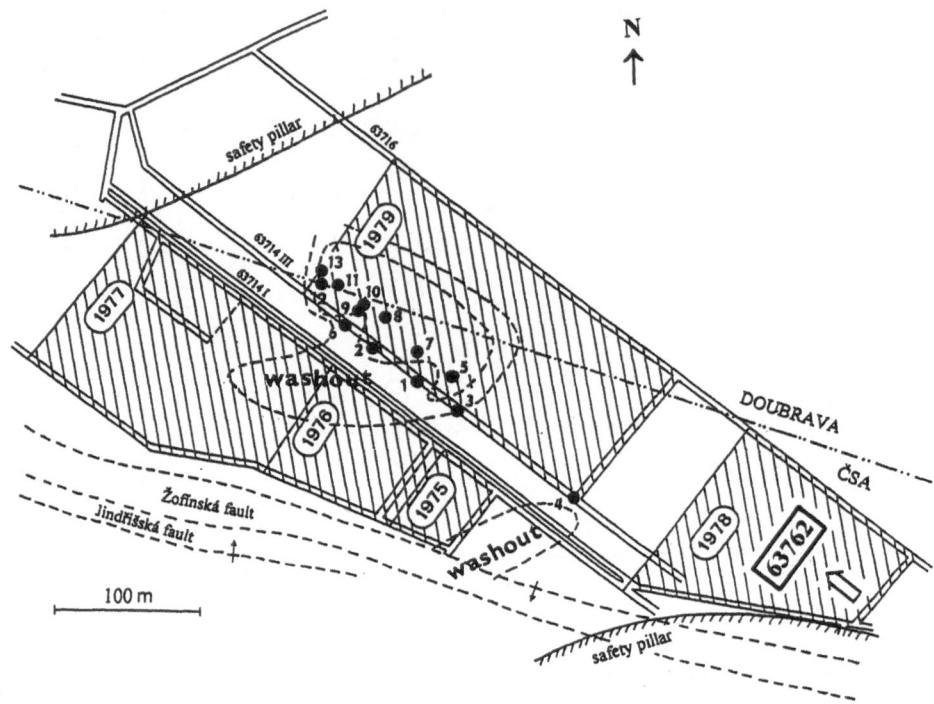

Figure 4
Plan view of rockburst positions defined according to the maximum damage in workings before and
during longwall face no. 63762 operation in the Doubrava Mine in 1978–1979 (after ZAMARSKÁ, 1981).

3.3. Red Beds

Another type of area predisposed to the creation of a focal zone in the rock
mass, which is also typical for the Ostrava-Karviná coal mines, is an area with red
beds which are associated with the paleorelief of the Carboniferous in this region.
Red beds are deposited subhorizontally, having different degrees of thermal alternation, according to which they can be classified in more detail (MARTINEC, 1994).
From a geomechanical viewpoint, the properties of these variegated rocks substantially differ from the properties of surrounding coal seams, which creates a marked
mechanical discontinuity between both rock bodies.

From the history of mining in the 2nd tectonic block in the Dukla Mine it
follows that the whole area of this block was relatively active; this fact was
documented by HOLUB (1995). In this block the longwall face no. 16212 was mined
in the coal seam no. XVI in its full thickness ($h = 4.5$ m) during 1980–1981, using
the caving method. According to the plan view in Figure 5 and the vertical cross
section in Figure 6, it is obvious that the longwall face is demarcated eastward by
the shaft safety pillar, while the westward limitation is created by a contact zone of
red beds. As mentioned above, these altered rocks display quite different physical
and mechanical properties than surrounding rocks and coal seams. That is why the

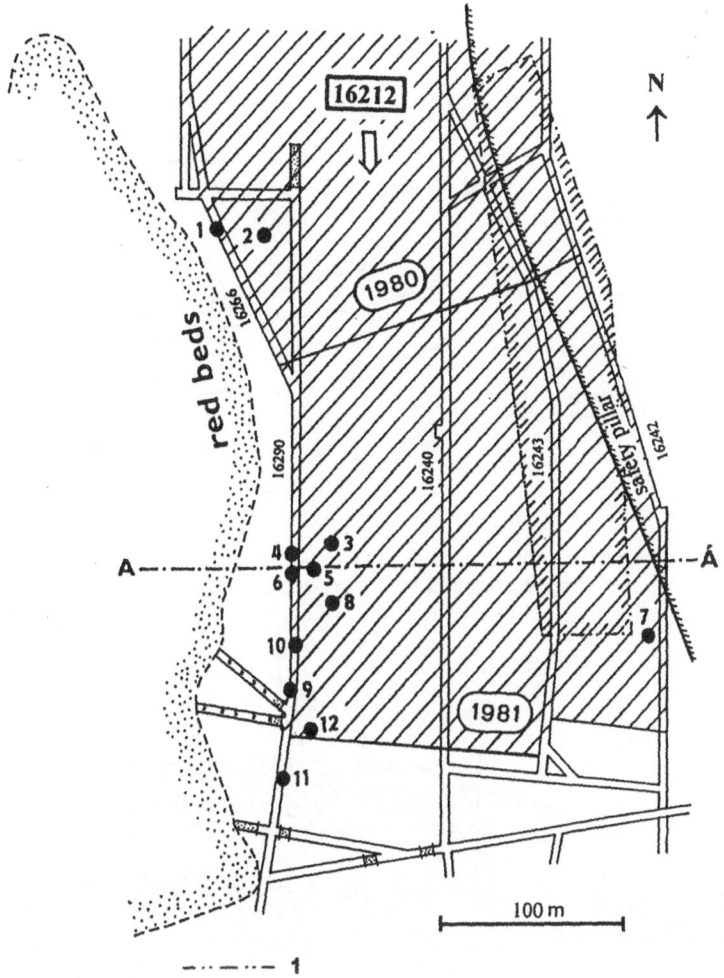

Figure 5

Plan view of rockburst positions defined according to the maximum damage in workings during longwall face no. 16212 operation in the Dukla Mine in 1980–1981. 1—edges of the mined out area in seam no. XII, A-Á line of the vertical cross section (see Fig. 6).

stability of the rock mass could be affected and, therefore, the adjacent area of their contact can be characterized as a rockburst prone zone.

These considerations became a reality once two strong rockbursts (nos. 1 and 2 in Figure 5) occurred near the tail gate no. 16266 in September 1980, both having released energy on the order of about 10^6 J. A series of rockbursts were observed during the time interval June–October 1981 (nos. 3–10), with energy being within the range of 1.5×10^3 to 2.5×10^4 J. The maximum damage was established near coal face no. 16212 and in adjacent tail gate no. 16290 following the contact zone, except for rockburst no. 7 which occurred in the area of the shaft safety pillar. The

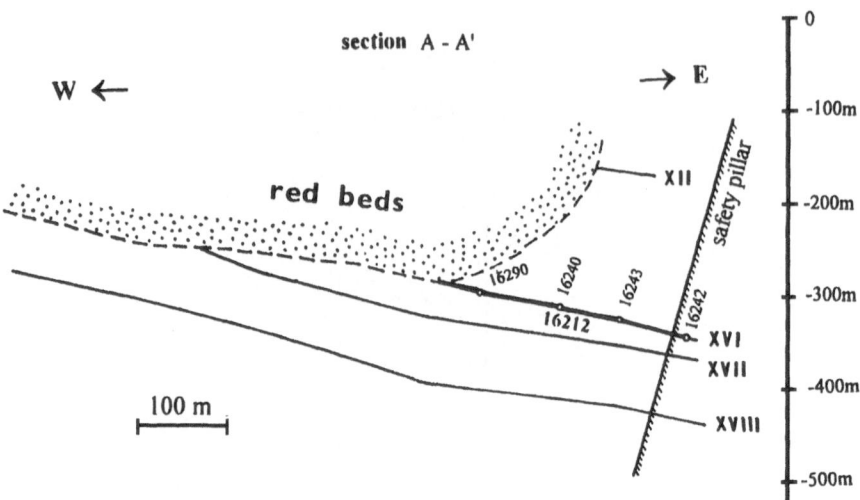

Figure 6
Vertical cross section in the region of longwall face no. 16212 along A-Á line. XII through XVIII
represents the Karviná coal seams numbering.

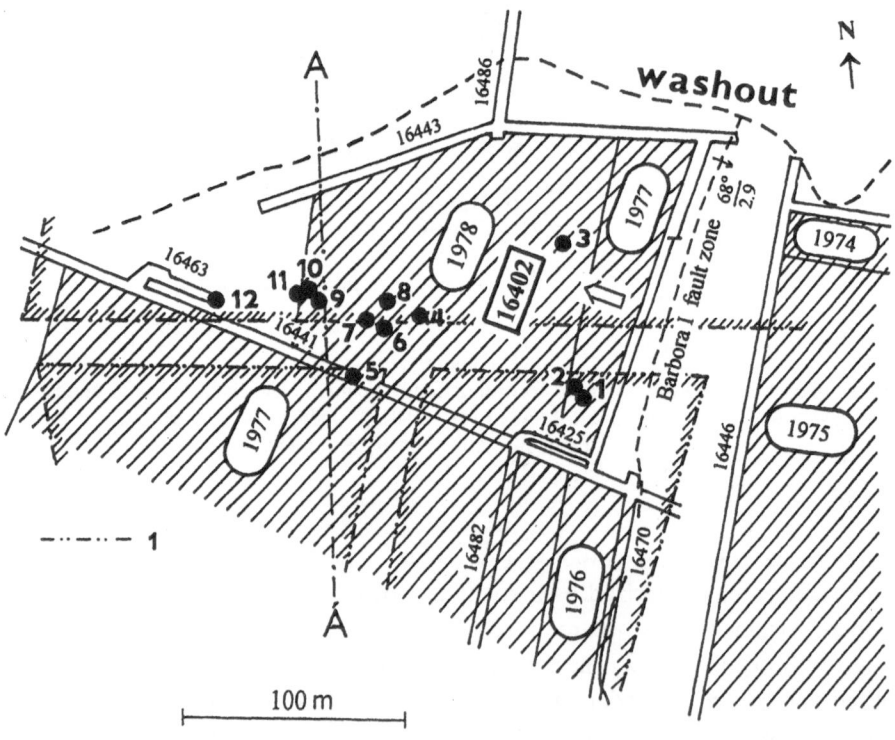

Figure 7
Plan view of rockburst positions defined according to the maximum damage in workings during longwall
face no. 16402 operation in the Dukla Mine in 1977–1978. 1—edges of the mined out area in seam no.
XII, A-Á line of the vertical cross section (see Fig. 8).

rockburst series in the area of longwall face no. 16212 given in Figure 7 terminated in November 1981 by two relatively strong rockbursts (nos. 11 and 12) releasing energy approximately in the range of 10^5–10^6 J.

Generally, all analyzed rockbursts were accompanied by an acoustic wave as well as by a shock wave which was often felt on the surface. It is noteworthy that beside a great influence of altered (red) beds, the influence of the edges of remnants in seam no. XII could be also taken into account.

The contact zone of red beds, like the washouts, represents a zone of weakness which affects the stability of the rock mass and its stress-strain behavior, usually resulting in the formation of focal zone.

4. Mining Conditions

4.1. Shaft and Crosscut Safety Pillars

The other group of factors influencing focal zone formation is represented by factors which appear as a consequence of man-made activities in the mines in the past and at the present time.

As a rule, for the shaft stability a safety pillar is created around it, the dimensions and the shape of which must be chosen very carefully due to expected damage to the construction. To this effect, in coal mines the shaft safety pillar is usually of conical shape. We sometimes find in the Ostrava-Karviná coal mines that a combination of conical and cylindrical shapes was accepted (RASZKA et al., 1990).

Crosscuts as well as other horizontal workings must also be effectively protected against undesirable stress state influences and, therefore, they are situated usually in safety pillars outside the stress-strain exposed regions within the mine field and/or inside the destressed zones created before their drivage. From the viewpoint of safety, all efforts leading to the creation and maintaining of safety pillars represent a necessary strategical measure.

In order to obtain an accurate space-time development of seismic activity in the 3rd tectonic block at the ČSA Mine for the time period May 1979 through February 1987, continuous seismological observations described by HOLUB et al. (1988) were used. During the course of the analysis of location plots, it was found that foci were clustering within several smaller regions as shown in Figure 3. It is evident that foci are either mostly concentrated along the nearest zone of faults or linked up with evident man-made boundaries. The energy scale of all recorded seismic events, denoted by open circles, is given on the right-hand side of the same figure.

In the group of seismic events originating near the man-made boundaries, two subsets of foci can be distinguished as seen in the respective foci plot. On the other hand, there exist foci clustering along the present shaft safety pillars Jan and Jindřich which, in principle, represent vertical compact bodies. The stability of these bodies is usually influenced by mining activities in their adjacent area, especially due to the irregular coal extraction of individual seams sometimes penetrating the proposed shape of the safety pillar. Thus, the contact of these vertical pillars with the surrounding rock mass containing unmined or mined out seams behaves, on the whole, as a spatial vertically formed discontinuity.

In contrast to that, one can see in Figure 3 foci locations (solid circles) which were depicted in the map according to the assessment of macroseismic observations *in situ* reported by ZAMARSKÁ (1981). Since the released seismic energy could not be objectively estimated, an equal size of solid circles was used. These foci follow again either the tectonic fault Hlubinská and Nepojmenovaná or the contours of the abandoned shaft safety pillars Hlubina, Františka and partly Hoheneger, the greater part of which is situated at the Darkov Mine. The contours of all three abandoned shaft safety pillars are depicted by thick dashed lines in Figure 3. Unlike the shape of safety pillars Jan and Jindřich, which are preserved as a continuous compact body extending to the present level of mining, i.e., approximately -500 up to -600 m m.s.l., the three old pillars were preserved to the level of about -300 up to -400 m m.s.l. being almost unmined. That is why their compact bodies are able to influence the stress-strain state in underlying layers of the coal deposit where coal seams are excavated throughout the area with no safety pillars.

Relative to the investigation of the influence of old mine workings on the present level of seismic activity and seismic event clustering, the effects of safety crosscut pillar in the 3rd tectonic block in the ČSA Mine using foci plot in Figure 3 was proved as well.

This finding was supported by the existence of foci concentration along the line passing near the eastern boundary of the abandoned shaft safety pillars Hlubina and Františka, where a crosscut was driven at the level of the coal seam no. 33 (approximately -325 m m.s.l.). Later, when mining operations were being performed at the level of the coal seam no. 37 (approximately -500 m m.s.l.), the influence of this pillar manifested itself as a zone of increased foci clustering.

Thus, the abandoned safety pillar become a source of additional loading on the workings operated beneath them and, therefore, remnant blocks in the rock mass create regions of stress concentration which can be anticipated as rockburst prone regions.

4.2. Remnant Pillars

Beside the safety pillars, the role of which is to protect workings, remnant pillars are sometimes formed within the course of mining. Unmined or unexploitable parts of the respective seams which have the shape of a remnant block, represent one of the most severe sources of additional stresses. The influence of these stresses

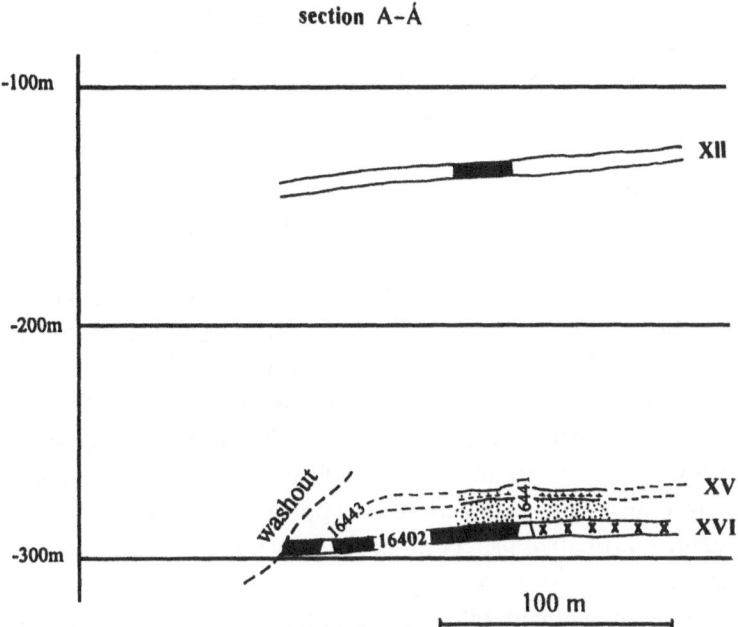

Figure 8
Vertical cross section in the region of longwall face no. 16402 along A-Á line. XII through XVI
represents the Karviná coal seams numbering.

manifests itself not only in the seam where the pillar was left, but also in seams
located in its neighborhood (above and/or below the mined seam), as was investigated (for example: BORISOW, 1980; KALENDA and SLAVÍK, 1992; KONEČNÝ,
1995). The extent of the zone of influence of remnant edges was also determined by
experimental seismic measurements carried out in Polish mines and reported by
DUBINSKI and WIERZCHOWSKA (1976). In the Ostrava-Karviná coal mines, the
existence of remnant pillars due to tectonic faults and the frequent reduction of the
thickness of seams and/or due to subcritical (diminished) blocks between neighboring longwall faces, cannot be omitted.

A special geomechanical problem is pillars left in those workings which are not
accessible at present. The position and impact of old workings must be considered
in any plan of further workings in an area which could be influenced by these
remnant pillars.

Figure 7 shows a complicated *in situ* situation which occurred within the course
of the mining of longwall panel no. 16402 in the Dukla Mine. It displays the
generation of a focal zone influenced by the existence of remnant pillar edges in seam
no. XII in the overburden, by a washout in the adjacent area and by a goaf of the
neighboring longwall panel. It can be seen from the section of the investigated
area which is given in Figure 8 that the interlayer distance between seams nos.
XII and XVI is approximately 150 m. Unfortunately, even after a reinterpreta-

tion of all the available data, we were unable to define the factors predominantly affecting the induced seismicity of this region. Moreover, the starting heading of the panel was situated in the proximity of the Barbora I fault zone and was parallel to it and, therefore, the influence of this fault on the seismic events origin could not be omitted.

The time of occurrence of induced events (nos. 1–12) and an approximate estimate of their seismic energy as shown in Figure 9, were obtained from observations made at a solitary seismic station situated roughly 1.1 km away. As evidence of the quantity of the energy released, most of the rockbursts were also recorded at the seismic stations Průhonice (PRU) and Kašperské Hory (KHC) distanced about 280 km and 370 km, respectively.

It was clear from an overall analysis of the situation that the drivage of roadway no. 16463 acted as a triggered mechanism for the occurrence of seismic event no. 12. This roadway was driven after longwall face no. 16402 was stopped; the drivage proceeded from an opposite direction to a previous focal zone. Since considerable potential energy probably has been always accumulated here, a sudden energy release occurred.

Another type of remnant pillar is the pillar left in the overburden around the former demarcation line between the Doubrava and ČSA Mines. This pillar

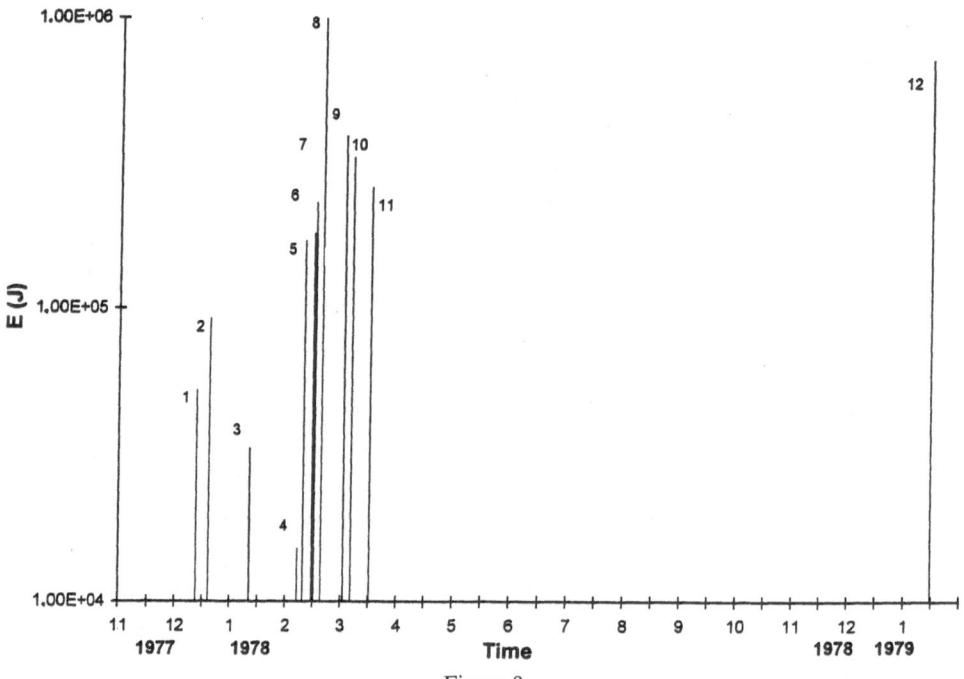

Figure 9
Time of occurrence of the rockbursts which originated in the area of longwall face no. 16212, numbering of events corresponds to Figure 7.

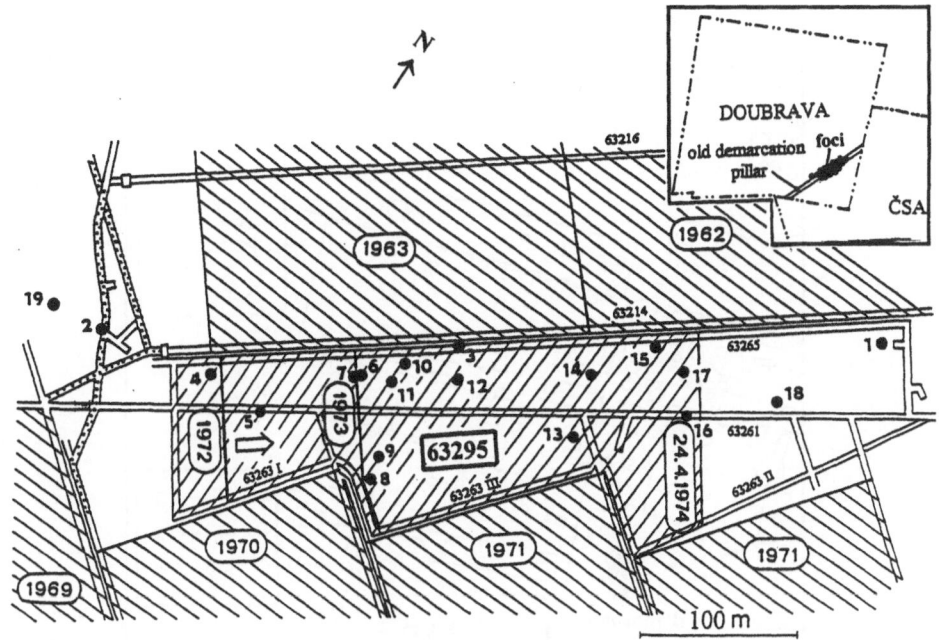

Figure 10

Plan view of rockburst positions defined according to the maximum damage in workings before and during the longwall face no. 63295 operation in the Doubrava Mine during 1971–1974. A schematic diagram of an "old demarcation" and rockburst foci is given in the upper right-hand corner.

induced seismic activity in the underlying workings in the Doubrava Mine. The diagram of the mining condition under investigation is given in Figure 10 where longwall panel no. 63295 in coal seam no. 32 N was mined during the period 1971–1974. Rockbursts nos. 1–3 occurred during the drivage of roadways, the remainder except rockburst no. 19, occurred during the longwall panel operation. All the above-mentioned rockbursts were assessed on the basis of macroseismic observations according to their consequences in the mine; several of them were felt on the surface. The strongest rockburst (no. 18) of an energy of about 10^9 J, which took place April 24, 1974, was also recorded at several European seismic stations, ranging in distance up to 2,000 km. A rockburst (no. 19) occurred beneath longwall panel no. 63295 two years later, on December 15, 1976, in coal seam no. 34, situated approximately 50 m below coal seam no. 32 N.

According to a detailed analysis of all the available geomechanical data, it was proved that the focal zone in coal seams nos. 32 N and 34 was induced by the influence of the primary demarcation pillar. This pillar was preserved up to coal seam no. 28, i.e. extending to a depth of about −290 m m.s.l. and, therefore, its pronounced influence on the stress-strain state in the underlying strata (coal seams) led to the creation of a focal zone in the area of longwall panel no. 63295 and in seam no. 34 beneath it. The interlayer distances between coal seams nos. 28–32 N

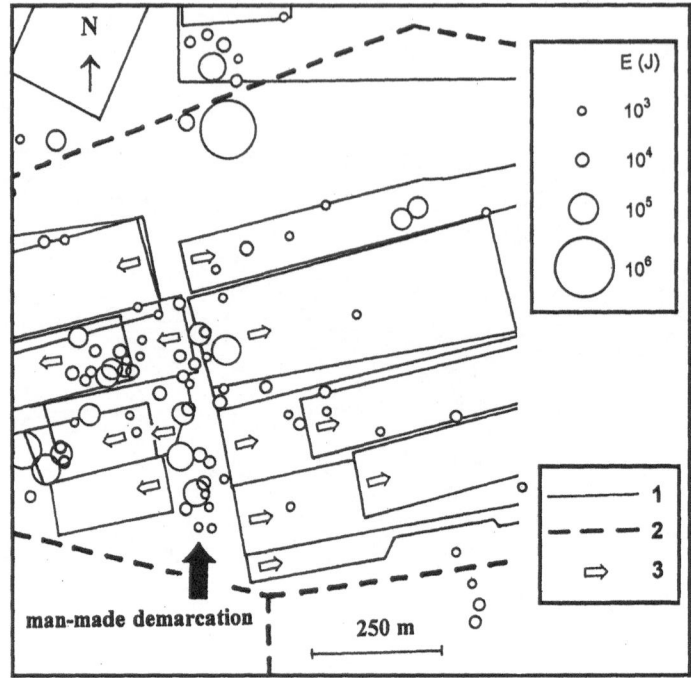

Figure 11

Location plot of seismic events recorded within 1992–1995 related to the vicinity of man-made demarcation in the 3rd tectonic block of the ČSA Mine created by mining of individual longwall faces. 1—contours of longwall faces, 2—demarcation of the 3rd tectonic block and 3—coal face advance.

and 32 N–34 were approximately 160 m and 35 m, respectively. As seen in the right-hand upper corner of Figure 10, all observed foci follow the line of primary demarcation, i.e., they are located within a narrow strip, the length of which was about 550 m and the width approximately 50 m.

On the basis of an investigation into the foci distribution in the 3rd tectonic (mining) block in the ČSA Mine, evidence of an unfavorable influence of a remnant pillar was obtained. As seen in Figure 11, one of the foci clusters follows a narrow strip which was artificially created by an accumulation of edges of non-mined-out areas in coal seams within the overburden. In principle, this artificial demarcation divides the 3rd tectonic block into two parts and, simultaneously, it creates a basis, from which individual longwall faces are developed westwards as well as eastwards. The approximate course of this artificial boundary between both parts of the tectonic block is also depicted in Figure 3 by a dotted line.

The examples of the influence of remnant pillars discussed above are not exhaustive because such pillars very often occur in our mines under various *in situ* conditions. Since all mining operations proceed to deeper levels, it can be generally stated that undermining the area of the existing remnant pillars is accompanied, due to the failure of the existing residual stress field, by an increase of the frequency of seismic events.

Conclusions

All geomechanical and mining conditions discussed here are documented prevailingly by foci plots conducted either on the basis of an assessment of earlier macroseismic observations *in situ* or of reliable instrumental data which have been assembled since the 1980s.

Studies of the above case histories were performed to comprehend the influence of geological and mining parameters on the induced seismicity associated with longwall mining in specified Czech coal mines. Our findings have proven that there exist in the rock mass predisposed zones of anomaly and/or weakness which manifest themselves by a pronounced concentration of foci of seismic events with various energy sizes. These zones were found in areas adjacent to tectonic faults, washouts, red beds and in the vicinity of areas in which the influence of various types of safety and remnant pillars is expected. However, it has been shown that in all cases discussed, the occurrence of most mining-induced events is very closely linked in time and space with regions of active mining. In principle, induced seismicity is surmised to be a result of the interaction of natural conditions, geomechanical conditions in mining, residual stress field left by previous workings and advancing coal faces, as also was reported by SATO and FUJII (1988). These current mining activities, acting as a triggering mechanism, seem to be an important factor in generating of rockbursts and rockburst-like phenomena.

Acknowledgments

I wish to thank the Institute of Geonics AS CR which funded this work as part of the internal research programme on rock mass failure. Critical and valuable suggestions given by anonymous reviewers are also highly appreciated.

REFERENCES

BORISOW, A. A., *Mechanics of Rocks and Solid Rock Masses* (Nedra, Moscow 1980) (in Russian).

DUBINSKI, J., and WIERZCHOWSKA, Z. (1976), *The Measurements of Influence Zone Range of Exploitation Edges and Heels of Coal in Upper Silesian Coal Basin*, Acta Montana *38*, 151–161.

GAY, N. C., *Mining in the vicinity of geological structures—an analysis of mining-induced seismicity and associated rockburst in two South African mines.* In *Rockbursts and Seismicity in Mines* (ed. Young, R. P.) (Balkema, A. A., Rotterdam/Brookfield 1993) pp. 57–62.

GILL, D. E., AUBERTIN, M., and SIMON, R., *A practical engineering approach to the evaluation of rockburst potential.* In *Rockbursts and Seismicity in Mines* (ed. Young, R. P.) (Balkema, A. A., Rotterdam/Brookfield 1993) pp. 63–73.

GRYGAR, R. (1987), *Regional Shear Deformation and Geotectonical Evolution of the Upper Silesian Basin*, Report VŠB-Technical University, Ostrava (in Czech, unpublished).

HOLUB, K., *Space and time patterns of induced seismicity*. In *Mechanics of Jointed and Faulted Rock* (ed. Rossmanith, H. P.) (Balkema, A. A., Rotterdam/Brookfield 1995) pp. 657–662.

HOLUB, K., *Origin and dynamics of some of induced seismicity focal regions*. In *Tectonophysics of Mining Areas* (ed. Idziak, A.) (Publishing House of the Silesian University, Katowice 1996) pp. 110–121.

HOLUB, K., KNOTEK, S., and VAJTER, Z. (1988), *Microseismology and Its Application in the Prevention of Rockbursts in the Ostrava-Karviná Coal Field, Czechoslovakia*, Publs. Inst. Geophys. Pol. Acad. Sc., M-10 (213), 203–215.

HOLUB, K., SLAVÍK, J., and KALENDA, P. (1995), *Monitoring and Analysis of Seismicity in the Ostrava-Karviná Coal Mining District*, Acta Geophys. Pol. *XLIII* (1), 11–31.

KALENDA, P., and SLAVÍK, J. (1992), *Analysis of the Seismoacoustic Emission during the Coal Extraction*, Acta Montana *A3* (89), 41–52.

KANEKO, K., SUGAWARA, K., and OBARA, Y., *Rock stress and microseismicity in a coal burst district*. In *Rockbursts and Seismicity in Mines* (ed. Fairhurst, Ch.) (Balkema, A. A., Rotterdam/Brookfield 1990) pp. 183–188.

KNEJZLÍK, J., GRUNTORÁD, B., and ZAMAZAL, R. (1992), *Experimental Local Seismic Network in the A. Zápotocký Mine of the Ostrava-Karviná Coal Field*, Acta Montana *84*, 97–104.

KNOLL, P., and KUHNT, W., *Seismological and technical investigations of the mechanics or rockbursts*. In *Rockbursts and Seismicity in Mines* (ed. Fairhurst, Ch.) (Balkema, A. A., Rotterdam/Brookfield 1990) pp. 129–138.

KONEČNÝ, P. (1989), *Mining-induced Seismicity (Rockbursts) in the Ostrava-Karviná Coal Basin, Czechoslovakia*, Gerlands Beitr. Geophys., Leipzig *89* (6), 523–547.

KONEČNÝ, P., *Influence of abandoned coal pillars on seismicity induced by their undermining*. In *Proceedings and Activity Report ESC 1992–1994 Vol. III* (eds. Makropoulos, K., and Suhadolc, P.) (Athens 1995) pp. 1356–1363.

MARTINEC, P., *Geotechnical classification of carboniferous rock mass with alternation of the "Variegated Beds" type*. In *Geomechanics 93* (ed. Rakowski, Z.) (Balkema, A. A., Rotterdam/Brookfield 1994) pp. 105–109.

PTÁČEK, J., and TRÁVNÍČEK, L., *Some views on the influence of tectonics on the occurrence of rockbursts*. In *Geomechanics 93* (ed. Rakowski, Z.) (Balkema, A. A., Rotterdam/Brookfield 1994) pp. 123–126.

RASZKA, T., KRZAN, S., JANAS, P., and ŽENČ, M. (1990), *The Experience of Operating in the Conditions of a Diminished Shaft Safety Pillar*, Uhlí *38* (2), 57–63 (in Czech).

SATO, K., and FUJII, Y. (1988), *Induced Seismicity Associated with Longwall Coal Mining*, Int. J. Rock Mech. Min. Sci. and Geomech. Abstr. *25* (5), 253–262.

TAKLA, G., and PTÁČEK, J. (1990), *Experience of Rockbursts Occurred in Ostrava-Karviná Coal District and its Application in Building the Rockburst Control System*, Uhlí *38* (8–9), 343–349 (in Czech).

ZAMARSKÁ, T. (1981), *Catalogue of Rockbursts Observed in the Ostrava-Karviná Coal Basin*, Final Report, Res. Min. Inst., Ostrava-Radvanice (in Czech, unpublished).

(Received June 18, 1996, accepted March 10, 1997)

Pure appl. geophys. 150 (1997) 451–459
0033–4553/97/040451–09 $ 1.50 + 0.20/0

Pure and Applied Geophysics

Spatial Distribution of Mining Tremors and the Relationship to Rockburst Hazard

G. Senfaute,[1] C. Chambon,[2] P. Bigarré,[1] Y. Guise,[3] and J. P. Josien[1]

Abstract—Within the Provence colliery (France), seismic remote monitoring is integrated in a research strategy on areas subject to rockburst. The purpose of this study is to characterise and analyse the induced microseismic events during the mining operations, in order to identify or establish criteria for the zones likely to generate rockburst. The detailed study of the space distribution of microseismic events enables consistent correlations to be demonstrated between the location of events, the changes in the state of stress in the massif during the mining operations and the configuration of exploitation. The results have opened interesting perspectives into the analysis of the spatial distribution of mining tremors and their relations to rockburst hazard.

Key words: Mining tremors, rockburst, seismic network, mining operations.

1. Introduction

The Provence lignite deposit is located in the south of France, close to the city of Marseilles. It is a sub-horizontal deposit, corresponding to a basin defined by the Maastrichtian phase (Gaviglio *et al.*, 1985). The exploited coal layer offers a thickness of 2 to 3 metres depending on the sectors. The foot wall and hanging wall, over a thickness of 250 metres, are limestone. The underlaying series consist of interbedded clays, marls and limestones. Mining is carried out according to the long-wall caving method, currently at a depth of approximately 1200 metres.

The Provence colliery has been affected by rockbursts for about ten years. These phenomena show great diversity regarding the natural and mining conditions present in the sectors concerned. Nevertheless, there is one common feature: these phenomena are associated with generally considerable seismic activity (Ben Sliman *et al.*, 1990). The recording and interpretation of this activity represent an important field of study for the understanding and prediction of such events. The purpose of this study is to characterise and analyse the induced seismic events during mining operations, in order to identify the relations between mining tremors and rockburst hazard.

[1] INERIS—Laboratoire de Mécanique de Terrains (Ecole des Mines), 54042 Nancy, France.
[2] Laboratoire de Mécanique de Terrains (Ecole des Mines), 54042 Nancy, France.
[3] Houillères de Bassin du Centre et du Midi, B.P. 534, 42007 St-Etienne, France.

2. Seismic Instrumentation

2.1 Data Acquisition System

The seismic network is comprised of eight vertical stations of the uniaxial geophone type as well as one triaxial station (natural frequency = 1 Hz). Four vertical stations and one triaxial station are on the surface. Data transmission from the surface network to the central site mainframe is by radio, while four underground stations, the depth of which is 1200 metres, are connected to the central site by cables (Fig. 1).

The central acquisition system is composed of two PC compatible microcomputers connected in parallel. The PCs are used for alternate data acquisition according to a master/slave logic, at a sampling frequency of 155 Hz. The triggering event is submitted to a combination of criteria to be validated. The master PC then acquires signals for the set-up time period and then switches to slave mode, whereas the slave PC switches to master mode (BIGARRÉ *et al.*, 1995a) (Fig. 2).

2.2 Data Processing

The automated processing run by a multitask machine in the network enables an automated bulletin to be edited for each event. The automated calculation of the wave arrival time is firstly based upon the neuromimetric network technology which determines a probable wave beginning presence range (BIGARRÉ *et al.*, 1995b) and upon the STA/LTA method (ALLEN, 1978) which involves a precise pointing of the wave arrival time. An interactive treatment can be chained a few moments after the automated processing of the event, in order to validate the calculated parameters. Finally, a database type file system is updated permanently in order to allow a further statistic study of the fundamental parameters (date, time, energy, magnitude, localisation) which characterise the microseismic activity generated by the various operated sites.

2.3 Event Location Accuracy

The location of tremors is an essential parameter in the analysis of the microseismic activity. The accuracy of the hypocentre calculation depends on various criteria: the geometric configuration of the network, the knowledge of the wave velocity and paths, the sampling frequency of the acquisition system as well as the pointing operation reliability. Experience demonstrates that the knowledge of a velocity model as close as possible to the real situation is essential in order to obtain proper accuracy of the coordinates (X, Y, Z, T) for each event. SENFAUTE *et al.* (1994) and SENFAUTE (1995) developed a suitable wave velocity model for locating events in the Provence area. It is a slanted multilayer model, which allows locations to be obtained with an accuracy of approximately ± 50 metres.

3. Space Distribution Study of Microseismic Events

Although all rockbursts generate seismic events, all seismic events are not due to rockbursts. The intrinsic causes, if any, for such particular seismic events, are not well known and their quick prediction raises intricate problems. Therefore, an abrupt energy release may occur at various points with respect to the working face, for example:
- in the operation layer, ahead of the working face,
- behind the working face,
- in the pillars located next to the mining,
- in two or more of the above places, simultaneously or almost simultaneously.

Therefore, the study of the space distribution of the seismic events is particularly significant in the analysis of the microseismic criteria for evaluating a hazard. A detailed study on event distribution for working face 07 of the Estaque district at the Provence colliery has been carried out with this purpose.

Working face 07 at the Estaque district is located to the south of the deposit (see Fig. 1). It is north-east oriented, with an dip of approximately 10° westward. The study period is between October 1992 and July 1993; it concerns 720 metres of operated working face, for which a total of 2223 microseismic events were recorded. The decision to study the space distribution of events associated with this working face was made according to the following criteria:
- this working face was operated following two different configurations, which allows the behaviour of the microseismic activity to be studied in two separate situations (Fig. 3);
- the operation of this working face was affected by 5 rockbursts;

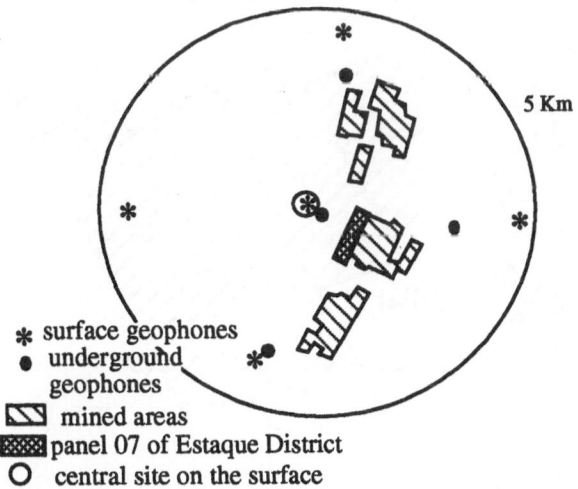

Figure 1. Geometry of the local Provence colliery seismic network.

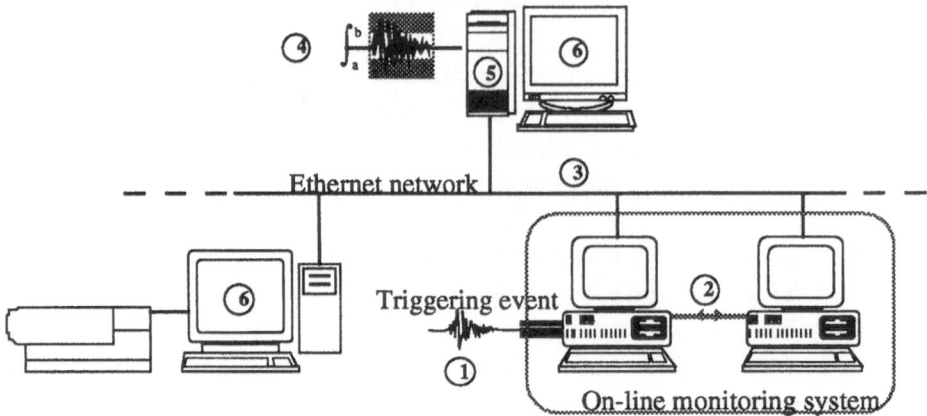

Figure 2. Seismic data acquisition and processing system. 1) Event detection; 2) data acquisition on the master PC, switching of the PC to slave mode and pre-processing of the signal; 3) data transmission to multitask machine; 4) complete signal processing; 5) data storage and database type file system updating; 6) interactive processing of the events in order to validate the calculated parameters.

– its central position with respect to the geometry of the seismic network allowed for good accuracy in the location of the events.

3.1 Location of Microseismic Events with Respect to the Working Face

For each event, the distance between the working face and the event location was determined, that is to say, to the rear or ahead of the working face. The results show that microseismic events strongly depend upon the excavation geometry:

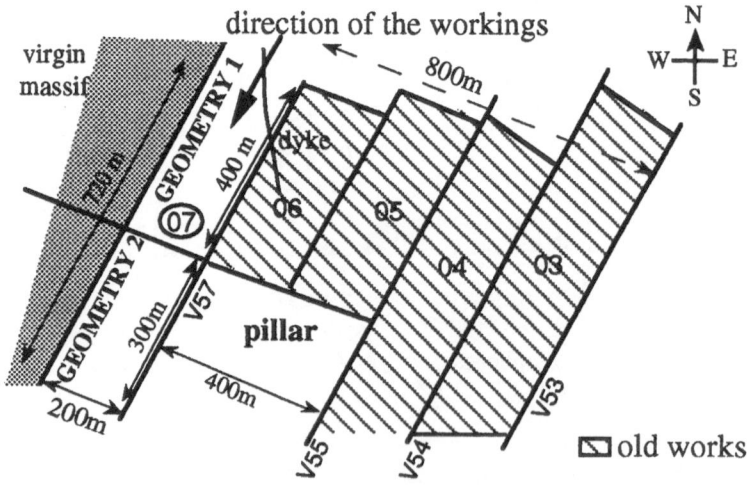

Figure 3. Schematic view of the operation configurations (geometry 1 and geometry 2) for working face 07 in the North Estaque district.

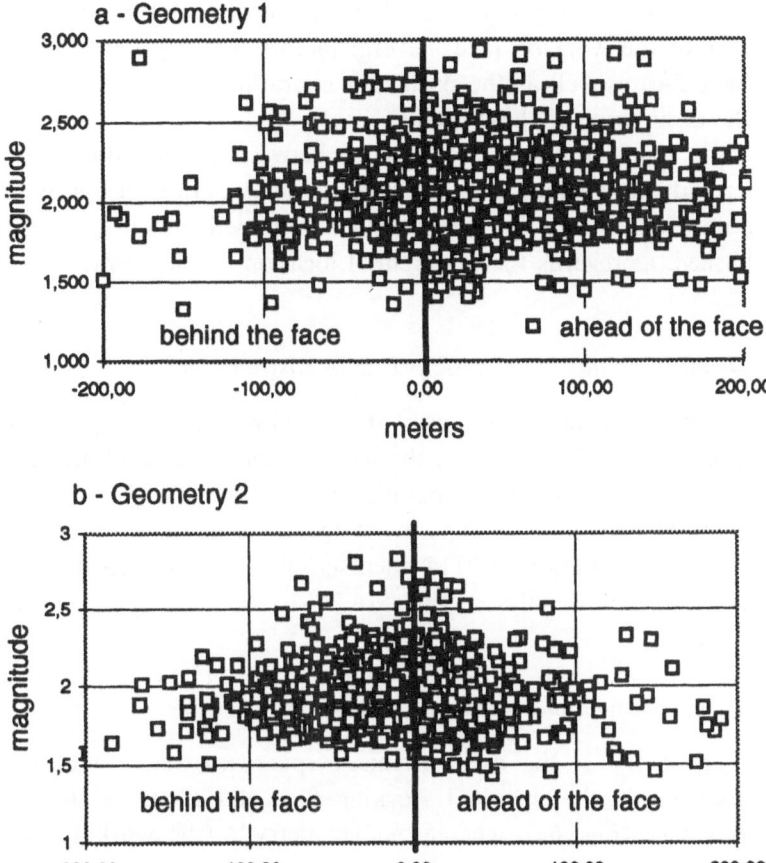

Figure 4. Location of seismic events with respect to the working face 07 in the North Estaque district.

— in geometry 1 (the working face runs along the old mined areas), the events are localised through the complete range of magnitudes ahead of the front and behind the face, but they are more numerous ahead of the face, between 0 and 50 metres from the working face (Fig. 4a);
— in geometry 2 (the working face runs along a pillar) the seismic events are also localised to the rear and ahead of the working face, but contrary to geometry 1, they are more numerous to the rear, between 0 and 80 metres from the working face (Fig. 4b).
These results provide a first classification of the seismic events:
— Class A events: a population of events located ahead of the working face, which follows the progress of the work. This corresponds to an area of high dynamic stresses. The load application and the fracturing ahead of the working face would be the generating cause of these events;

— Class B events: a population of events located between 0 and 80 metres to the rear of the working face. Several factors, among which the caving-in of the back face as well as the particular operation configuration would account for the occurrence of these events.

These results outline a consistency between the location of the microseismic events and the changes that occur in the sites during the progress of mining. A quantitative study of the event distribution will enable concrete criteria to be established, as regards the preferential location of the events.

3.2 Quantifying the Microseismic Events Distribution

In order to quantify the event distribution, the working face is broken down into areas. In this break-down, the working face is divided into 16 different areas. Each area contains the corresponding events, classified by magnitude classes and geometry of exploitations. The event locations in these 16 areas will enable the preferential distribution of 2114 microseismic events recorded during working face 07 operation to be analysed (Fig. 5).

Events with Magnitudes < 2.0

In geometry 1, the microseismic events with magnitudes less than 2.0 are localised within the operated working face and about 25 metres ahead of the working face (Fig. 6a), whereas in geometry 2 (the working face runs along a

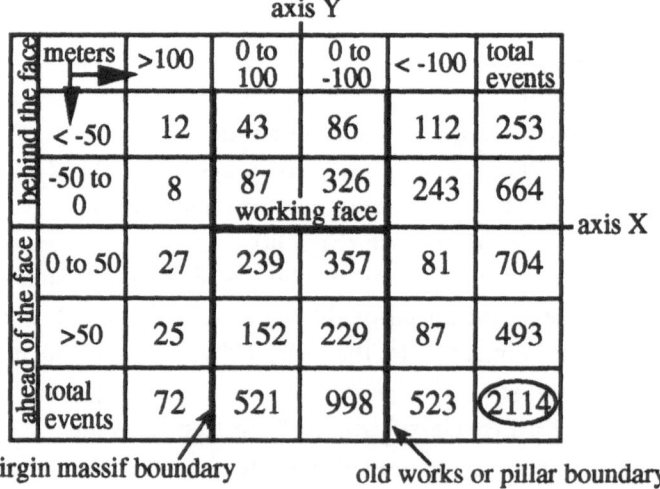

Figure 5. Graphic illustration of working face 07 break-down into areas, in order to quantify the seismic focuses distribution. The X axis, at the centre of the square, corresponds to the working face, whereas the Y axis, at the centre, corresponds to the central axis of the working face.

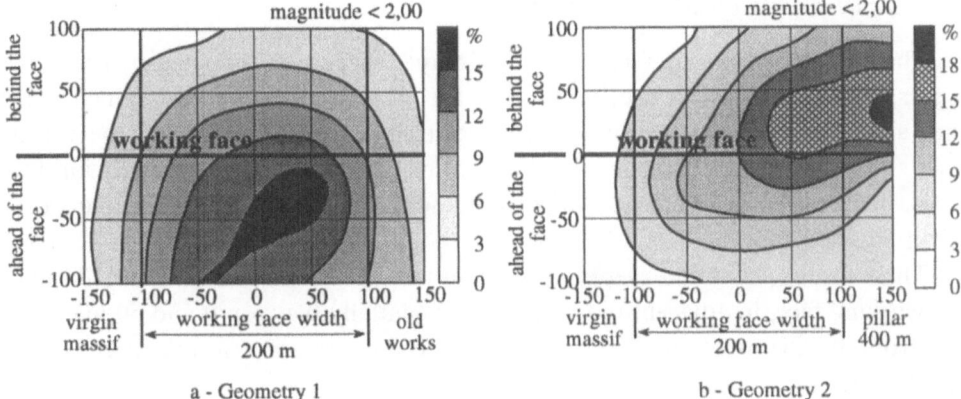

Figure 6. Preferential location of small seismic events (magnitude <2.0) induced by the mining of face 07 at the North Estaque district, according to two different operation configurations.

pillar), the events are localised about 25 metres to the rear of the working face and to the outside of the working face, that is, in the pillar area (Fig. 6b).

The preferential distribution of this category of events highlights the bearing areas due to the voids created behind the working face. In geometry 2, the pillar adjacent to the operation constitutes a significant bearing area due to the void in the back face. The pillars which bound the old operations are overloaded areas (REVALOR et al., 1985). The working faces reaching this type of area create an overload already overloaded, whcih results in microfractures accounting for the occurrence of the small seismic events. In an operation adjacent to old works (geometry 1), the small events are preferably localised ahead of the working face. This characterises the main bearing area due to the voids created at the back of the face.

Events with Magnitudes ≥2.5

In geometry 1, the microseismic events are localised within the working face and preferably ahead of the face, between 25 and 100 metres of the front (Fig. 7a), whereas in geometry 2, the events are also grouped within the panel and preferably at the front of the working face, between 20 and 50 metres from the face (Fig. 7b).

These tremors with magnitudes ≥2.5 are representative of the dynamic loading and fracturing at the front of the working face. Part of this category of events is also localised in the old works adjacent to the face. They are associated with the opening or slipping of former fractures revived during the mining of the new panel.

SYREK and KIJKO (1988) obtained similar results in the Polish coal mines. These authors demonstrated that the most energetic events are preferably localised ahead of the working face.

3.3 Seismic Events Location and Rockbursts

The mining of working face 07 at the Estaque district was affected by rockburst events. These phenomena are localised in a configuration where the mined panel runs along the old works of an adjacent face (geometry 1, Fig. 7a). All the seismic events associated with damage (rockbursts) are localised next to the working face front, between 0 and 50 metres to the front and in the closest part of the old works of an adjacent face.

These results show that events with magnitude ≥ 2.5 localised next and ahead of the working face, in adjacent old works and essentially between 0 and 50 metres to the front of the working face, are seismic events creating a potential risk of rockbursts.

4. Conclusion

The study of the space distribution of 2114 microseismic events highlights consistent correlations between the location of the focuses and the load application areas during mining. The results demonstrate that the mining configuration appreciably influences the preferential location of microseismic events.

In terms of the evaluation of a hazard based upon the location of the focuses, these results seem to indicate that when the focuses of events with magnitudes ≥ 2.5 are very close to the working face front (essentially ahead of the face and on the side of old works in an adjacent face, geometry 1) a potential risk of rockburst exists.

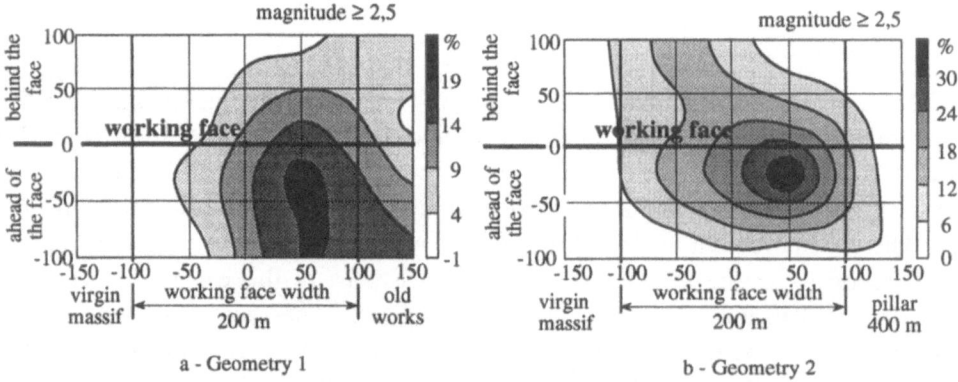

Figure 7. Preferential location of seismic events with magnitudes ≥ 2.5 induced by the mining of face 07 at the North Estaque district, according to two different operation configurations.

REFERENCES

ALLEN V. (1978), *Automatic Earthquake Recognition and Timing from Single Traces*, Bull. Seismol. Soc. Am. *68*(5), 1521–1532.

BEN SLIMAN, K., BESSON, J. C., MANDEREAU, G., and CHAMBON, C. (1990), *Seismic Monitoring: A Tool for Planning Mining Sites Submitted to Dynamic Phenomena*. 3rd Franco-Polish Symposium, Wroclaw. Vol: XI, Nos. 1 and 2, pp. 55–68.

BIGARRÉ, P., LIZEUR, A., BENNANI, M., and FELT, T. (1995a), *SYTMIS: Software for Real-time Microseismic Monitoring Systems*. FMGM95 4th International Symposium Bergamo-Italy.

BIGARRÉ, P., LABLÉE, S., DODO AMADOU, A., PIGUET, J. P., and JOSIEN, J. P. (1995b), *The Use of Neuromimetric Networks in the Automatic Location of Microseismic Events*, International Congress on Rock Mechanics, Tokyo.

GAVIGLIO, P., REVALOR, R., PIGUET, J. P., and DEJEAN, M., *Tectonic structure, strata properties and rockburst occurrence in a French coal mine*, Proc. 2nd Int. Symp. *Rockbursts and Seismicity in Mines* (C. Fairhurst, Balkema, Minneapolis 1985).

REVALOR, R., ARCAMONE J., and JOSIEN, J. P. (1985), *In situ Rock Stress Measurements in French Coal Mines: Relations between Virgin Stresses and Rockbursts*, 26th U.S. Symposium on Rock Mechanics. Vol. 2.

SENFAUTE, G., BIGARRÉ, P., JOSIEN, J. P., and MATHIEU, E., *Real-time microseismic monitoring; Automatic wave processing and multilayered velocity model for accurate event location*. In *Rock Mechanics in Petroleum Engineering*, EUROCK '94 (Balkema, Delf/Netherlands 1994) pp. 631–638.

SENFAUTE, G. (1995), *Microseismic Monitoring of Underground Coal Mining at the Provence Coal Mines*, Thesis (Ph.D.) defended before the French National General Engineering Institute of Lorraine, 321 pages.

SYREK, B., and KIJARO, C. (1988), *Energy and Frequency Distributions of Mining Tremors and their Relations to Rockburst Hazard in the Wujek Coal Mine, Poland*, Acta Geophy. Pol. *36*.

(Received June 18, 1996, accepted May 15, 1997)

Pure appl. geophys. 150 (1997) 461–472
0033–4553/97/040461–12 $ 1.50 + 0.20/0

Pure and Applied Geophysics

Induced Seismicity in Liaoning Province, China

Yi Zhang Zhong[1], Changbo Gao[1] and Bai Yun[1]

Abstract—We describe three types of induced seismicity observed in Liaoning Province, China: reservoir-induced seismicity, mine-induced seismicity and collapse earthquakes. A shock with magnitude $M = 5.2$ took place on December 22, 1974 at Shenwo Reservoir and some smaller earthquakes caused by impoundment also took place near other reservoirs. Numerous earthquakes associated with mining activity occurred in some coal mines. 56 collapse earthquakes with magnitude of $M > 1.8$ occurred at Binggou coal mine in Jianchang county. An analysis of the cause and some features of these three categories of the induced earthquakes are described in this paper.

Key words: Reservoir-induced seismicity, mine-induced seismicity, collapse earthquakes, Liaoning Province.

1. Introduction

Strong seismic activity has brought about tremendous casualties and economic losses in heavily populated and industrial areas in northeastern China, such as Liaoning Province. Historically, there have been tens of destructive earthquakes with magnitude $M \geq 4.7$ in this Province. In the last three decades, besides tectonic earthquakes, induced earthquakes have become more frequent. This paper presents a comprehensive analysis of the induced earthquakes and their genesis.

2. The Observed Induced Earthquakes

Liaoning Province in northeastern China is an earthquake-prone part of the country frequently hit by induced earthquakes (Fig. 1). Three types of induced earthquakes have been observed. These are associated with reservoirs, mines and collapse in excavated areas of mines (Zhong *et al.*, 1991).

---

[1] Institute of Seismology, Liaoning Province, Shenyang 110031, China.

2.1 Reservoir-induced Earthquakes

According to the catalogues of historical earthquakes, the area in which the Shenwo Reservoir is located is considered to be an aseismic region where no significant seismic activity has been observed before reservoir impoundment (Fig. 1). In March 1973 seismicity started to be felt in this area due to reservoir impoundment. The largest earthquake at Shenwo Reservoir occurred on December 22, 1974 and registered a magnitude of $M = 5.2$. This event resulted in damage to buildings in the meizoseismal area including the cracking of walls, twisting of chimneys, slipping of house tiles and a slight cracking of dam.

The characteristics of the seismic activity in the Shenwo Reservoir area are as follows:

—The foreshock-main shock-aftershock sequence of seismic activity is similar to that of a typical reservoir-induced seismic sequence. The duration of the aftershock activity is long, enduring for a ten-year period. The attenuation factor (P) of Omori's law of aftershock of tectonic earthquakes is greater than 1.3 (LI, 1995; ZHAO *et al.*, 1988), P value of induced earthquake is less than 1.0 and for the Shenwo earthquake $P = 0.98$.

—Most epicenters were concentrated in a small area, within about 12 km from the dam. This fact strongly suggests that the seismic activity is closely related to the impoundment of the reservoir (Fig. 2).

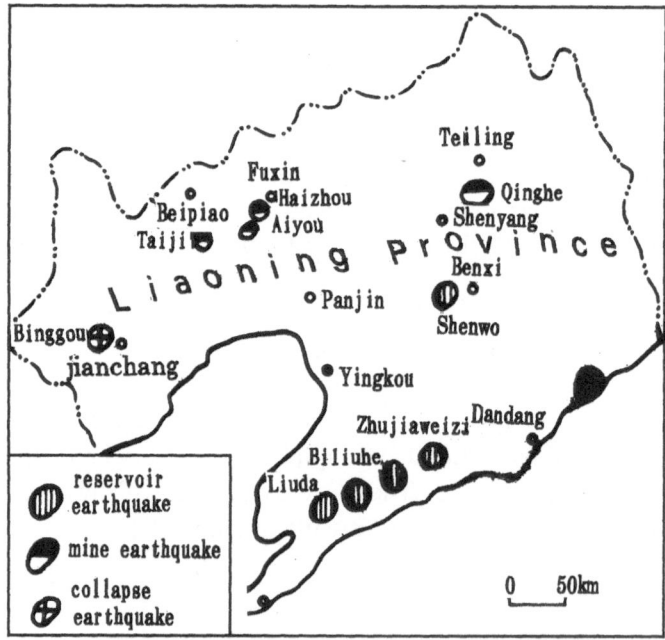

Figure 1
Distribution of induced earthquakes in the Liaoning Province.

Figure 2
Epicenters of earthquakes at the Shenwo Reservoir and the major faults in the area.

—Some parameters of seismicity are close to that of induced earthquakes. The magnitude difference between the main shock of Shenwo earthquake and the largest aftershock $(M - M_1)$ is 0.8. An equivalent magnitude difference was observed for the seismic sequence at Xinfenjian reservoir in Guang Dong Province. The ratio of the magnitude of the aftershock to that of the main shock (M_1/M) is 0.84 similar to that observed for the Koyna reservoir earthquakes in India (GUPTA and RASTOGI, 1976).

—The tectonic earthquakes near the reservoir area range in depth from 10–15 km, but all the reservoir-induced seismic events near the Shenwo Reservoir have shallow depths, usually in the range of 4–8 km. However, the shallowest events were at a depth of about 1 km and the depth of Haichen earthquake, which was located 60 km from the Shenwo Reservoir was 12 km.

—From the correlation of seismicity with water levels (Fig. 3) it can be seen that in March 1973, three months after impoundment started, a seismic event with a magnitude 2.5 occurred in the reservoir area. After the reservoir was fully impounded, i.e., up to its maximum of water level (elevation of 97.4 m) on 30 November 1974, the main shock took place on 22, December 1974. The fact that the origin time of the main shock coincided with the time when the water level attained its maximum elevation, strongly suggests that the seismic activity is closely related to the change of water level in the reservoir area.

In summary, the available data suggest that the earthquakes observed at Shenwo Reservoir are induced by the impoundment of the reservoir (ZHONG *et al.*, 1981; LI *et al.*, 1989). There are over twenty large reservoirs in the Liaoning Province. Some seismicity has been observed in other reservoirs and their neighboring areas such as Qinghe and Zhujiaweizi reservoirs (Fig. 1) with the largest event reaching a magnitude $M = 2.9$. Because of insufficient data we have not yet determined whether these earthquakes are also induced by reservoir impoundment.

2.2 Earthquakes Associated with Coal Mines

The seismic activity caused by the excavation of mines first appeared at Taiji coal mine (Fig. 1), Beipiao County, Liaoning Province (ZHONG and CUI, 1984; CHEN *et al.*, 1987). In 1970 when the excavation of the shaft reached a depth of 500 m, a series of small earthquakes occurred. Until 1989 a total of over 1300 earthquakes were recorded by local seismological stations, among them there were 140 events with a magnitude $M \geq 1.8$. In general, a shock with a magnitude of about $M = 2.5$ can cause damage in mining areas. Table 1 summarizes the damage caused by this type of seismicity. An earthquake with a magnitude of $M = 4.3$ took place on April 28, 1977 causing damage in an area of 2 km². Over 100 houses were seriously damaged. The walls of rock tunnel showed rock splitting and the steel tracks used for transportation in the tunnels were twisted.

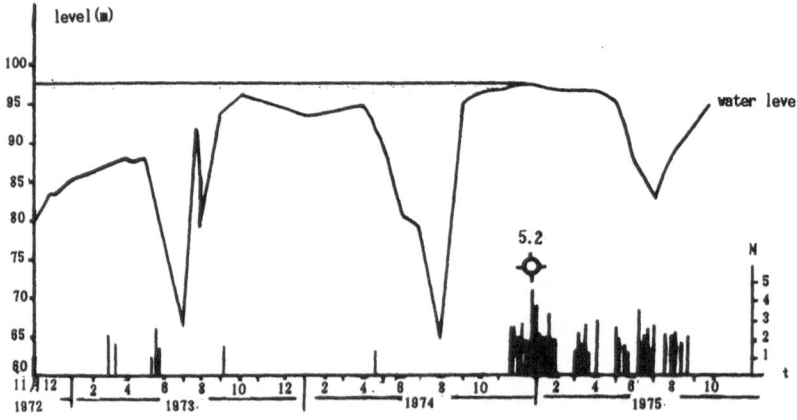

Figure 3
Correlation of earthquakes with water level at the Shenwo Reservoir. The horizontal line shows the maximum of water level (elevation of 97.4 m).

Table 1

Damage caused by earthquakes associated with coal mines

No.	Date Y.M.D.	Magnitude (M)	Damage	Observed fault activity
1	1977.04.28	4.3	Houses and tunnels seriously damaged. Intensity is VII	Movement along No. 10 fault
2	1981.04.07	2.9	Rock mass fall	Dislocation along No. 13 fault
3	1981.08.22	3.3	Old houses crack	Offset along No. 10 fault
4	1981.12.13	3.2	Tunnels partly collapse	Movement along No. 12 fault
5	1982.03.13	3.1	Chimneys of houses fall	Movement along No. 12 fault
6	1983.04.14	3.0	Walls split	Movement along No. 12 fault
7	1983.04.27	2.7	Ground subsidence	Movement along No. 8 fault
8	1983.08.09	3.2	Buildings crack	Movement along No. 12 fault
9	1990.02.27	4.1	Houses and tunnels damaged. Intensity is VI	Movement along No. 10 fault

The common characteristics of earthquakes associated with the excavation of the coal mine are as follows:

—The seismic activity seems to alternate between an active phase and a calm phase. The majority of the earthquakes associated with the excavation of the coal mine occur in spring and autumn when the extraction rate is high.

—The focal depth of these earthquakes is in the range of 0.8–1.3 km. These shallow earthquakes have simple wave patterns, well developed surface waves, high-frequency components, long periods and a slow decrease of wave amplitude. Because of their shallow foci the corresponding intensities are higher. The main distinction between quarry blasts and the earthquakes associated with the excavation of the coal mine is the different polarities of the first arrival of P waves observed at the vertical components of the sensors: the motion is upwards in the case of the quarry blasts.

—These earthquakes can be correlated to rockburst and outburst activity. In space, the areas where earthquakes occur frequently coincide with the zones experiencing a high density of rockburst and outburst activity. As the excavation depth increased, the number of earthquakes associated with the excavation of the coal mine increased as well (CHEN, 1987).

—The seismicity is closely related to active faults. The majority of the earthquakes with $M \geq 2.5$ took place on No. 10, 8 and 12 fault (Fig. 4). The largest shock with $M = 4.3$ occurred on No. 10 fault (ZHONG and CUI, 1984).

—There is a correlation between the occurrence of the seismicity associated with the excavation of the coal mine and that of the natural earthquakes. As the number of natural earthquakes increased, the number of earthquakes associated with the excavation of the coal mine also increased. For example, an increase of natural earthquakes in the western part of the Liaoning Province, correlated with

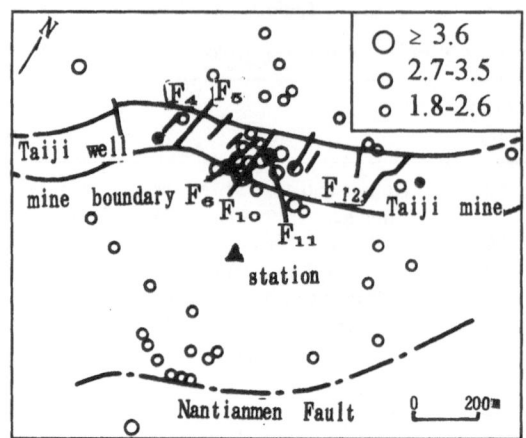

Figure 4
Plan view of the area of Taiji mine showing the faulting in the area and the distribution of the
earthquakes associated with mining.

increased earthquakes associated with the excavation (Fig. 5). Both types of
seismicity are strongly affected by the tectonic stress field. This observation suggests
that many of the earthquakes associated with mining at Taiji coal mine are
triggered by natural earthquakes.

Besides the earthquakes occurring at Taiji coal mine, since 1970 a series of
earthquakes have occurred in other coal mines, such as Xilin, Haizhou, Wulong,
and Aiyou mines (Fig. 1), with the largest earthquake registering a magnitude
$M = 3.8$. A specific type of seismic activity occurring in the coal mines is which is
closely related to coal extraction. These earthquakes have characteristic features

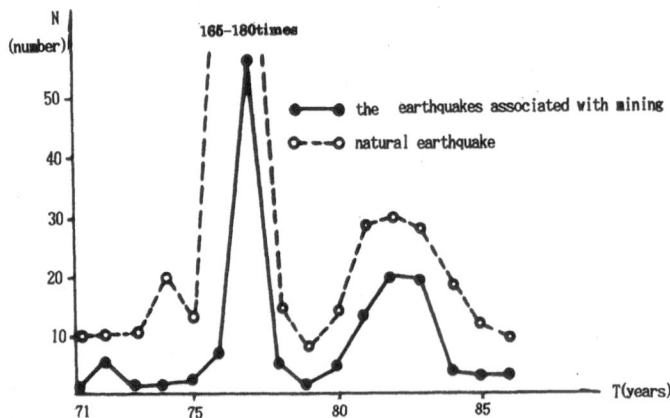

Figure 5
Correlation between the natural earthquakes in the western part of the Liaoning Province and the
earthquakes associated with mining at the Taiji coal mine.

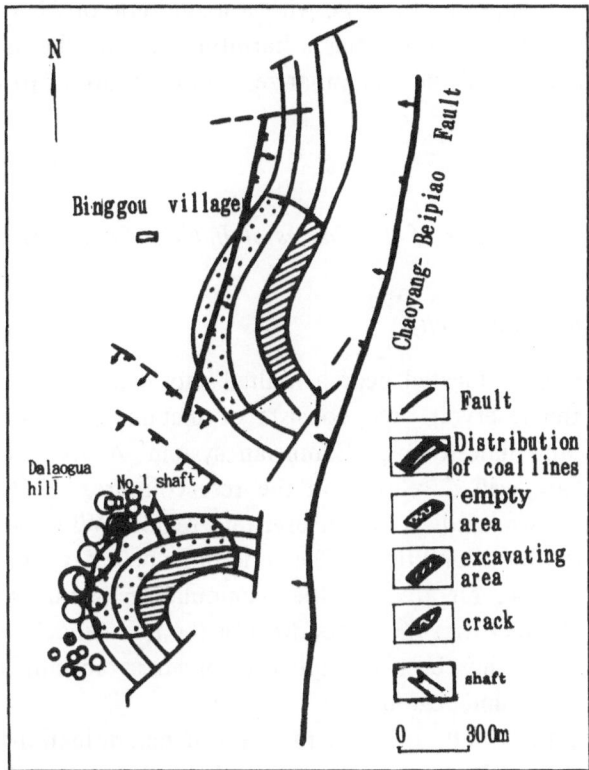

Figure 6
Geological map of the area of collapse earthquakes at the Binggou coal mine.

such as shallower depths and smaller magnitudes than previous events. However, they present a large threat to the safe production of coal.

2.3 Collapse Earthquakes

The cases of collapse earthquakes caused by excavating mines were only found at Binggou coal mine, Jianchang county (Fig. 1). The Binggou coal field consists of Jurassic coal deposits where there are four faults. This coal mine has an excavation history exceeding 200 years, and there are many empty areas left by the excavation. Many collapse earthquakes with a maximum magnitude $M = 3.9$ have been recorded here. In particular, fifty-six collapse earthquakes with a magnitude of $M \geq 1.8$ occurred between 1976 and 1985. From hypocenter distribution (Fig. 6) it can be seen that the collapse earthquakes are concentrated at Dalaogua hill near the No. 1 shaft. Because of a gradual expansion in the empty area and as a result of ground vibration associated with mining activity, a ground fissure zone striking NNE has been formed near Dalaogua hill. This fissure zone is 150 km long and 3 m wide and the height difference between the two sides of the feature has reached

4 m. In general, a collapse earthquake with a magnitude of $M = 2.5$ is felt by the people living in the mine area and has a harmful effect on the mining production (ZHONG and CUI, 1984). Table 2 summarizes major collapse earthquakes detected at Binggou mine.

3. Analysis of the Causes of Induced Seismicity

3.1 Reservoir-induced Seismicity

Shenwo Reservoir is located near a faulted subsidence zone of Mesozoic era. The stratum at the reservoir area consists of metamorphic rocks, quartzite of Precambrian era and limestone of Cambrian system. A great number of minor faults and joints are well developed in the reservoir area. Historically, Shenwo Reservoir area is considered to be an aseismic region. Why can induced earthquakes occur under these conditions? The conditions necessary for causing induced earthquakes in the reservoir area are the accumulation of initial stresses and the action of water. The fact that the stress drop of the main shock was as high as 60 bars reveals that there has been a high level of stress accumulation at Shenwo Reservoir and its surround region.

Table 3 shows the results of focal mechanism determinations for four major earthquakes at Shenwo Reservoir. The azimuth of the two nodal planes (A&B) obtained from the focal mechanism solution of the foreshock, the main shock and the two aftershocks are NE and NW, respectively. The azimuth of the nodal planes A, the orientation of the longer axis of isoseismic line of epicentral area and the

Table 2

List of major collapse earthquakes at the Binggou mine

No.	Date Y.D.M.	Magnitude (M)	Remarks
1	1976.5.19	3.0	Felt
2	1976.5.22	3.2	Felt
3	1978.2.16	3.3	Collapse
4	1978.3.30	2.5	Felt
5	1978.4.14	3.2	Felt
6	1979.2.13	2.9	Felt
7	1979.4.14	2.7	Felt
8	1979.10.13	3.9	Strongly felt
9	1980.7.16	3.6	Felt
10	1980.12.20	3.2	Felt
11	1981.12.8	3.0	Felt
12	1984.3.23	3.2	Felt
13	1984.4.20	3.5	Felt

Table 3

Fault-plane solutions for Shenwo Reservoir earthquakes

Date	Origin time	Magnitude (M)	Location Lat.	Location Long.	Depth (km)	Nodal plane A Az	Nodal plane A α°	Nodal plane B Az	Nodal plane B α	P axis Az	P axis α	T axis Az	T axis α°
1974.12.22	06-59-19	3.6	123° 35'	41° 16'	7.0	34°	70°	98°	40°	84°	52°	330°	18°
1974.12.22	12-46-19	5.2	123° 35'	41° 16'	5.8	34°	70°	98°	40°	84°	52°	330°	18°
1974.12.30	19-27-06	3.9	123° 35'	41° 16'	6.9	26°	49°	126°	80°	250°	20°	80°	36°
1975.6.9	09-41-03	4.4	123° 34'	41° 17'	8.0	34°	60°	114°	75°	251°	32°	347°	10°

Table 4

Fault-plane solutions for two natural earthquakes and the M = 4.3 earthquake associated with the coal mine

Date	Magnitude (M)	Location Lat.	Location Long.	Nodal plane A Az	Nodal plane A α°	Nodal plane B Az	Nodal plane B α	P axis Az	T axis Az
1975.2.4	7.3	122° 48'	40° 35'	292°	75°	23°	81°	66°	157°
1977.06.05	5.1	121° 18'	41° 57'	279°	90°	9°	90°	54°	144°
1977.04.28	4.3	120° 41'	41° 50'	296°	90°	28°	90°	72°	162°

strike of the main fault are all in the NE quadrant. These observations indicate that tectonic stresses before the main shock have been concentrated on the structure belt striking NE. In addition to initial stress accumulation, the action of water in inducing earthquakes should not be neglected. The water can decrease the strength of rocks by increasing the pore pressure. BELL and NUR (1987) defined the change in strength ΔS by the following equation:

$$\Delta S = \mu_f(\Delta\sigma_n - \Delta P) - \Delta\tau \tag{1}$$

where $\Delta\tau$ and $\Delta\sigma_n$ are changes in the shear stress on the fault in the direction of slip and in the compressive normal stress across the fault respectively; μ_f is the coefficient of friction and ΔP is the change in pore pressure. Failure occurs when ΔS decreases below a threshold level. From the above equation we note that a decrease in ΔS can be brought about by a decrease in $\Delta\sigma_n$ (unloading) or an increase in pore pressure. The pre-existing fractures, joints and karsts, which are well developed in the reservoir, have provided favorable conditions for the permeation of the lake water and the increase of pore pressure. When water permeates at depth along fracture planes or planes of weakness, it can contribute to the expansion of the fracture. Indeed, the increase in pore pressure and permeation of water occur slowly. Initially, water permeates to very shallow depths. The initial stresses are released by tremors. The tremor activity would have the fractures further expanded and water further penetrated in the rock mass. In particular, the existing carbonate strata is favorable to the permeation of water in Shenwo Reservoir. The action of water decreases the strength of the fault plane triggering the inducement of earthquakes. The above-mentioned processes would continue until the main shock occurs.

3.2 Earthquakes Associated with Coal Mines

There have been many earthquakes associated with the excavation of coal mines in the western part of Liaoning Province as described previously. The frequency and strength of the earthquakes associated with mining observed in Taiji coal mine coincided with that of regional tectonic seismic activity. Focal mechanisms are also consistent with each other (Table 4). This observation suggests that the earthquakes associated with excavation and tectonic earthquakes are responses to the same stress field. The first type of activity is closely related to the producing mines. The deep and large empty areas left by the excavation of coal mines can change the initial stress field. Under the influence of the change of the regional tectonic stress field, the associated excavation-induced stresses and the lithostatic stress field, a high stress concentration area is formed in the rock mass near the excavation and a high level of strain energy is accumulated in the rock mass. Under certain conditions, this energy is on the verge of being released rapidly and violently, causing strong ground vibrations (ZHANG *et al.*, 1996). The deeper the excavation, the greater the probability of the occurrence of earthquakes associated with the excavation (CHEN et al., 1987).

3.3 Collapse Earthquakes

The creation of empty areas of mines left after the excavation of mines for a prolonged period changes the initial state of equilibrium and stress field in the rock mass. These empty areas are influenced by vibrations and other phenomena over a long period of time and the strata collapse under the imbalance. Collapse earthquakes would be created by a sudden breakdown of rock in areas of high stress concentration. Therefore, these kind of collapse earthquakes are caused by the action of the local stress field.

4. Conclusion

1. There were three types of induced earthquakes in the Liaoning Province: reservoir-induced earthquakes, earthquakes associated with the excavation of coal mines and collapse earthquakes.

2. Induced seismicity results from the perturbation of the initial tectonic stress field by stress changes associated with human activity. These changes result from the impoundment of reservoirs and excavation of mines. Other factors influencing induced earthquakes are lithology, the structure and geomorphology of faults and their response to human activity. The seismicity after a main shock decreases in a reservoir area. However, the number of earthquakes associated with the excavation of mines increases as the excavation depth increases.

3. Induced earthquakes tend to correspond to natural earthquakes temporarily, although generating sites are not identical. This suggests that many of the observed earthquakes have been triggered by natural earthquakes. Induced earthquakes always take place in specific areas while natural seismicity is rather diffuse.

4. Other features of induced earthquakes are their shallow foci, high intensity and destructive power. For example, an event with a magnitude of about 2.5 can be felt in the reservoir area. A shock with a magnitude of about $M = 1.8$ in a mine is felt and a shock with a magnitude larger than $M = 2.5$ can cause minor damage.

REFERENCES

BELL, M. L., and NUR, A. (1978), *Strength Changes Due to Reservoir-induced Pore Pressure and Application to Lake Oroville*, J. Geophys. Res. *83*, 4469–4483.

CHEN, Z. J., XU, S. J., and YU, Z. C. (1987), *A Study of the Mine-shocks in Beipiao*, Northeastern Seismol. Res. *3*, 41–50.

GUPTA, H. K., and RASTOGI, B. K., *Dams and Earthquakes* (Elsevier, Amsterdam 1976).

LI, Z. A., *Induced seismicity in Danjiangkou reservoir*. In *Proceedings of the International Symposium on Reservoir-induced Seismicity* (ed. Chinese Society of Hydroelectric Engineering and Westcan Environmental Inc., Canada; Beijing 1995) pp. 142–150.

LI, A. R., XU, Y. J., and HAN, X. G., *Discussion on the environmental factors of the induced earthquakes*. In *Crust Deformation and Earthquakes* (ed. Institute of Seismology, S.S.B.) (Seismological Press, Beijing 1989) pp. 140–147.

ZHANG, S. Q., GUO, J. M., and ZHANG, L. C. (1996), *The Idea and Project of the Medium-scale Experiment Field for Earthquake Prediction*, Acta Seismological Sinica *4*, 679–690.

ZHAO, Z., JIANG, X. Q., and FAN, Y. P. (1988), *The Characteristics of the Shenwo Earthquake*, Earthq. Res. in China *4*, 116–124.

ZHONG, Y. Z., JIANG, X. Q., and CHEN, A. P., *Geological Disasters in Liaoning Province* (Seismological Press, Beijing 1991).

ZHONG, Y. Z., and CUI, R. S., *Seismic activity induced by excavating mine at Taiji coal mine, Beipiao county, Liaoning Province*. In *The Induced Earthquakes in China* (ed. Institute of Seismology, S.S.B.) (Seismological Press, Beijing 1984) pp. 162–167.

ZHONG, Y. Z., JIANG, Y. Q., and HAN, D. Z. (1981), *Discussion on the Induced Seismicity in Shenwo Reservoir Area*, Seismology and Geology *4*, 59–69.

ZHONG, Y. Z., and JIANG, X. Q., *Study of the generating conditions of earthquakes at Shenwo Reservoir, Liaoning Province*. In *Proceedings of the International Symposium on Reservoir-induced Seismicity* (ed. Chinese Society of Hydroelectric Engineering and Westcan Environmental Inc., Canada; Beijing 1995) pp. 171–178.

(Received May 15, 1996, accepted September 2, 1997)

Pure appl. geophys. 150 (1997) 473–492
0033–4553/97/040473–20 $ 1.50 + 0.20/0

On the Nature of Reservoir-induced Seismicity

Pradeep Talwani[1]

Abstract—In most cases of reservoir-induced seismicity, seismicity follows the impoundment, large lake-level changes, or filling at a later time above the highest water level achieved until then. We classify this as initial seismicity. This "initial seismicity" is ascribable to the coupled poroelastic response of the reservoir to initial filling or water level changes. It is characterized by an increase in seismicity above preimpoundment levels, large event(s), general stabilization and (usually) a lack of seismicity beneath the deepest part of the reservoir, widespread seismicity on the periphery, migrating outwards in one or more directions. With time, there is a decrease in both the number and magnitudes of earthquakes, with the seismicity returning to preimpoundment levels. However, after several years some reservoirs continue to be active; whereas, there is no seismicity at others. Preliminary results of two-dimensional (similar to those by Roeloffs, 1988) calculations suggest that, this "protracted seismicity" depends on the frequency and amplitude of lake-level changes, reservoir dimensions and hydromechanical properties of the substratum. Strength changes show delays with respect to lake-level changes. Longer period water level changes (∼1 year) are more likely to cause deeper and larger earthquakes than short period water level changes. Earthquakes occur at reservoirs where the lake-level changes are comparable or a large fraction of the least depth of water. The seismicity is likely to be more widespread and deeper for a larger reservoir than for a smaller one. The induced seismicity is observed both beneath the deepest part of the reservoir and in the surrounding areas. The location of the seismicity is governed by the nature of faulting below and near the reservoir.

Key words: Mechanism of reservoir-induced seismicity, Koyna, Monticello Reservoir, Lake Mead.

Introduction

Since the identification of a causal association of seismicity with the impoundment of Lake Mead in the early 1940s (Carder, 1945) reservoir-induced seismicity (RIS) has been observed at over seventy locations worldwide (Simpson, 1976, 1986; Gupta, 1992). Following damaging reservoir-induced earthquakes in the 1960s at Koyna, India; Hsingfengkiang, China; Kariba, Zimbabwe and Kremasta, Greece, there was great improvement in seismic monitoring. Local networks were deployed in the vicinity of several reservoirs in the 1970s. These resulted in lower detection thresholds and improved locations of recorded seismicity. Complementary field studies led to the identification of factors that control the observed RIS. These

---

[1] Department of Geological Sciences, University of South Carolina, Columbia, South Carolina 29208, U.S.A.

factors include ambient stress field conditions, availability of fractures, hydromechanical properties of the underlying rocks, geology of the area, together with
dimensions of the reservoir and the nature of lake-level fluctuations.

As case histories of RIS accumulated, the effect of reservoir loading on the
existing stress field has been the subject of several studies (SNOW, 1972; BELL and
NUR, 1978; TALWANI and ACREE, 1984; SIMPSON, 1976, 1986; SIMPSON et al.,
1988; ROELOFFS, 1988; RAJENDRAN and TALWANI, 1992). Except for ROELOFFS'
(1988) study, all of them addressed seismicity associated with the initial impoundment of the reservoir. However, there are other cases in which protracted and
"significant" RIS has been observed several years after initial impoundment.
"Significant" here implies both a larger number and a higher magnitude of
seismicity than at preimpoundment levels. The seismicity in such cases appears to
be related to water level fluctuations. The ongoing seismicity at Koyna, India, three
decades after impoundment; the observed seismicity at Lake Mead for over three
decades after the observation of the initial seismicity and to a lesser extent, the
current seismicity at Lake Jocassee and Monticello Reservoir South Carolina are
examples of this protracted RIS.

Initial and Protracted Seismicity

We classify the seismic response of a reservoir into two temporal categories. The
first, which is widely observed, is associated with the initial impoundment or large
lake-level changes. This category also applies to seismicity associated with lake-level
increases above the highest level attained thus far. We call this category of RIS,
"initial seismicity." The second category of seismicity, which is observed in rare
cases, occurs after the effect of initial filling has diminished. It persists for many
years without a decrease in frequency and magnitude. We call it "protracted
seismicity."

The initial seismicity results from the instantaneous effect of loading (or
unloading) and the delayed effect of pore pressure diffusion. Following this initial
activity, there is an increase in the frequency and magnitude of earthquakes. The
largest associated event usually occurs after completion of the reservoir impoundment and the attainment of maximum water level. The delay between the start of
filling and the larger events varies from months to years and is associated with the
reservoir and local site characteristics. Spatially there is a general stabilization and
(usually) an absence of seismicity beneath the deepest part of the reservoir and
widespread seismicity on the periphery, migrating outwards in one or more
directions. This period of increased seismicity is followed by a gradual decay in
activity (over months to years) to preimpoundment levels, indicating the cessation
of the coupled poroelastic response to the impoundment.

In the case of protracted seismicity, modeling suggests that the pore pressure increase that causes the seismicity, is related to the frequency and amplitude of lake-level changes (ROELOFFS, 1988). Peak changes in pore pressures occur directly beneath the lake and decrease away from it. Strength changes show delays with respect to lake levels. In this category, earthquakes are associated with reservoirs with large and/or rapid lake-level rises and longer periods (lower frequencies) of water level changes. Seismicity is observed both beneath the deepest part of the reservoir and in surrounding areas. The seismicity continues for decades and does not appear to die out.

The factors controlling the spatial and temporal patterns of these two categories of RIS are different. In this paper I address the nature of these two categories of RIS.

Poroelastic Response to Reservoir Impoundment

BELL and NUR (1978) defined the change in strength ΔS by the following equation

$$\Delta S = \mu_f(\Delta\sigma_n - \Delta p) - \Delta\tau \qquad (1)$$

where $\Delta\tau$ and $\Delta\sigma_n$ are changes in shear stress on the fault in the direction of slip and compressive normal stress across the fault respectively. μ_f and Δp are the coefficient of friction and change in pore pressure respectively. Failure occurs when ΔS decreases below a threshold level. From equation (1) we note that a decrease in ΔS can be brought about by a decrease in $\Delta\sigma_n$ (unloading) or an increase in pore pressure. The temporal effect of impoundment can be divided into two parts, instantaneous and delayed. (We use the poroelastic approach of RICE and CLEARY (1976) in which both the solid and fluid phases are assumed to be compressible.)

The instantaneous effect is due to the elastic and undrained response to loading. The delayed effect is due to the drained response and pore pressure changes by diffusion. The net result is a *coupled* response of the different responses mentioned above. These responses are shown schematically in Figure 1. The various effects have been reviewed by RAJENDRAN and TALWANI (1992). In order to present a comprehensive review here, and compare with field observations, the next section has been extracted from that paper and illustrated with Figure 1.

Elastic response: The elastic response of the subsurface to loading causes changes in normal and shear stresses on the fault plane. Under assumptions of isotropic conditions, $\Delta\sigma_n$ (Figure 1b) mimics the reservoir loading curve (Figure 1a), the change being instantaneous. In general, increased normal stress tends to stabilize (increase ΔS) the region, especially under the reservoir. An example of this is provided by comparing the RIS observed at Lake Mead with elevation changes found by releveling. Intense seismicity was observed following the impoundment of

Lake Mead behind the Hoover Dam in the late 1930s and early 1940s. CARDER and SMALL (1948) found that the epicenters were located near the periphery and were not associated with the region of maximum crustal load due to the lake.

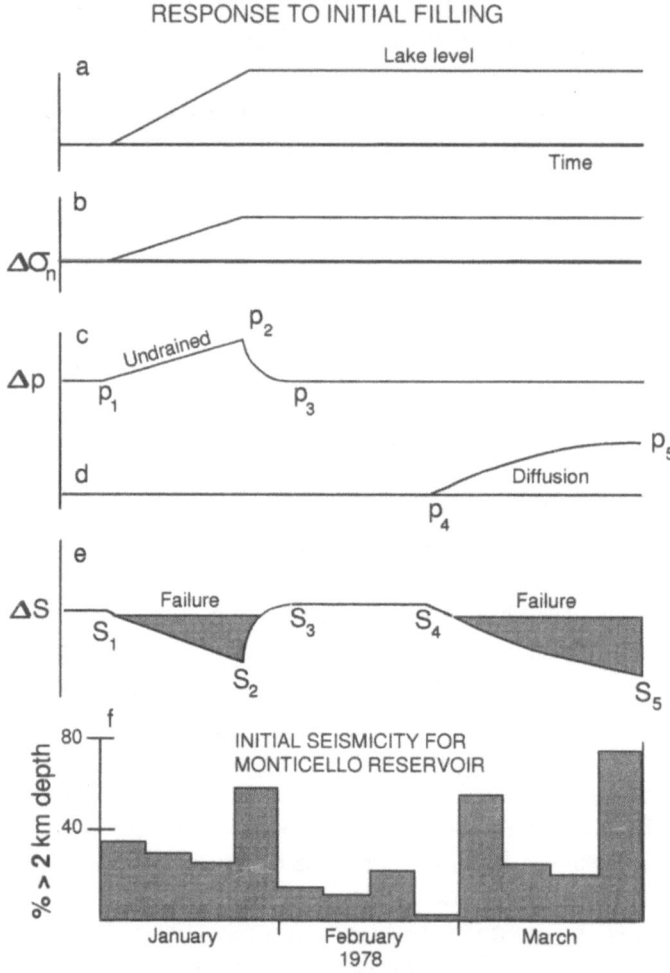

Figure 1

Schematic figure to illustrate the processes observed in initial seismicity. (a) shows the filling curve of the reservoir, it is associated with an increase in $\Delta\sigma_n$ (b) due to the load. The undrained response in a clogged pore causes an increase in the pore pressure (p_1 to p_2) (c) and a corresponding decrease in strength (S_1 to S_2) (e). When the pore is unclogged, the increased pore pressure dissipates (p_2 to p_3) and the strength increases (S_2 to S_3). When the pore pressure front due to reservoir loading arrives, there is an increase in pore pressure (p_4 to p_5, (d)) and a corresponding decrease in strength (S_4 to S_5, (e)). When the strength decreases below a critical threshold (marked FAILURE) seismicity occurs (shaded pattern). Panel (f) shows the percentage of "deep" events associated with the initial filling of Monticello Reservoir.

Undrained response: We borrow the nomenclature from soil mechanics, according to which undrained conditions prevail if the rock sample is subjected to a change in confining pressure and pore fluid is prevented from escaping or entering. In the case of reservoir impoundment, there will be an instantaneous increase in pore pressure in the substratum due to the additional load at the surface. If no fluid is allowed to flow, for example in the case of clay filled fractures, there will be an increase in pore pressure, Δp_u. This pore pressure increase, due to undrained response, Δp_u, will persist until it dissipates into the surrounding fractures. The undrained response is given by

$$\Delta p_u = B\sigma_{\kappa\kappa}/3 \tag{2}$$

where B is the Skempton's constant and $\sigma_{\kappa\kappa}$ is the mean stress. If clogged fractures are present, Δp_u can increase with the loading (p_1 to p_2 in Figure 1c) and be sustained until flow occurs. In such a case, there is a corresponding decrease in the strength, (S_1 to S_2), which can lead to failure, when the strength decreases below a threshold value (labeled FAILURE in Figure 1e).

Drained response occurs when the pore fluid is enabled to enter or leave and the pore pressure decreases to the original value. In the case of a clogged fracture, the drained response occurs when the fluid leaves it and Δp_u decreases to zero (p_2 to p_3 in Figure 1c). The drained response is delayed with respect to the initial impoundment and the delay depends on the hydromechanical properties, chemical composition of fluids (for stress corrosion), nature of clays, etc. The drained response results in a decrease in pore pressure and an increase in ΔS (S_2 to S_3 in Figure 1e).

Pore pressure diffusion from the surface to the substratum also causes an increase in pore pressure. Pressure flow is governed by the diffusion equation (JAEGER and COOK, 1969). In one dimension it is

$$\delta^2 p/\delta z^2 = 1/C(\delta p/\delta t) \tag{3}$$

where p is the pore pressure at depth z, t is time and C is the coefficient of diffusivity

$$C = k/\eta\beta \tag{4}$$

where k is the permeability of the rock, η is the viscosity of the pore fluid and β is the bulk compressibility of fluid-filled rocks. The pore pressure increase following impoundment is delayed, the lag depending on hydraulic diffusivity C (and hence permeability, k) and the distance. Equation (3) has a solution of the form

$$p(z, t)/p(0, 0) = 1 - \operatorname{erf}[z^2/4ct]^{1/2}. \tag{5}$$

The pore pressure increase due to diffusion (p_4 to p_5 in Figure 1d) may occur after the increase in $\Delta p_u (p_1$ to $p_2)$ has already dissipated. This pore pressure increase is associated with a decrease in strength (S_4 to S_5 in Figure 1e) and earthquakes occur when the strength decreases below a threshold value (labeled FAILURE in Figure 1e).

Coupled response: Actually what we observe is the coupled response of the different responses mentioned above. For isotropic fluid-saturated porous elastic medium, RICE and CLEARY (1976) calculated the coupled response

$$\sigma_{ij} = 2G\varepsilon_{ij} + \frac{v}{1+v}\,\sigma_{\kappa\kappa}\,\delta_{ij} - \frac{3(v_u - v)}{B(1+v)(1+v_u)}\,p\,\delta_{ij} \tag{6}$$

where v and v_u are drained and undrained Poisson's ratios and G is the shear modulus. Using RICE and CLEARY's (1976) results, ROELOFFS (1988) modified equation (5) to include the term due to the undrained response. For a unit step increase in pore pressure at the surface, $p(0, t) = H(t)$, she calculated the pore pressure at a depth z after time t, $p(z, t)$. For a one-dimensional case she found

$$p(z, t) = (1 - \alpha)\,\mathrm{erfc}\,[z^2/4ct]^{1/2} + \alpha(H(t)) \tag{7}$$

where erfc is the complementary error function, $H(t)$ is Heaviside unit step function and $\alpha = B(1 + v_u)/3(1 - v_u)$. Thus the coupled response may be dominated by the undrained response immediately on impoundment and be primarily due to diffusion later. At any depth, after enough time has elapsed, $p(z, t)$ approaches the load applied at the surface, there are slight changes in the pore pressure and the RIS decays to preimpoundment levels.

The above arguments are given for isotropic conditions. However, the presence of fractures on which RIS is usually observed, clearly suggests that anisotropic conditions prevail. In such a case, pore pressure increase can cause earthquakes on vertical fractures in a normal faulting environment and on horizontal fractures in a reverse faulting environment (CHEN and NUR, 1992).

An Example of Initial Seismicity

Figure 2 shows the filling curve at Monticello Reservoir, South Carolina superimposed on monthly seismicity. Impoundment (~ 32 m) occurred between December 3, 1977 and February 8, 1978. Thereafter the lake level was kept within 1.5 m of the mean water level. "Initial seismicity" persisted for a few years and then began to decay, approaching preimpoundment levels.

The spatial pattern of the earthquakes provides further insight into the nature of initial seismicity. Most of the seismicity was shallow ($z < 3$ km). We should anticipate loading to increase $\Delta\sigma_n$ and thus ΔS, the greatest increase occurring beneath the deepest part of the reservoir. The increased pore pressure (which tends to destabilize) thus has a greater effect on the periphery of the reservoir. In 1978, following impoundment, the initial seismicity surrounded the deepest part of the reservoir (Figure 3). The "deeper" seismicity ($z \geq 2.0$ km) associated with impoundment provides further insight into the relative roles played by undrained response and diffusion. For $C \sim 10^4$ cm^2/s the time for pore pressure to diffuse to depths, of,

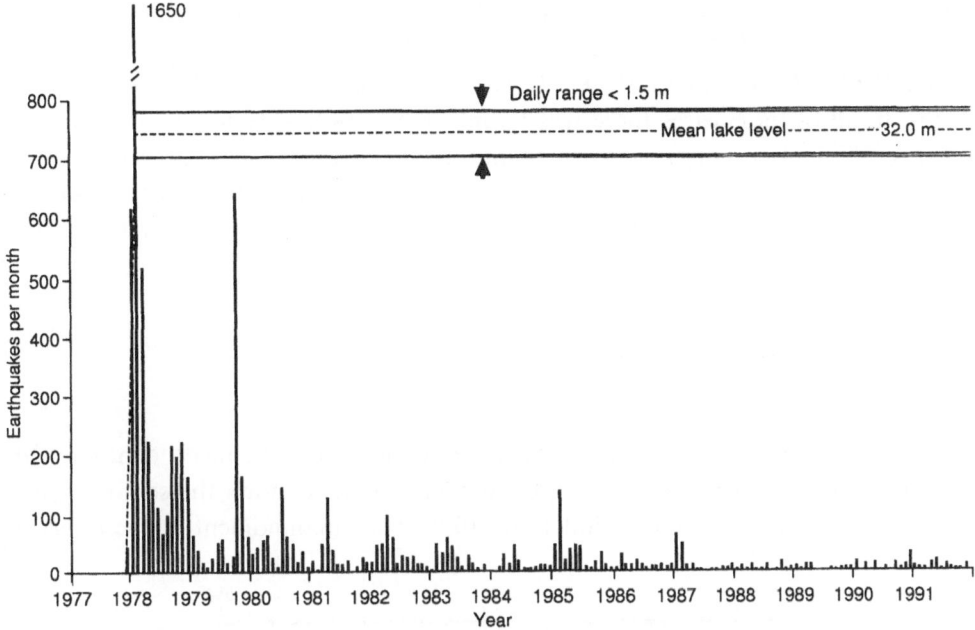

Figure 2
Lake level at Monticello Reservoir (dashed) compared with monthly seismicity for the period 1977 to 1991.

say 2 and 2.5 km is ∼46 and 72 days. In January 1978, over 30% of the seismicity was deeper than 2 km (Figure 1f). We attribute this seismicity to increased pore pressure due to the undrained response of the reservoir. There was a decrease in the "deeper" seismicity in February, 1978 (<10%). We attribute this to leaking of the increased pore pressure (drained response, corresponding to p_2 to p_3 in Figure 1c). The decrease in pore pressure causes strengthening, (increase in ΔS) and thereby a decrease in the "deeper" seismicity (Figure 1f). The increased "deeper" activity (>40%) in March, 1978 is attributed to increased pore pressure due to diffusion.

Thus Figure 1f shows the different aspects of initial seismicity. The seismicity in January 1978 was due to both the undrained response (deeper earthquakes) and to diffusion (shallower earthquakes). The initial undrained response dissipated by February, 1978 and the ensuing seismicity was primarily due to diffusion. The initial seismicity was located outside the deepest part of the reservoir.

The burst of seismicity in 1985 was located beneath deeper parts of the reservoir. The nature of this protracted seismicity is different from the initial seismicity and is described in the next section.

Initial seismicity is the most widely observed category of RIS. Most case histories where adequate data are available, can be explained by the coupled response to initial loading, loading at a later time above the highest water level achieved, thus far and due to rapid water level changes. Examples of these include

the observed seismicity at Lake Mead (CARDER, 1945); Nurek (SIMPSON and
NEGMATULLAEV, 1981); Manic-3 (LEBLANC and ANGLIN, 1978); Kariba (SIMPSON
et al., 1988); Hsinfengkiang (SHEN et al., 1974); Lake Jocassee (TALWANI et al.,
1976). In all of these cases there was an increase in seismicity above preimpoundment levels; large event(s) followed filling and there was a decay in seismicity to
preimpoundment levels. Where accurate data are available, the initial seismicity
appears to occur away from the deepest part of the reservoir and migrates
outwards.

Protracted Seismicity

Unlike the case of initial seismicity, at some reservoirs seismicity continues for
several years, even decades after impoundment. Figure 4 shows the seismicity and
lake levels at Koyna for the period 1961–1995. The impoundment of the reservoir

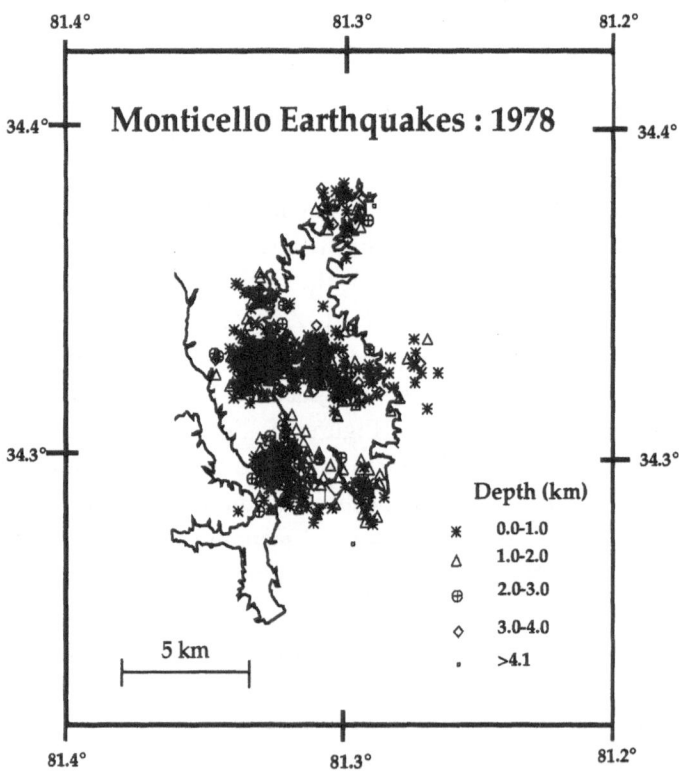

Figure 3

Seismicity observed near Monticello Reservoir in 1978. Note that most of the earthquakes lie in two
bands—in the middle of the reservoir where the water is relatively shallow and on the southwestern and
southern banks of the reservoir. There is a general absence of seismicity below the deepest (and
southernmost) part of the reservoir.

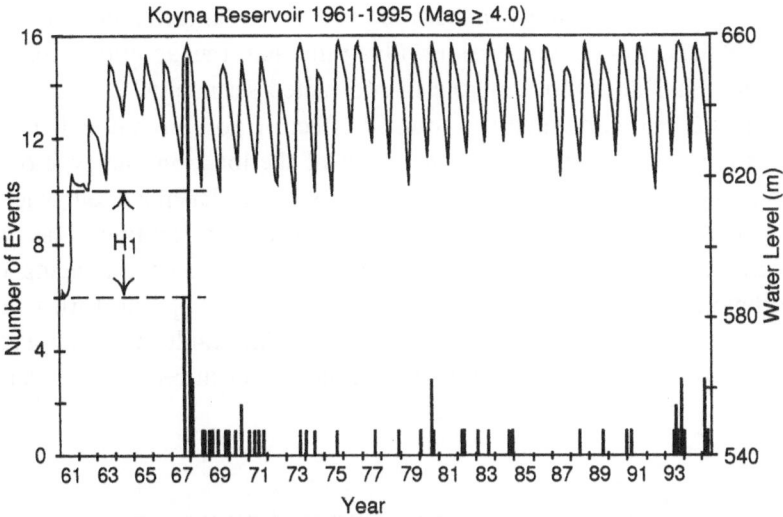

Figure 4

Monthly seismicity ($M \geq 4.0$) at Koyna Reservoir compared with lake levels for the period 1961 to 1995. (Figure taken from TALWANI et al., 1996.)

occurred between 1962 and 1964. Earthquakes continue to occur there; the latest $M > 5.0$ event occurring in February, 1994.

This seismicity, occurring long after the strength changes associated with the coupled response to initial filling have stabilized, is relatable to the large water level changes. These are shown schematically in Figure 5. The daily/weekly/annual change H_2 is usually considerably less than the least depth of water, H_1 for the case of initial seismicity. This was the case at Monticello Reservoir where $H_1 \sim 31.8$ m and $H_2 \sim 1.5$ m (Figure 2). When the change in water level (weekly/monthly/annual) H_2 is comparable or a large fraction of the least depth of water, H_1, the seismicity is governed by the frequency lake-level changes. For Koyna H_2 is 20 to 40 m (with largest value ~ 47 m) and $H_1 \sim 30$ m.

This effect is illustrated by observations from Lake Mead impounded by the Hoover Dam. Figure 6 shows the lake level and the time of occurrence of larger earthquakes ($4.0 \leq M \leq 5.0$). (The revised magnitudes and lake levels are taken from a report by ANDERSON and O'CONNELL (1993).) Two large events (M 5.0 and 4.4) occurred in May and June 1939 following initial impoundment ($H_1 > 100$ m). Ten other events with $4.0 \leq M \leq 4.9$ took place in the period between 1942 and 1963. These events happened following large annual changes in the lake levels (15 m $\leq H_2 \leq 30$ m) (Figure 6). Four events with $M \geq 4.0$ occurred between 1963 and 1965. These were related to elastic unloading (decrease in $\Delta\sigma_n$ and hence ΔS). The average magnitude of the larger events between 1939 and 1963 was 4.3. Lake-level changes (H_2) were less than 10 m following the construction in 1965 of Glen Canyon Dam located upstream of Lake Mead. No earthquake with $M > 3.7$ has

occurred since 1965. The mean magnitude of the 13 largest events in the period
1965–1992 is 3.3. These observations illustrate that the amplitude of lake-level
changes (H_2) plays an important role in protracted seismicity.

ROELOFFS (1988) computed the coupled effect of pore pressure changes due to
a cyclic load. She also incorporated the effect of the location and type of faulting.
She found that the effect of an oscillating reservoir depended on where it was
located with respect to the fault and the nature of the faulting. The oscillating
reservoir maintained a stabilizing effect if it was located on the hanging wall of a
steeply dipping reverse fault, directly above a shallowly dipping thrust fault or if
there was a shallow vertical strike-slip fault or normal fault at the reservoirs edge
(Figure 7a). However, destabilization (earthquakes) occurred if the reservoir was

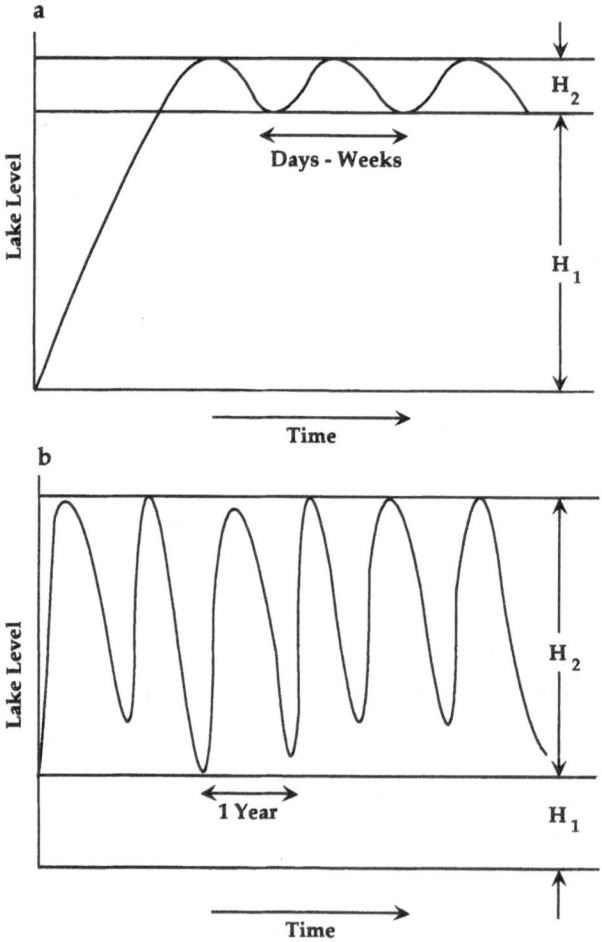

Figure 5
Comparison of the amplitude of cyclic lake level changes (H_2) with the least water level (H_1). For initial
seismicity $H_2 \lll H_1$ and for *protracted* seismicity H_2 is comparable to H_1.

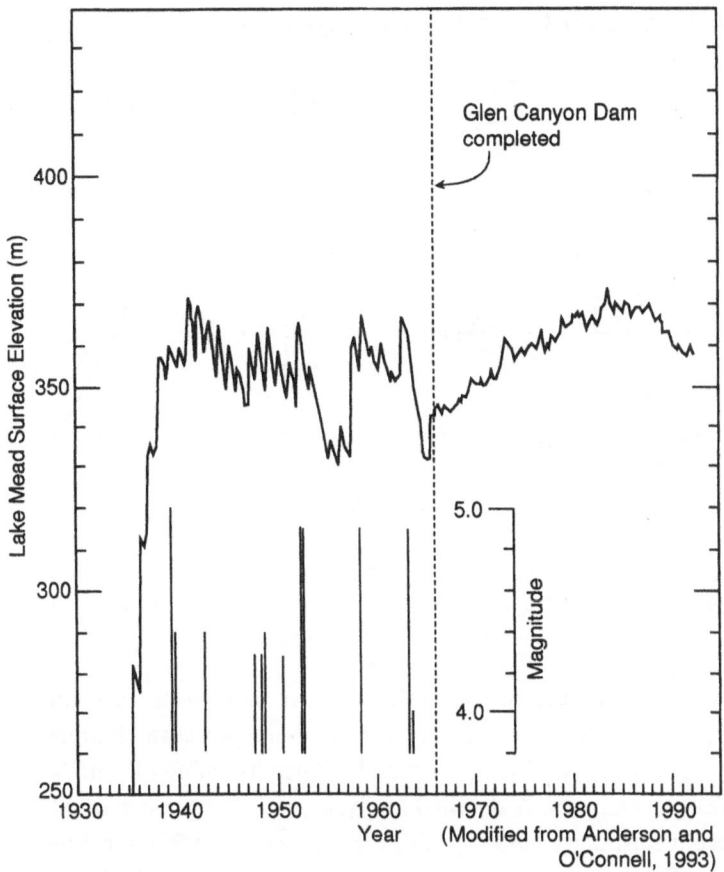

Figure 6

Larger earthquakes ($4.0 \leq M \leq 5.0$) at Lake Mead compared with lake levels for the period 1935–1992. Note a decrease in larger events after construction of Glen Canyon Dam in 1965. (Modified from ANDERSON and O'CONNELL, 1993.)

located on the foot wall of a steeply dipping reverse fault or on the hanging wall of a shallowly dipping thrust. Seismicity also occurred below the reservoir if there was a vertical strike-slip fault or a normal fault located there (Figure 7b).

Her calculations for a simple 2-D reservoir suggested that the changes in stress and pore pressure fields produced by reservoir loads are governed by a dimensionless frequency Ω, given by

$$\Omega = \omega L^2 / 2C, \tag{8}$$

where ω is loading frequency (1/year for Koyna and 1/day for Monticello Reservoir) and L is the width of the reservoir. The depth, z^*, below which pore pressure changes were negligible is given by

$$z^* = \Pi (2C/\omega)^{1/2}. \tag{9}$$

Effect of Oscillating Reservoir Load (⬆⬇)

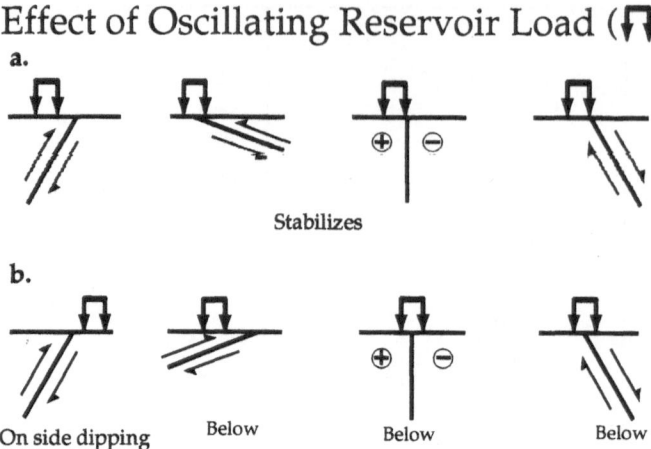

a.

Stabilizes

b.

On side dipping
away from reservoir
Below   Below   Below

Destabilizes

Figure 7
Schematic figure to show the effect of an oscillating reservoir load. It results in stabilization or seismicity, depending on the location and orientation of faults *vis à vis* the reservoir and on the stress field, based on ROELOFFS (1988).

Table 1 illustrates that the deepest effects occur in regions with higher diffusivity and longer periods. The model would predict pore pressure changes to a depth of about 25 km below Koyna for an assumed diffusivity value of 1 m²/s and an annual cycle of lake-level changes. A detailed analysis provided good estimates of the depth extent of recent seismicity (1993–95) (TALWANI *et al.*, 1996). Current seismicity lies to the south of the reservoir and between depths of 5 and 16 km.

However, as ROELOFFS (1988) demonstrated, the location of maximum destabilization also depended on the nature of faulting. For a strike-slip and a reverse fault below the reservoir, we calculated the maximum change in strength over the entire cycle of loading (TALWANI *et al.*, 1992). The results are shown in Figures 8 and 9.

Table 1

z* for various periods and diffusivities

Diffusivity m²/s	Peroid (days)		
	1	30	360
	z* km	z* km	z* km
0.1	0.4	2.3	7.8
1.0	1.3	7.2	24.8
10.0	4.1	22.6	78.4

z* Depth below the surface, below which the coupled pore pressure changes due to lake-level changes are negligible.

Change of Strength in Strike Slip Environment

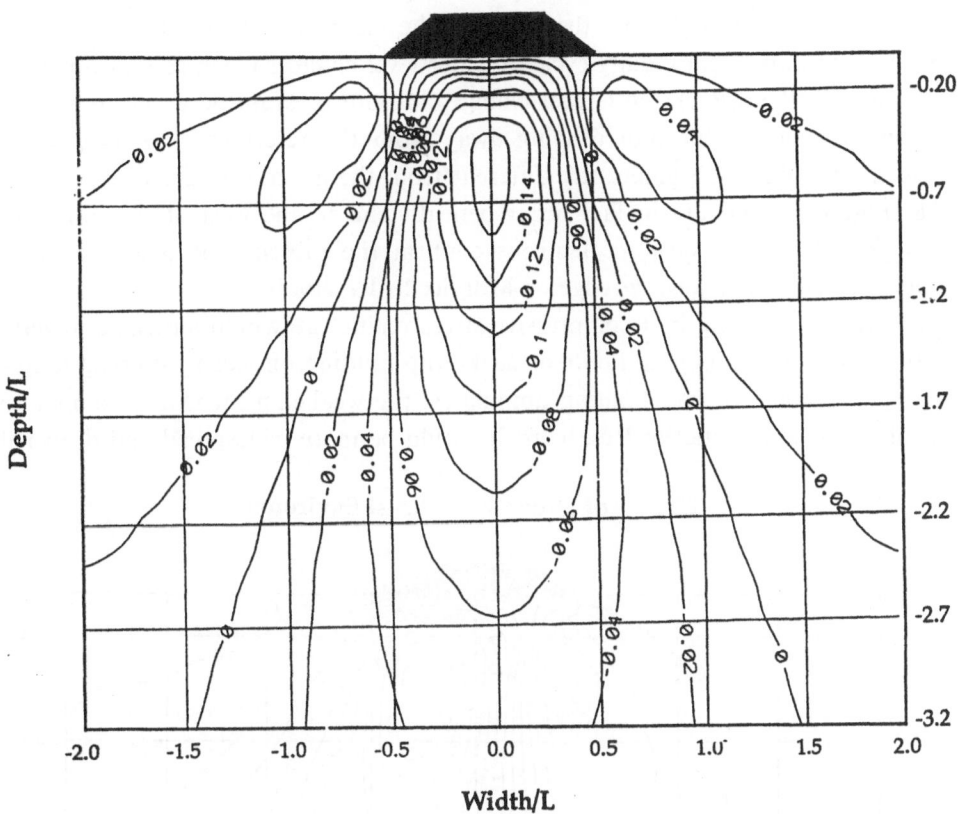

Figure 8

Maximum change in strength (in bars) over the entire cycle of lake-level change. A vertical fault is assumed directly below the reservoir (shaded). The 2-D reservoir and fault extend into the plane of the paper. The cyclic load varies from +1 bar to −1 bar over the 360 days cycle. The depth and width are normalized with respect to the width of the reservoir.

To calculate the strengths we modified Equation (7) and assumed parameters given in Table 2.

For the 2-D model we calculated the changes in strength (in bars) corresponding to a cyclic load (p_0 varying from +1 bar to −1 bar) (+0.1 MPa to −0.1 MPa). The depth and horizontal distances are normalized with respect to the width of the reservoir. For a cyclic load corresponding to p_0 of say ±5 bars (0.5 MPa) (corresponding to water level changes of ±50 m) the changes in strength will be multiplied by 5. Negative values of changes in strength correspond to weakening. The change in strength occurs over the entire cycle. The maximum change in strength over the entire cycle is plotted. In the case of the strike-slip fault, the largest changes occur under the reservoir (Figure 8), and for a reverse fault dipping 60° to the left weakening occurs on the left bank of the reservoir (Figure 9). To obtain actual depths and distances in Figures 8 and 9 multiply by the width of the reservoir. For a strike-slip fault the largest changes in pore pressure occur at depths between about 0.5 to 0.7 times the width of the reservoir, whereas for the reverse fault dipping at 60° and a daily cycle in water level changes (such as that observed at Monticello Reservoir), the largest changes in pore pressure occur in the middle and on the side of the fault dipping away from the reservoir. The location of seismicity in 1985 was significantly different from the initial seismicity observed in 1978 (Figure 3), and did in fact occur beneath and to the west of the reservoir. Geological data and focal mechanisms confirm the presence of steep westward dipping faults on the western edge of Monticello Reservoir.

For reservoirs lying in a compressive stress regime we would anticipate reverse faulting on shallow, dipping faults or strike-slip faulting on steeply dipping faults. For regions where reverse faulting dominates, the weakening would be similar to that observed at Monticello Reservoir. It would be more widespread and deeper if

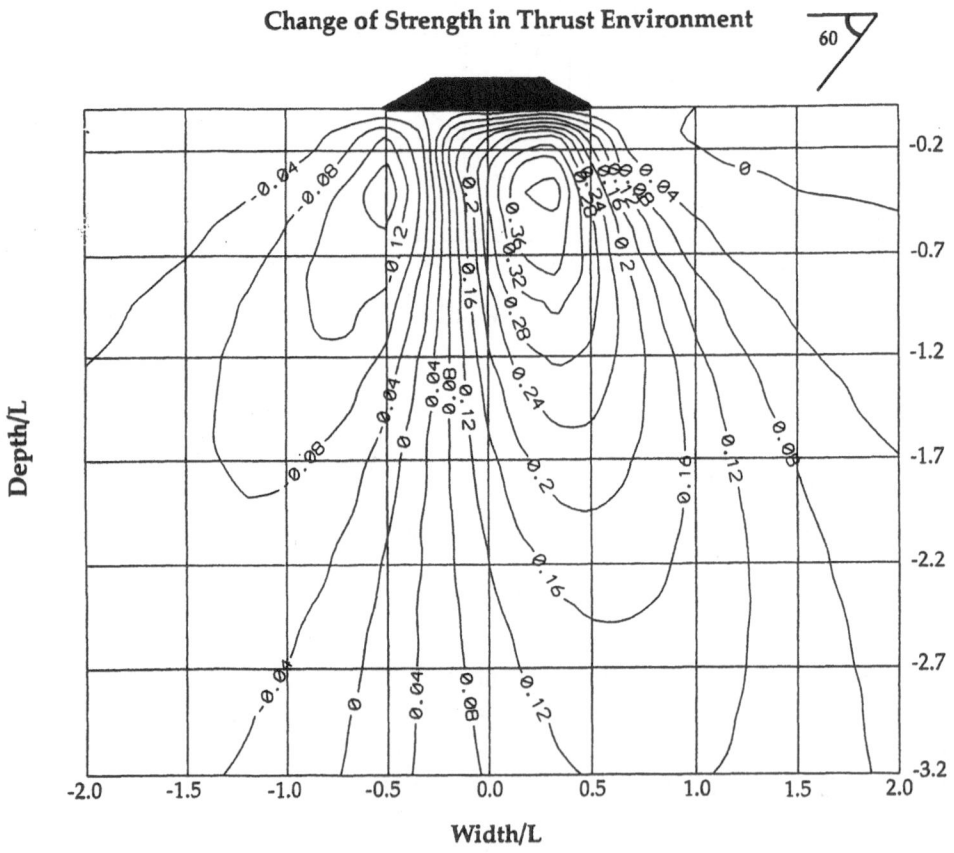

Figure 9
Maximum change in strength (in bars) over an entire cycle of lake-level change. A reverse fault below the reservoir dips 60° to the left. Weakening occurs below and to the left of the reservoir.

Table 2

Model parameters

	Strike-slip fault (Koyna)	Reverse fault (Monticello)
B	0.7	0.7
C	1 m^2/s	1 m^2/s
v_u	0.3	0.3
v	0.25	0.25
Period $(1/\omega)$	360 days	1 day

it is associated with longer periods of water level cycles, compared to Monticello Reservoir (1 day) (Table 1). The temporal changes in pore pressure and strength to the left, below and to the right of the reservoir at depths equal to half the width of the reservoir, over half a cycle are shown in Figure 10. Weakening occurs when the

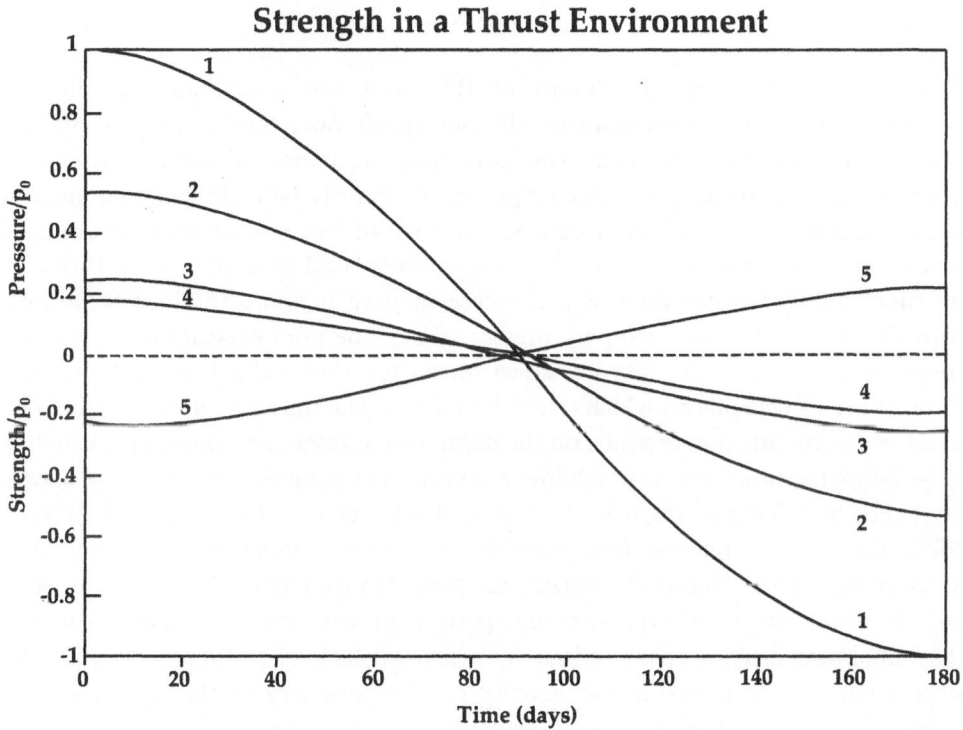

Figure 10

Temporal changes associated with 1/2 cycle of reservoir water level change. The water level change (curve 1) (corresponding to $p_0 = +1$ to -1) results in a change in pore pressure (curve 2). Pore pressure changes on the left bank, center and right bank (curves 5, 4 and 3) are calculated at depths equal to half the width of the reservoir.

ratio of strength to pore pressure becomes negative, and strengthening takes place when this ratio is positive. The fault is assumed to lie underneath the reservoir and dip 60° to the left. The change in pore pressure (curve 2) follows the lake level (load, curve 1), with a lag. At full reservoir there is weakening on the left side of the reservoir (decrease in ΔS, curve 5) and a strengthening on the right side of the reservoir (curve 3). The change in strength below the middle of the reservoir (curve 4) is similar to that on the right side. At minimum water level (day 180), there is a maximum weakening on the right bank of the reservoir and strengthening below the left bank. The weakening below the right bank at minimum load is comparable to the weakening below the left bank at maximum load. Thus, depending on the *in situ* conditions prevailing below the reservoir, we can have seismicity on the left side or on the right side. Also note that the changes in strength below different parts of the reservoir are out of phase with the water level curve. The delays depend on the hydraulic diffusivity, Skempton's constant, geometry of the reservoir, frequency of water level changes and fault geometry.

Discussion

We divide the temporal pattern of RIS into two categories. The first is associated with initial impoundment, the raising of water level above the highest water level achieved until then. The poroelastic response of the reservoir is a coupled response. Initially and occurring simultaneously with the impoundment is the undrained response. This occurs because of an increase in pore pressure in closed pores (by fault gouge and clay). As the increased pore pressure diffuses to the surrounding regions, there is a decrease in pore pressure (drained response). With the arrival of a diffusive pore pressure front, the pore pressure increases and causes seismicity. In reality all three effects occur together and the coupled response of the reservoir depends on which effect dominates. The time for an increase in pore pressure due to diffusion depends on the depth of the reservoir, geometry, availability of faults/fractures, etc. For shallow reservoirs the coupled response may take a few weeks to a few months (e.g., at Monticello Reservoir (TALWANI and ACREE, 1987), whereas for large and deep reservoirs it may take years (e.g., Hsingfengkiang SHEN *et al.*, 1974), Nurek (SIMPSON and NEGAMATULLAEV, 1981), etc. In both cases however we classify the temporal pattern of seismicity as initial seismicity. The initial seismicity is characterized by a general lack of seismicity beneath the deepest part of the reservoir and activity on the periphery of the reservoir. The seismicity increases after the impoundment is completed (or highest water level is achieved) and the largest earthquake usually occurs after that. Then there is a decay in seismicity (over 5–10 years) to preimpoundment levels.

SIMPSON *et al.* (1988) suggested that the temporal distribution of induced seismicity following the filling of large reservoirs exhibits two types of response;

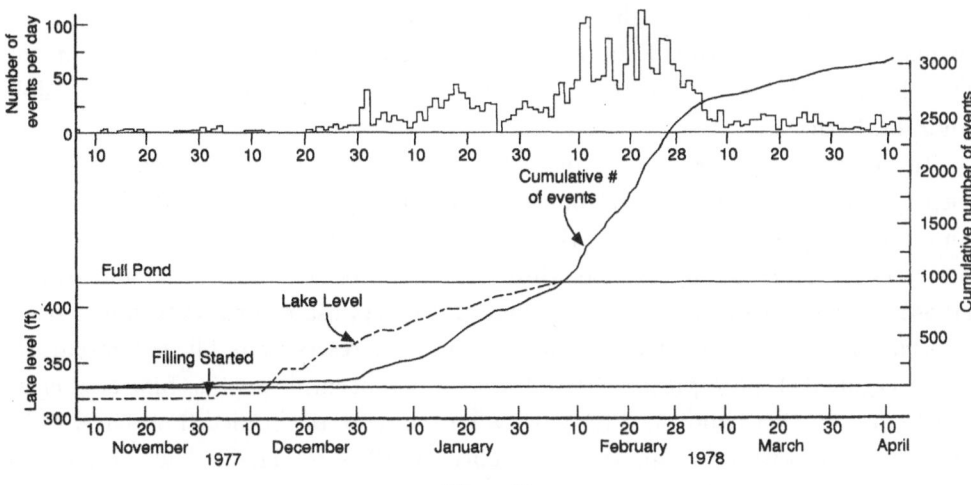

Figure 11
Seismicity associated with the filling of the Monticello Reservoir. (From TALWANI and ACREE, 1984.)

instantaneous and delayed. They suggested that the seismicity that began almost immediately following initial impoundment was due to the instantaneous elastic response and the undrained response, resulting in an increase in pore pressure. The delayed seismicity e.g., at Koyna, was attributed to an increase in pore pressure due to the diffusion of pore pressure to hypocentral depths. However, in our classification, both types of RIS alluded to by SIMPSON et al. (1988), are integral components of initial seismicity.

This is illustrated by thoroughly examining the observed seismicity at Monticello Reservoir in 1978. Filling occurred between December 3, 1977 and February 8, 1978 (Figure 11) (TALWANI and ACREE, 1984, 1987). The curve showing the cumulative seismicity is the most instructive. There are three distinct breaks in the slope of the curve. The first occurs about three weeks after the beginning of reservoir filling. It is associated with the start of RIS. The increased seismicity continues until February 8, 1978 (the date the reservoir is filled) when there is a second break in the cumulative seismicity curve. This third leg of the curve is associated with an increased rate of seismicity, which persists until about March 7, 1978. A lower rate of seismicity, last leg, follows thereafter. The seismicity between about the last week of December and February 8, 1978 is a result of three effects. These are the increasing stabilizing effect due to the reservoir load (especially below the deepest part of the reservoir), the destabilizing effect due to an instantaneous increase in pore pressure (undrained effect) and a delayed increase in pore pressure due to diffusion. The persistent three week lag between the bends in the filling curve and the cumulative seismicity curve attests to the role of diffusion in the observed seismicity. In the second phase of seismicity, between about February 8, 1978 and March 7, 1978, there is no longer an increase in the stabilizing effect (the reservoir has already been filled) although the effect of diffusion associated with increasing lake levels during the last three weeks of filling dominates, and results in the increased seismicity. After February 8, 1978, there are few (or no) changes in the lake levels and the seismicity is predominantly due to diffusion and possible undrained effects.

In summary, the seismicity following impoundment at Monticello Reservoir displays a coupled poroelastic response that includes pore pressure diffusion and an undrained response, and not ONLY an undrained, instantaneous response that SIMPSON et al. (1988) suggest. The difference arises because TALWANI and ACREE (1984) assume that diffusion and the undrained response begins when filling starts (December 3, 1977) and therefore by the time the reservoir is filled, February 8, 1978 (Figure 11), pore presure has diffused to hypocentral depths. SIMPSON et al. (1988) on the other hand, start the clock after the reservoir is filled (February, 1978) and thus the seismicity observed in February 1978 is classified as "rapid" or "instantaneous," and is *not* attributed to diffusion.

It is clear that "instantaneous" elastic and diffusion related responses are functions of the reservoir size and depths of earthquakes. The coupled, poroelastic, response at a large reservoir would take considerably longer to be fully manifested than for a small shallow reservoir like Monticello Reservoir. The coupled poroelastic response due to impounding at Monticello Reservoir thus appears to be instantaneous when compared with the response of larger reservoirs, although both cases display the same coupled poroelastic response.

SIMPSON et al. (1988) classify the seismicity at Koyna as "delayed response," and note that the major burst of activity did not occur until a number of annual filling cycles had passed. They attribute the seismicity as being dominated by diffusion of pore pressure. Our analysis suggests that the ongoing seismicity at Koyna is a case of protracted seismicity. The seismicity is the coupled poroelastic response to annual lake-level fluctuations.

For reservoirs where the lake-level changes are large, typically a large fraction of the least water depth, and have a longer period (~ 1 year), seismicity continues long after the initial coupled response due to impoundment is over. This protracted seismicity is rare; the two best cases being the seismicity observed at Lake Mead, U.S.A. and Koyna Reservoir, India. Protracted seismicity depends on the frequency of lake-level changes, the reservoir dimensions, hydraulic diffusivity and the geometry of faults *vis à vis* the reservoir. The longer the period of water level changes, the deeper and more pronounced are the effects, often over 10 km, for an annual cycle, and 100 s of m to a few km for daily or weekly cycles.

Thus in cases where the lake level depends on the annual rainfall, it is difficult to control the annual cycle of seismicity. However, for pumped storage dams, where the lake level is controlled by pumping water from the lower reservoir at times of low power demand, the seismicity is restricted to shallow depths and low magnitudes.

Acknowledgements

The ideas in the paper have developed over the years and the research has been supported by grants from the U.S. Geological Survey.

I thank an anonymous reviewer for an insightful review, and Ron Marple, Bob Trenkamp and Lynn Hubbard for their support with the figures and word processing.

REFERENCES

ANDERSON, R. E., and O'CONNELL, D. R. (1993), *Seismotectonic Study of the Northern Lower Colarado River—Arizona, California and Nevada for Hoover, Davis and Parker Dams*, U.S. Department of the Interior, Bureau of Reclamation, Denver, Colorado.

BELL, M. L., and NUR, A. (1978), *Strength Changes Due to Reservoir-induced Pore Pressure and Application to Lake Oroville*, J. Geophys. Res. *83*, 4469–4483.

CARDER, D. S. (1945), *Seismic Investigations in the Boulder Dam Area, 1940–44 and the Influence of Reservoir Loading on Local Earthquake Activity*, Bull. Seismol. Soc. Am. *35*(4), 175–192.

CARDER, D. S., and SMALL, J. B. (1948), *Level Divergences, Seismic Activity and Reservoir Loading in the Lake Mead Area, Nevada and Arizona*, Trans. of the Am. Geophys. Union *29*, 767–771.

CHEN, Q., and NUR, A. (1992), *Pore Fluid Pressure Effects in Anisotropic Rocks: Mechanisms of Induced Seismicity and Weak Faults*, Pure appl. geophys. *139*, 463–479.

GUPTA, H. K., *Reservoir-induced Earthquakes, Developments in Geotechnical Engineering* (Elsevier 1992).

JAEGER, J. C., and COOK, N. G. W., *Fundamentals of Rock Mechanics* (Methuen, London 1969).

LEBLANC, G., and ANGLIN, F. (1978), *Induced Seismicity at the Manic-3 Reservoir, Quebec*, Bull. Seismol. Soc. Am. *68*, 1469–1485.

RAJENDRAN, K., and TALWANI, P. (1992), *The Role of Elastic, Undrained and Drained Responses in Triggering Earthquakes at Monticello Reservoir, South Carolina*, Bull. Seismol. Soc. Am. *82*, 1867–1888.

RICE, J. R., and CLEARY, M. P. (1976), *Some Basic Stress Diffusion Solutions for Fluid-saturated Elastic Porous Media with Compressible Constituents*, Rev. Geophys. *14*, 227–241.

ROELOFFS, E. A. (1988), *Fault Stability Changes Induced Beneath a Reservoir with Cyclic Variations in Water Level*, J. Geophys. Res. *93*, 2107–2124.

SHEN, C., CHANG, C., CHEN, H., LI, T., HUENG, L., WANG, T., YANG, C., and LU, H. (1974), *Earthquakes Induced by Reservoir Impounding and their Effect on the Hsinfengkiang Dam*, Sci. Sinica *17*, 232–272.

SIMPSON, D. W. (1976), *Seismicity Changes Associated with Reservoir Impounding*, Eng. Geol. *10*, 371–385.

SIMPSON, D. W. (1986), *Triggered Earthquakes*, Ann. Rev. Earth Planet. Sci. *14*, 21–42.

SIMPSON, D. W., and NEGMATULLAEV, S. K. (1981), *Induced Seismicity at Nurek Reservoir*, Bull. Seismol. Soc. Am. *71*, 1561–1586.

SIMPSON, D. W., LEITH, W. S., and SCHOLZ, C. H. (1988), *Two Types of Reservoir-induced Seismicity*, Bull. Seismol. Soc. Am. *78*, 2025–2040.

SNOW, D. T. (1972), *Geodynamics of Seismic Reservoirs*, Proc. Sym. Percolation Through Fissured Rock, Stuttgart, Ges. Erd und Grundbau, *T2J*, 1–19.

TALWANI, P. (1997), *Seismotectonics of Koyna-Warna Region*, Pure appl. geophys. (this volume).

TALWANI, P., and ACREE, S. (1984), *Pore Pressure Diffusion and the Mechanism of Reservoir-induced Seismicity*, Pure appl. geophys. *122*, 947–965.

TALWANI, P., and ACREE, S. (1987), *Induced Seismicity at Monticello Reservoir, A Case Study*, Final Technical Report, Contract No. 14-08-0001-21229,22010, U.S. Geological Survey, 271 pp.

TALWANI, P., STEVENSON, D., CHIANG, J., and AMICK, D. (1976), *The Jocassee Earthquakes—A Preliminary Report*, Third Technical Report, U.S. Geological Survey, Reston, Virginia, 127 pp.
TALWANI, P., RUIZ, R., DICKERSON, J., and RAJENDRAN, K. (1992), *Temporal Pattern of Reservoir-induced Seismicity*, EOS, Transactions, AGU, 1992 Fall Meeting *73*, *No. 43*, 405 pp.
TALWANI, P., KUMARA SWAMY, S. V., and SAWALWADE, C. B. (1996), *Koyna Revisited: The Revaluation of Seismicity Data in the Koyna-Warna Area*, 1963–1995, 343 pp.

(Received October 30, 1996, accepted April 10, 1997)

Pure appl. geophys. 150 (1997) 493–509
0033–4553/97/040493–17 $ 1.50 + 0.20/0

Pure and Applied Geophysics

Seismicity around Dhamni Dam, Maharashtra, India

B. K. RASTOGI,[1] P. MANDAL[1] and N. KUMAR[1]

Abstract—Dhamni Dam (height 59 m, capacity 285 Mm^3) was constructed about 100 km north of Mumbai (Bombay), India over the Deccan flood basalt and across the Surya River. The filling of the reservoir started in 1983. Construction of the dam was completed in 1990. However for want of environmental clearance, the maximum water column height in the reservoir since 1988 has been 45 m (8 m short of the maximum possible, 53 m) with the volume of water in the reservoir being 175 Mm^3. The first phase of seismicity started in August 1984, soon after the reservoir reached a 22.5 m depth over the river bed level, and the increased level of seismicity continued for two years, when 605 shocks of $M \geq -1.7$ to 2.5 were recorded. During 1987–93, there were only a few shocks. Seismicity rejuvenated in 1994 when over 2000 shocks of $M \geq -1.7$ to 3.8 occurred, including 20 shocks of $M \geq 3.0$ which occurred during the months of January–February and August–September. Seismicity has continued at a low level during 1995 and 1996. The hypocenters are located in a volume of $10 \times 10 \times 10$ km^3 situated just south of the reservoir along the NW trending Kalu-Surya fault. Correlation of a space-time pattern of seismicity with reservoir filling and the seismic characteristics like b value, foreshock-aftershock pattern and decay rate of aftershocks indicate that the seismicity is reservoir induced.

Key words: Reservoir-induced seismicity, Dhamni Dam, seismicity of India.

Introduction

In the Western Ghats (mountains) of Peninsular India, several reservoirs impounded behind dams have been associated with induced seismicity, viz. Koyna (height 103 m, start/increase of seismicity in 1962), Idukki (169 m, 1975), Mula (56 m, 1972) and Bhatsa (89 m, 1983) (GUPTA *et al.*, 1969, 1972a,b; RASTOGI *et al.*, 1995; RASTOGI, 1995 and RASTOGI *et al.*, 1986). When seismicity continued at Bhatsa, another dam of smaller size (height 59 m and capacity 285 Mm^3) some 50 km north of it started to be filled in 1983. It also began to show seismicity in 1984 with only 22.5 m of reservoir depth. The dam was raised subsequently and intense seismicity occurred in 1994.

This paper describes the geology of the area, particulars of the dam, correlation of water level and seismic characteristics which indicate the possibility that the mild seismicity was induced by reservoir filling with a reservoir depth of 22.5 m only, and intense seismicity followed a few years after the reservoir attained a depth of 45 m.

---

[1] National Geophysical Research Institute, Hyderabad-500 007, India.

Particulars of the Dam

Dhamni Dam (1/10 capacity of Koyna) is situated about 100 km north of Mumbai (Bombay) and 80 km west of Nasik on the Surya River (Fig. 1, Table 1). This dam was designed for power generation and irrigation. Further downstream is a smaller Kavdas Dam which is a pickup weir. The main canals start from both banks of the pickup weir. Both the dams and other pertinent structures are called the Surya Project which is a major river valley project. The project provides irrigation to a 270 km² crop area and will produce about 7 MW of power. The other particulars of the Dhamni Dam are given in Table 1.

Construction of the Kavdas Dam was started in 1975 and completed in 1979. Construction of the Dhamni Dam began in 1978 and was completed in 1990. Filling started in 1983. In 1984, storage was effected to a reservoir depth of 22.47 m and then raised subsequently (Fig. 2, Table 2).

Geology

The Dhamni area lies in the Deccan Traps (volcanics) region. The lava erupted in the late Mesozoic-Cenozoic period, covering an area of over 500,000 km². Its western slope is steep, nearly reaching sea level in a few kilometers while on the eastern side it has a gradual slope. There is a narrow coastal strip of about 40–50 km width between the sea and Western Ghats. The seismically affected areas are located on the steep slope side of the Western Ghats (mountains). The geomorphology of the area indicates a youthful topography characterized by deep gorges, narrow V-shaped valleys, rapids and waterfalls (PATIL *et al.*, 1984). This suggests that the area might be uplifting. The elevations in the area range from 100 to 600 m. Some 10 km east of the study area, the Western Ghats reach the maximum elevation of 1 km above the Mean Sea Level. The Deccan Trap thickness in the study area is of the order of 1 km or more and constitutes dominantly amygdaloidal basalts, exhibiting gentle slopes towards the west (KRISHNAN, 1960). The flows are fine to medium grained, dark gray to green and sparsely to coarsely porphyritic in nature. Generally, the individual flows consist of dense basalt changing into a vesicular type with or without a fragmentary top and are nearly horizontal.

The region is dissected by an intense drainage pattern, resulting in a terraced hilly appearance with moderate to steep slopes. As the Western Ghats in the region slope towards the west, the rivers flow westwards. The Surya River flows westwards commencing about 5 km east of the Dhamni Dam. The area south of the Surya Valley slopes towards the south, facilitating a flow of ground water in that direction. The elevation south of the valley extending to Kunj is around 150 m, falling to approximately 80 m near Vehelpada within a distance of about 5 km.

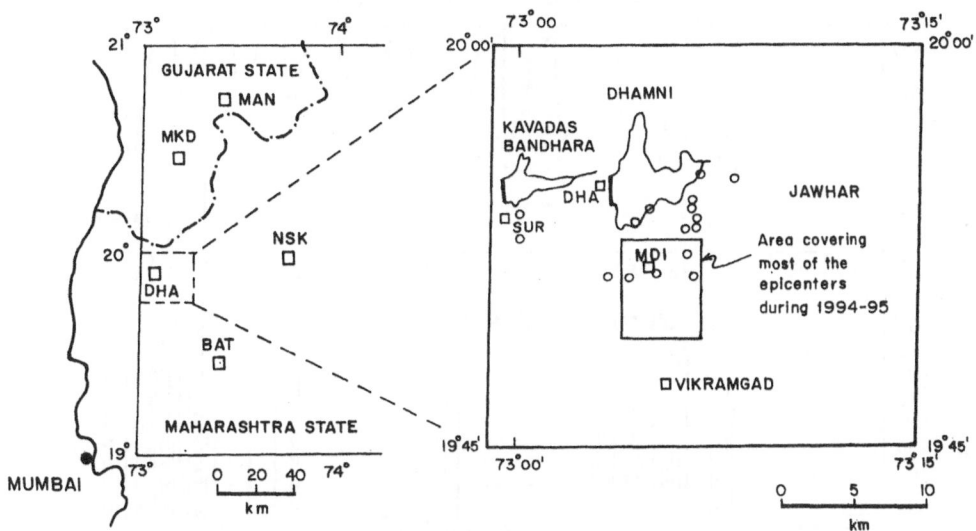

Figure 1

Location of the Dhamni Dam, reservoir boundary, and the epicenters during January 1994 determined by MERI using the data of seismic stations in the states of Maharashtra and Gujarat. The big rectangle indicates the area of maximum concentration of epicenters during 1994–95. Small squares indicate seismic stations.

The basalt flows have been profusely intruded by dolerite and basaltic dikes which continue for distances of 5–15 km. The general trend of the dikes in the area is N-S (Fig. 3).

The Ghod lineament (POWAR, 1981) also named the Kalu-Surya fault (PATIL *et al.*, 1984) is 300 km long. In its southern part it trends NW and then swings towards WNW. Its northern end comes to Surya (Dhamni). Some 50 km south of Dhamni, i.e. near the Bhatsa Dam, intense seismicity was noticed during 1983–84 (RASTOGI *et al.*, 1986) with a maximum magnitude of 4.9 along this fault.

Table 1

Particulars of the Dhamni Dam

1. Height	59.0 m
2. Capacity	285.0 M m³
3. River bed level	67.5 m
4. Crest level of spillway	110.6 m
5. Gates*	12 m × 8 m × 5 no.
6. Full reservoir level	118.6 m

* Note: The crest gates have been erected but kept open and there is no storage against the gates for want of clearance.

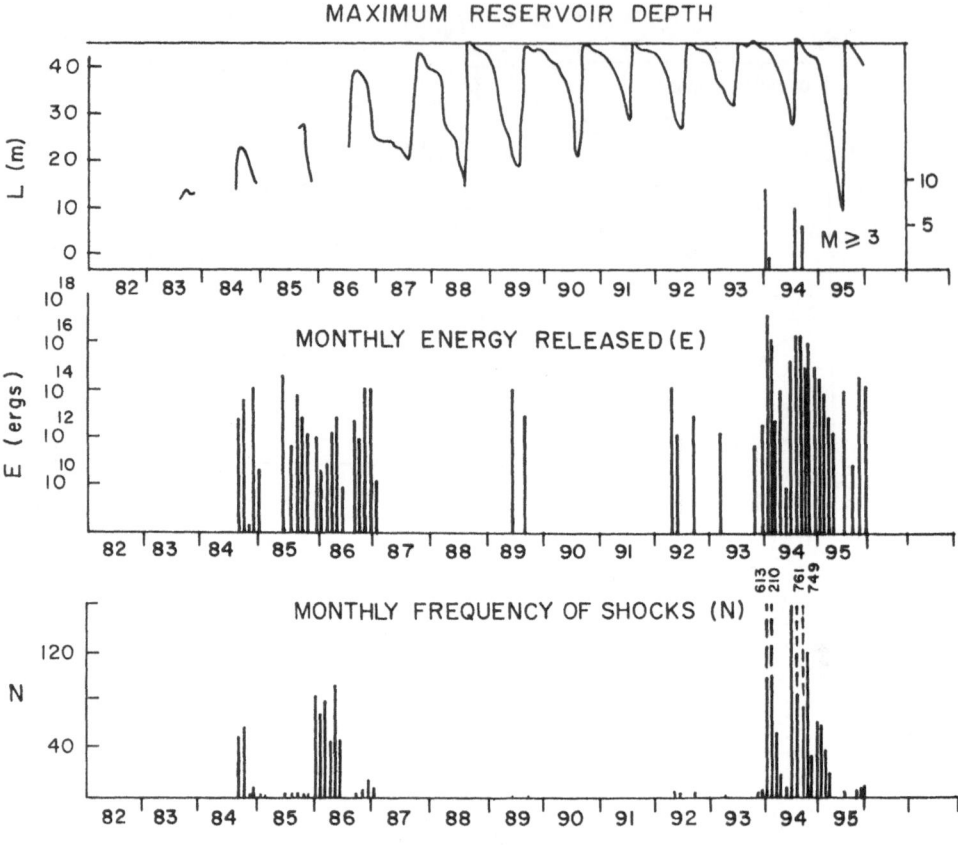

Figure 2
Dhamni (Surya) reservoir levels, monthly frequency of shocks and monthly energy released (1982–95).

Seismic Monitoring

One seismic station at Dhamni was established on June 16, 1982 prior to the start of seismicity and two more at Vikramgad and Surya were installed on November 14, 1984 and May 23, 1985, respectively (Figs. 1 and 3, Table 3). Vikramgad and Surya were withdrawn in 1987. A total of 5 stations were deployed during the period of rejuvenated seismicity in 1994 and 1995 with analog seismographs and vertical component short-period seismometers. Only one station is operating since 1996.

The Determination of Hypocenters

The hypocentral parameters of the earthquakes during July 1994 to December 1995 were estimated using the HYPO71PC program. Hypocenters of all the earthquakes of $M \geq 2.0$ and the majority of the $M \geq 1.0$ earthquakes could be determined. A 4-layered crustal velocity model (Table 4) constrained by DSS results for Koyna Profile-II (KAILA *et al.*, 1981a,b) and a *Vp/Vs* ratio of 1.73 have been used for hypocentral calculations.

The RMS is usually less than 0.05 and on average 0.1. The error in the epicentral location is less that 1 km, whereas the average error in depth estimation is 1.7 km. The estimated focal depths are found to be less than 10 km. The minimum station distance is 0.3 to 7 km. The quality of hypocentral parameters estimation is mainly of class B or C. The magnitude of these earthquakes was estimated from the recorded duration (d), using the relation,

$$M = -0.18 + 1.48 \log{(d)} \tag{1}$$

Hypocenters were determined for 507 shocks which are found to be located in a volume of $10 \times 10 \times 10$ km^3, situated just south of the reservoir. The epicentral locations of most of the shocks are within 5 km radius of the village of Medhi and about 10 km south of the Dhamni reservoir. The epicenters are shown in Figure 4 and the depth section along the profile AB in the NE direction is shown in Figure 5. The depth section in this direction shows the least scatter and shows a near 50° trend towards NE for depths >2 km.

Migration of Hypocenters in Space and Time

The migration of hypocenters in space and time from the start of seismicity in 1984 cannot be established because of inadequate monitoring. However, a migra-

Table 2

Yearly impounding of water in Dhamni Dam
(1984–1996)

Year	Max. water depth (m)
1983	18.77
1984	22.47
1985	38.17
1986	43.10
1987	45.00
1988	45.00
1989	44.00
1990	45.00
1991	45.00
1992	45.00
1993	45.17
1994	45.27
1995	45.27
1996	45.00

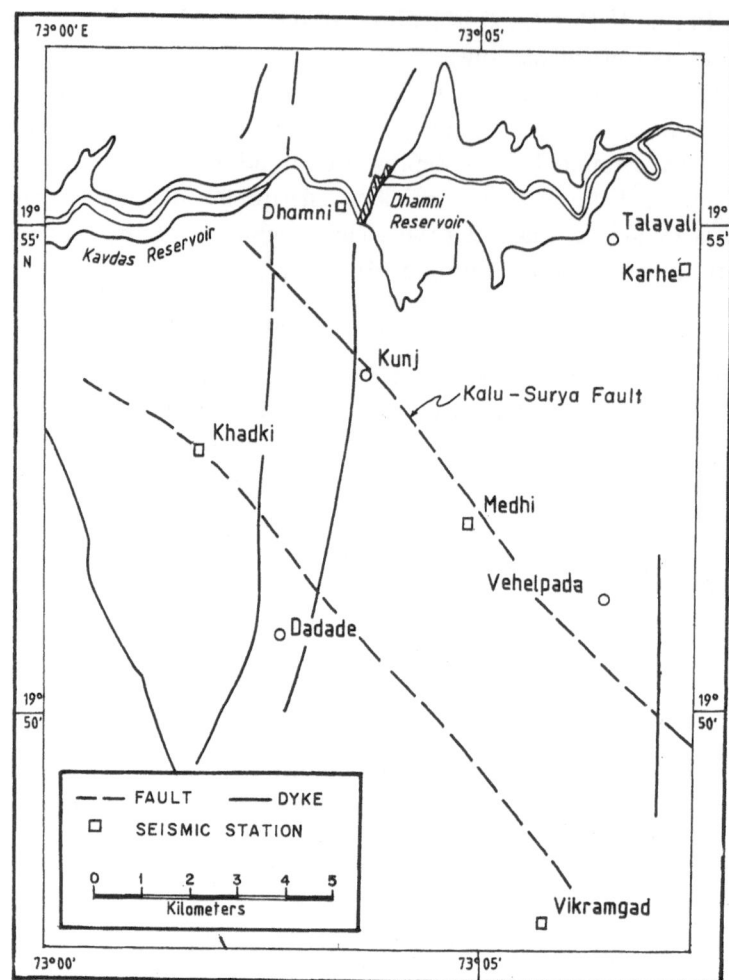

Figure 3
Faults and dikes near Dhamni Dam (after PESHWA and KALE, 1995).

tion towards the south and to deeper depths is indicated for some periods. Initially during 1984–86, the shocks were reported being felt at Surya and Dhamni i.e., near reservoir boundary. In 1994–95, the maximum number of shocks was felt further south. The epicenters during January 1994, located by MERI using the data of Dhamni and other seismic stations in the states of Maharashtra and Gujarat extend to Medhi in the south (latitude 19°52′N, Fig. 1). During August 1994, the epicenters

Table 3

List of seismic stations

Location		Recorder	Seismometer	Date of installation	Date of withdrawal
				D M Y	D M Y
1. Dhamni	(DHA)	MEQ 800	L4-C	16.06.82	
2. Vikramgad	(VIK)	MEQ 800	L4-C	14.11.84	1.11.87
				26.05.94	1.01.97
3. Surya	(SUR)	MEQ 800	L4-C	23.05.85	1.11.87
4. Khadki	(KDH)	PS2	Ranger	27.05.94	1.01.96
5. Karhe	(KRE)	PS2	Ranger	03.06.94	1.01.96
6. Medhi	(MDI)	RV320B	L4-C	10.08.94	1.01.96

migrated further south by 3–4 km to a latitude of 19°50′ (Fig. 6a) and in September 1994, the epicenters migrated further south to 19°48′ (Fig. 6b).

The migration to deeper depth is indicated by the depths of $M \geq 3.0$ shocks as given in Table 5. All the shocks except one until February 1994, i.e., serial numbers 3–8 have depths shallower than 4 km, while most of the shocks during August–September 1994 have depths 4–10 km. During August 1994, all but one of the recorded shocks had depths of <6.5 km, while during September 1994, the depths reach 9.6 km (Fig. 7).

Correlation of Seismicity to Water Levels

Seismic activity started in August 1984, soon after the reservoir reached a 22.5 m depth over the river bed level (Fig. 2) and the increased level of seismicity continued for two years in the first phase. During this phase (1984–86), a total of 143 shocks of $M = 0.0$ to 2.5 (605 shocks including negative magnitudes) were recorded.

The storage to spillway crest level was achieved in 1987, establishing a water column height of 43 m. Due to overflooding there has been an additional 2 m of

Table 4

Velocity model

Velocity (km/s)	Depth (km)
4.950	0.0
6.200	1.0
6.600	20.5
6.900	30.0
8.260	37.0

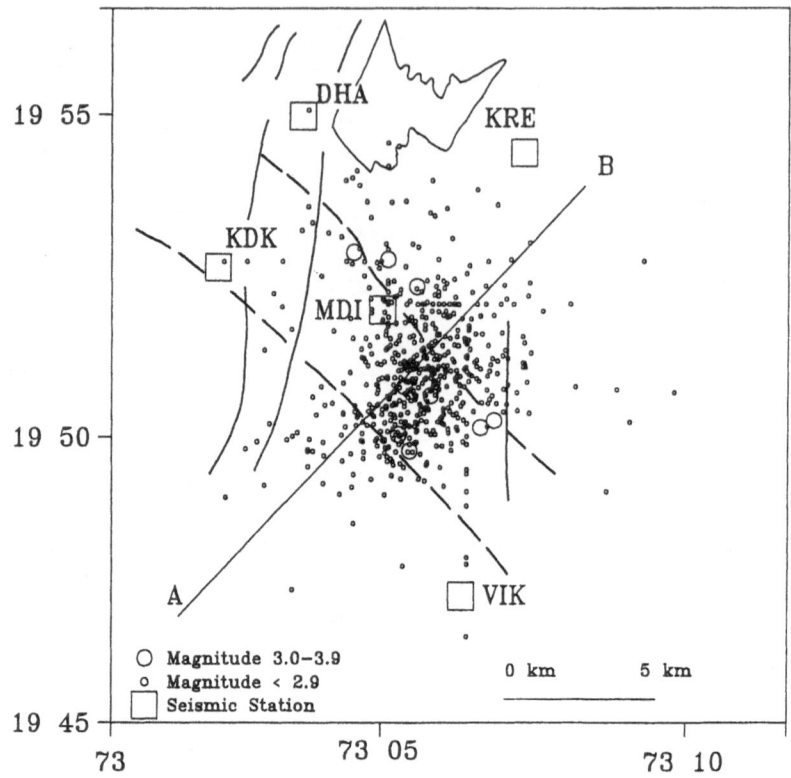

Figure 4
Dhamni reservoir and epicentral map (July 1994–December 1995).

water column, setting a maximum water column height of 45 m during 1988–96. However, the seismicity was negligible during 1987 to 1993. The gross storage at spillway crest level is 175.482 Mm³. When the 8 m high gates are closed the water depth would increase accordingly.

During 1987–93 though the water level was higher, only a few shocks were recorded (Table 6). Seismicity rejuvenated on January 1, 1994. Local shocks were felt in the area and particularly at Medhi and Kunj. A tremor of M3.8, the maximum magnitude to date, occurred on January 9, 1994. During January–February 1994, a total of 511 earthquakes of magnitude 0.0 to 3.8 were recorded (858 including negative magnitudes, from which 621 shocks were recorded in January and 237 in February). There were a total of 8 tremors of magnitude 3 or above during January and February. Seismicity declined during April–June 1994.

As the seismicity enhanced during 1994–95, we have examined the difference in water level or rate of loading during 1993–95 as compared to the previous years. It was observed that during 1993–95 the maximum water depth was about 0.25 m more than that of 1988–92. During 1993, the high water depth over 45 m was kept for considerably longer duration as compared to other years during 1991–95.

In view of the anticipated seismicity after the rains, NGRI established three seismic stations by June 1994 (Table 3). Seismicity subsequently rejuvenated from July 12 soon after the heavy rains with a tremor of magnitude 2.8. On August 3, 1994 a tremor of magnitude 3.2 occurred. On August 19 a repeat tremor of magnitude 3.5 was registered. About 10–20 shocks were being recorded every day. During Aug. 1 to Sep. 7 there were 12 tremors of $M \geq 3.0$ (Table 5). Earthquakes were felt to the maximum in the village of Medhi. During July–December 1994, a total of 1882 earthquakes of $M \geq 0.0$ was recorded.

There was a total of 20 shocks of magnitude ≥ 3.0 which have occurred during January–February and August–September 1994 (Table 5). The time of occurrence of these shocks is related to the maxima of water level. During the rainy season water level is high in July–August and the seismicity is high after some delay during August to November generally and sometimes up to February.

During 1995 a total of 158 shocks of $M = 0.0$ to 2.7 was recorded. Seismicity has continued at a low level during 1996.

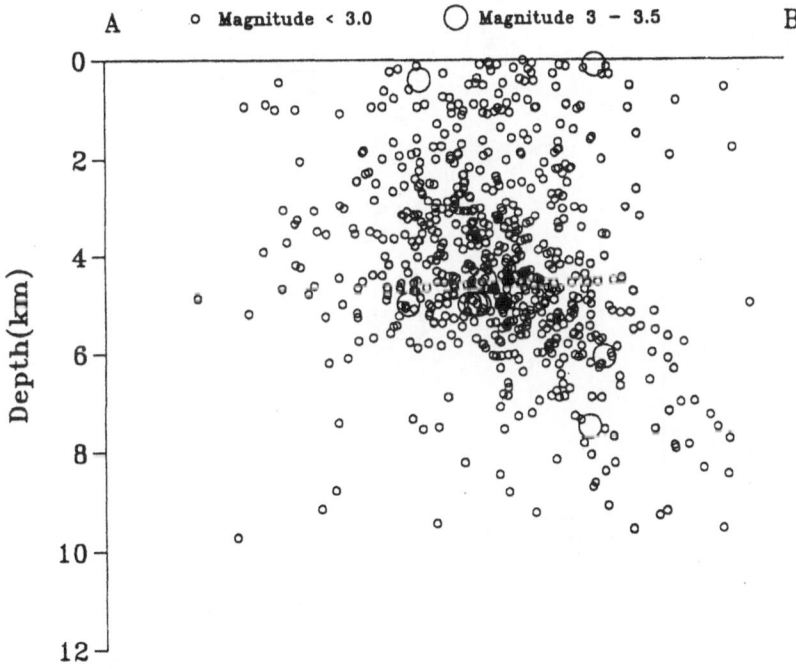

Figure 5
Depth section in NE direction. Both horizontal and vertical scales are the same.

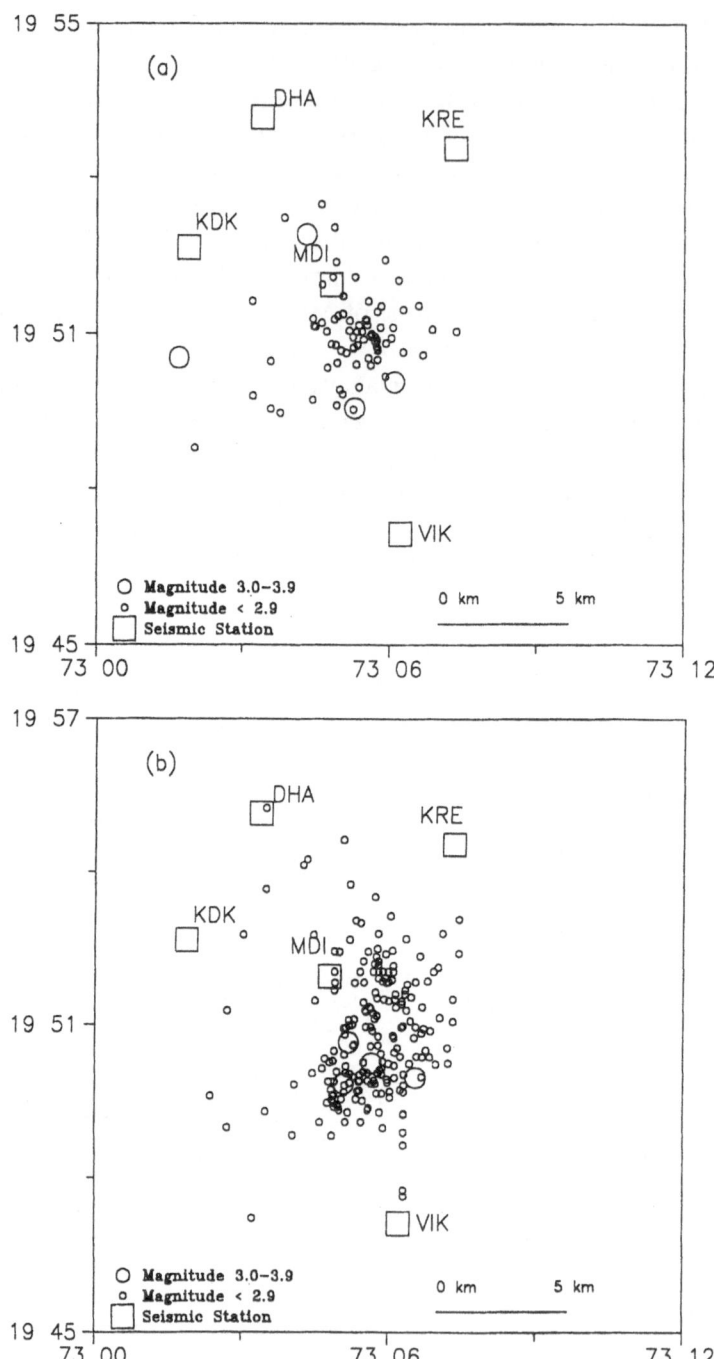

Figure 6
Epicentral locations during (a) August 1994, (b) September 1994.

Composite Fault Plane Solution

A composite fault plane solution has been determined from first motion directions for shocks of $M \geq 3.0$. The data has been plotted on the lower hemisphere of the equal area projection as shown in Figure 8. The figure indicates that the NW striking and steeply dipping nodal plane could be the fault plane which matches with the Kalu-Surya fault and indicates normal faulting. Nearly vertical dip is supported by the depth section.

The Discriminatory Characteristics

The Foreshock-Aftershock Pattern

Figure 9 shows the foreshock-aftershock pattern for the main earthquake on January 9, 1994. It indicates a swarm type pattern which is of Type III model of Mogi. The swarm nature of activity is typical of RIS when strength changes occur in a critically stressed area. Under the influence of a reservoir the rock strength

Table 5

List of $M \geq 3.0$ earthquakes

Sr. No.	Date	Origin time (GMT)			Location		Depth (km)	Mag.
		H	M	S	Lat. N	Long. E		
1.	940102	04	47					3.0
2.	940107	10	33					3.5
3.	940108	23	02	58.00	(19 47.30	73 04.10	5.0)	3.5
4.	940109	05	30		(19 52.80	73 05.70	1.5)	3.8
5.	940113	01	16		(19 49.20	73 03.00	1.2)	3.3
6.	940114	11	24	11.20	(19 49.90	73 05.70	1.1)	3.5
7.	940119	00	07		(19 50.00	73 06.20	3.0)	3.4
8.	940216	17	09	28.50	(19 53.00	73 07.00	1.7)	3.6
9.	940801	08	37		–	–	–	3.0
10.	940802	03	39		–	–	–	3.0
11.	940803	00	32	14.50	19 52.90	73 04.30	0.1	3.2
12.	940803	00	33		–	–	–	3.0
13.	940819	01	23	47.80	19 50.08	73 06.10	4.5	3.3
14.	940819	11	51	32.34	19 50.54	73 01.70	9.3	3.5
15.	940830	09	36	47.65	19 49.57	73 05.30	5.0	3.2
16.	940902	21	23		19 49.84	73 05.10	0.4	3.3
17.	940903	06	11		19 50.66	73 05.20	5.0	3.3
18.	940907	07	08	53.21	19 50.25	73 05.67	5.0	3.3
19.	940907	13	10		–	–	–	3.0
20.	940907	14	30	44.09	19 49.96	73 06.56	5.0	3.1

Note: Parenthesis indicates approximate locations with data of stations in Maharashtra state.

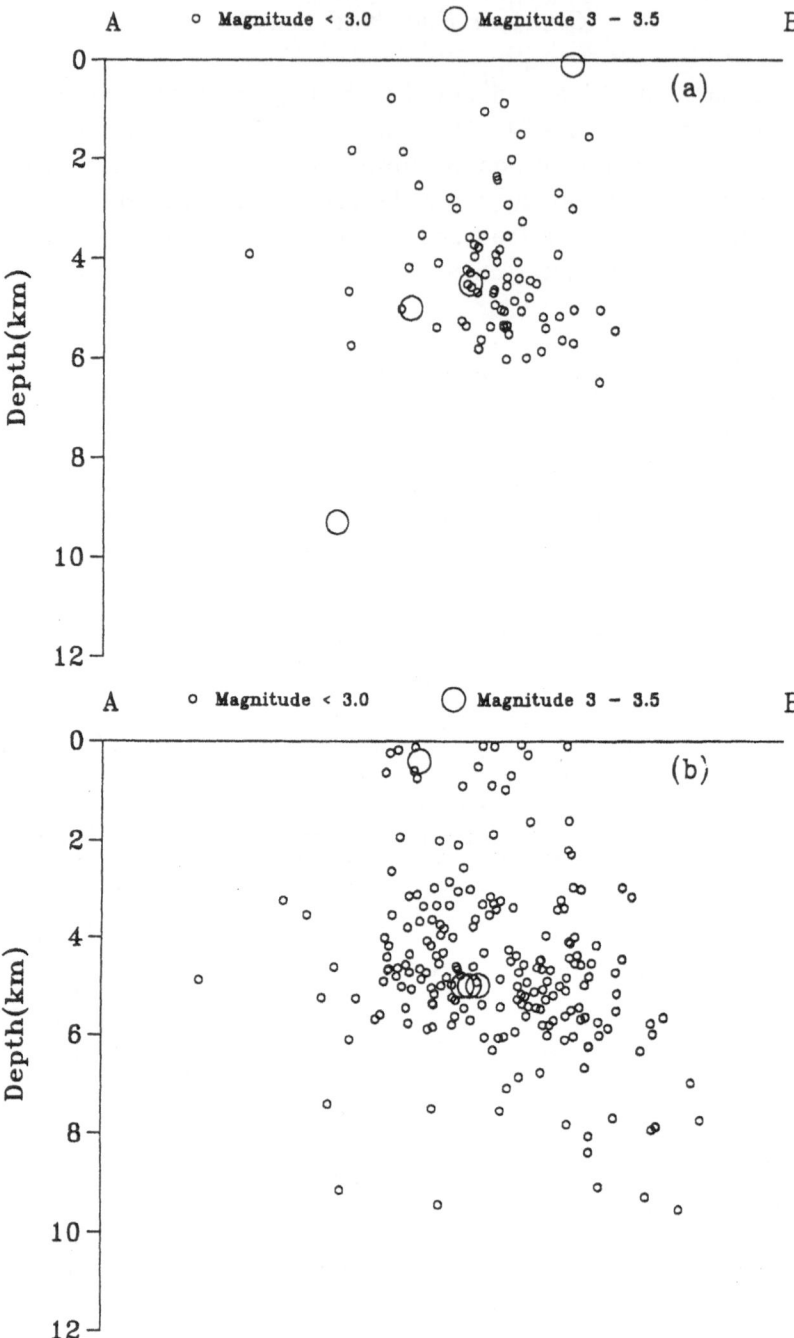

Figure 7
Depth sections during (a) August 1994, (b) September 1994.

decreases. In many cases of RIS the homogeneous rock condition has been inferred to change to slightly heterogeneous, resulting in a good number of foreshocks or Type II model of Mogi (GUPTA et al., 1972b). It can also change to quite heterogeneous conditions, giving rise to a swarm type of activity. The same type of pattern is seen for other independent earthquakes of $M > 3.5$ in the sequence such as the August 19, 1994 shock.

The decay rate of aftershocks was calculated to be 0.8 as shown in Figure 9. This rate is slow and identical to that for other cases of RIS (GUPTA et al., 1972b).

The b Value and M_1/M_0 Ratio

As only one station was operating at the time of the largest tremor of $M = 3.8$ on January 9, 1994, the b value for foreshocks and aftershocks was not estimated due to a slight uncertainty of magnitude. For the shock of $M = 3.5$ on August 19, 1994 when the network was operating, the foreshocks b value of 0.72 is comparable to the aftershocks b value of 0.83 (Fig. 10). These values verge on those for Koyna and are higher than the regional value of about 0.5 in Peninsular India (GUPTA et al., 1972a). The ratio of the magnitude of the largest aftershock (M 3.6) to that of the main shock (M 3.8) is 0.9 which is high.

Discussions

The region around the Dhamni Dam i.e., northern part of the west coast region of India is moderately seismic with a record of a few damaging earthquakes of magnitudes 5 and 6 and some 100 other felt earthquakes. The region is traversed by

Table 6

Annual number of shocks

Year	$M \geq -1.7$	$M \geq 0.0$	$M \geq 3.0$
1983	Nil		
1984	135	71	
1985	89	18	
1986	381	116	
1987	Nil		
1988	Nil		
1989	1	1	
1990	Nil		
1991	Nil		
1992	5	5	
1993	7	2	
1994	2740	2393	20
1995	158	158	

PHI	DIP	RAKE		TREND	PLUNGE
A: 121.0	88.0	−90.0		T: 211.0	43.0
B: 301.0	2.0	−90.0		P: 31.0	47.0

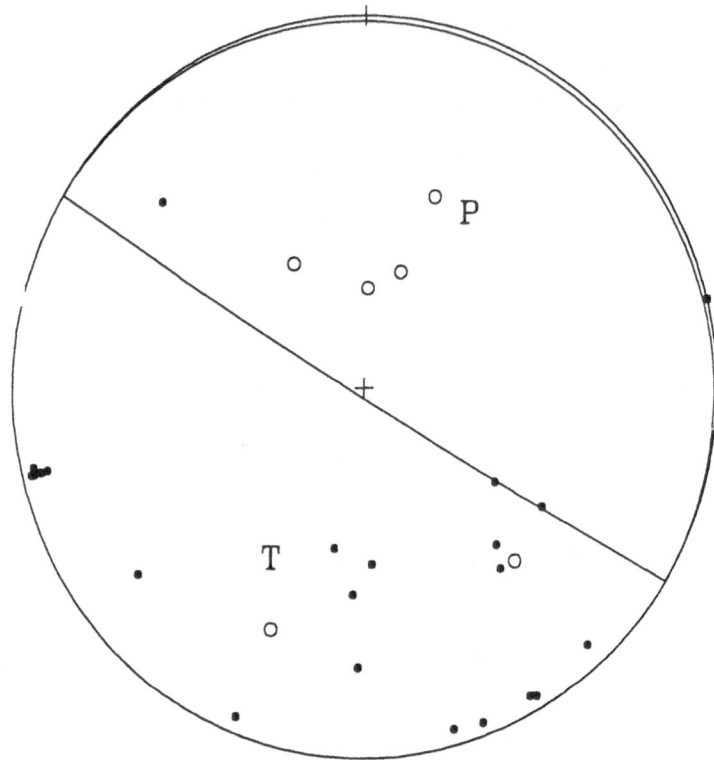

Figure 8

Composite fault plane solution (plotted on lower hemisphere) for the earthquakes of $M \geq 3.0$ during the period July 1994–December 1995 supports a NW trending near vertical fault. Filled circles are compressions and open circles are dilatations.

a number of faults, some of which may be critically stressed and prone to reservoir-induced seismicity. Hence, monitoring of seismicity around the Dhamni Dam was started prior to its filling. Filling began in 1983. Seismicity initiated in August 1984, a few days after the water column reached a maximum depth of 22.5 m. The seismicity nearly ceased after two years, though the filling continued. It rejuvenated after a gap of seven years in 1994, when 20 shocks of $M = 3.0$ to 3.8 and 2720 other smaller shocks occurred.

The shocks of magnitude ≥ 3.0 occurred during January–February and August–September, 1994. The time of occurrence of these shocks is related to the maxima of water level as also observed for the Koyna reservoir. During the rainy season the water level is high in July–August and the seismicity is high after some delay during

August to November usually and sometimes extending to February (GUPTA and RASTOGI, 1976; GUPTA, 1992).

Accurately determined hypocenters indicate that the seismically active zone of $10 \times 10 \times 10$ km^3 is situated adjacent to the reservoir. The spatial-temporal correlation of seismicity with the newly filled reservoir is rather strong as observed for other cases of reservoir-induced seismicity. The hypocenters appear to have migrated away from the reservoir boundary towards the south and have become deeper with time, indicating the role of pore pressure diffusion in triggering of

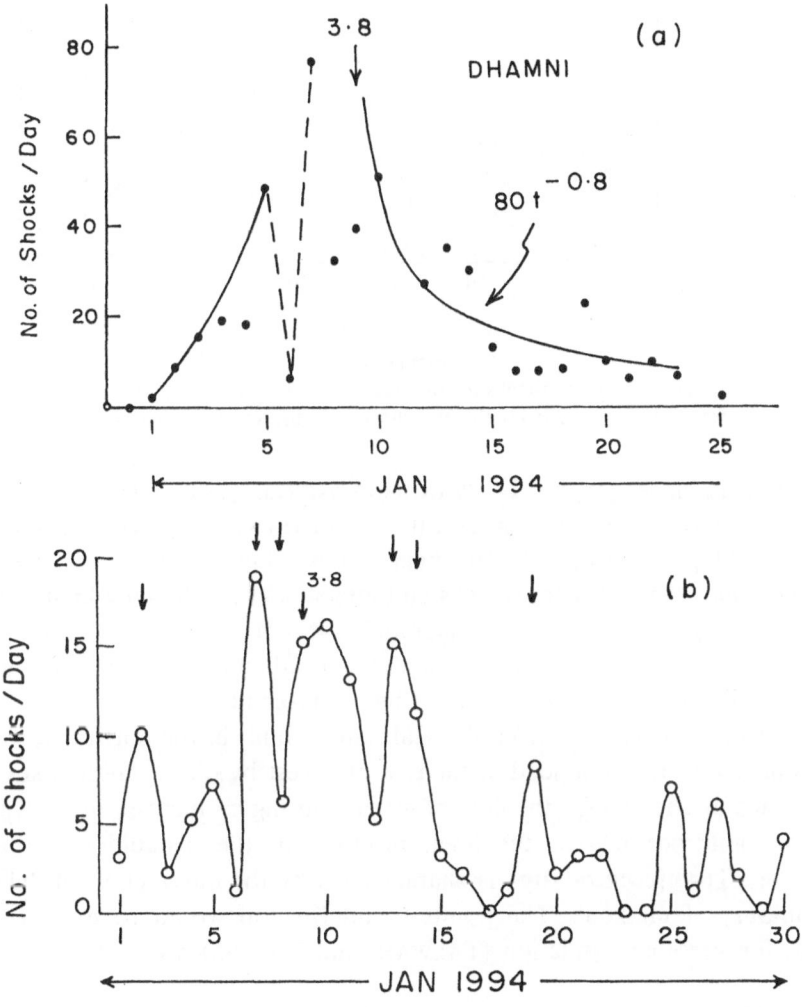

Figure 9
(a) Foreshock-aftershock pattern for M 3.8 shock on January, 1994 (daily number of all the recorded shocks) shows Mogi's Type III (swarm) pattern and decay of aftershock at a slow rate (h or $p = 0.8$) as observed for other cases of RIS. A precursory quiescence has also been observed. Part (b) shows the daily number of shocks of $M \geq 1.0$ and indicates a swarm pattern more clearly. Arrows indicate the time of occurrence of $M \geq 3.0$ earthquakes.

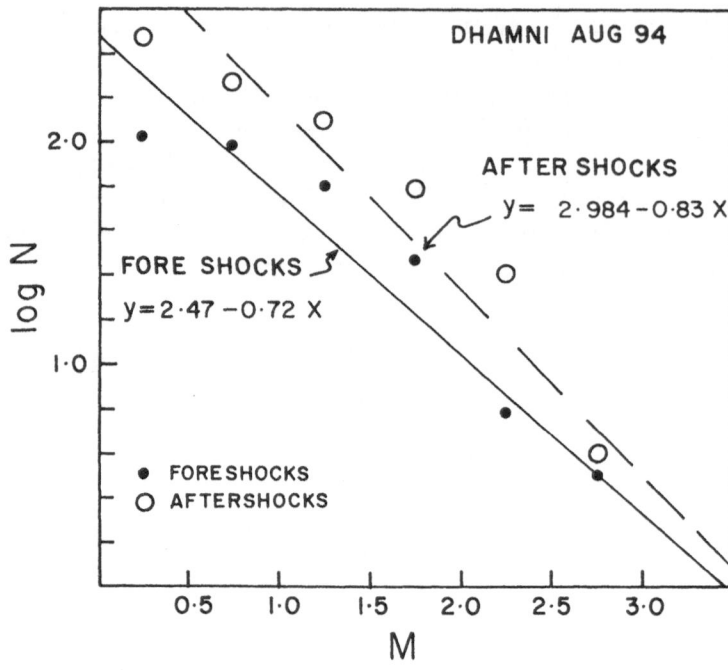

Figure 10
The *b* value for foreshocks and aftershocks for the shock of *M* 3.5 on August, 1994. The *b* value for the foreshocks is similar to that for the aftershocks as observed for other cases of RIS.

earthquakes as envisaged by TALWANI and ACREE (1985). The discriminatory seismic characteristics as seen for several cases of reservoir-induced seismicity such as the swarm type of foreshock-aftershock pattern, the slow decay of aftershocks, nearly the same *b* value for foreshocks and aftershocks (both being higher than the regional *b* value) along with a high magnitude ratio of the largest aftershock to the main shock are also observed in the case of Dhamni seismicity. Hence, this seismicity is likely to have been induced by reservoir filling.

The seismic zone extends from the Kalu-Surya fault in the north and a parallel fault in the south. It is bounded in the east and west by dikes. These dikes appear to have acted as barriers for the flow of water, causing an increase in pore pressure along the fault portions in the area, resulting in the instability and induced earthquakes. The epicentral area is characterized by the intersection of dykes with the Kalu-Surya fault zone. The points of intersection would be more favorable locations for stress accumulation (TALWANI and RAJENDRAN, 1991).

Conclusions

The 300-km long NW trending Kalu-Surya fault which extends to Surya has been activated for a length of about 12 km and depth of 10 km. The dikes have acted as barriers to the flow of water from the reservoir, causing zones of instability by increasing pore pressure along the faults bounded by them.

The spatial-temporal correlation of seismicity with the Dhamni reservoir filling and the discriminatory characteristics indicate that the Dhamni seismicity is reservoir-induced. A low level of seismicity with a magnitude reaching 2.5 started with a water column of 22.5 m and earthquakes of magnitude up to 3.8 occurred after a few years of water column reaching 45 m.

Acknowledgements

The authors are grateful to Dr. Harsh K. Gupta, Director, NGRI for his interest in the study, to MERI, Nasik for providing the data and to the Department of Science and Technology, New Delhi for support.

REFERENCES

GUPTA, H. K., NARAIN, H., RASTOGI, B. K., and MOHAN, I. (1969), *A Study of the Koyna Earthquake of December 10, 1967*, Bull. Seismol. Soc. Am. *59*, 1149–1162.

GUPTA, H. K., RASTOGI, B. K., and NARAIN, H. (1972a), *Common Features of the Reservoir Associated Seismic Activities*, Bull. Seismol. Soc. Am. *62*, 481–492.

GUPTA, H. K., RASTOGI, B. K., and NARAIN, H. (1972b), *Some Discriminatory Characteristics of Earthquakes near the Kariba, Kremasta and Koyna Artificial Lakes*, Bull. Seismol. Soc. Am. *62*, 493–507.

GUPTA, H. K., and RASTOGI, B. K., *Dams and Earthquakes* (Elsevier, Amsterdam 1976) 229 pp.

GUPTA, H. K., *Reservoir Induced Earthquakes* (Elsevier, Amsterdam 1992) 355 pp.

KAILA, K. L., MURTHY, P. R. K., RAO, V. K., and KHARETCHKO, G. E. (1981a), *Crustal Structure from Deep Seismic Sounding along the Koyna II (Kelsi-Loni) Profile in the Deccan Trap Area, India*, Tectonophys. *73*, 365–384.

KAILA, K. L., REDDY, P. R., DIXIT, M. M., and LAZARENKO, M. A. (1981b), *Deep Crustal Structure at Koyna, Maharashtra, Indicated by Deep Seismic Sounding*, J. Geol. Soc. India 22, 1–16.

KRISHNAN, M. S., *Geology of India and Burma* (Higginbothams, Madras 1960) 604 pp.

PATIL, D. N., BHOSALE, V. N., GUHA, S. K., and POWAR, K. B. (1984), *The Khardi Earthquakes, 1983*, Current Sci. *53*, 805–806.

PESHWA, V. V., and KALE, V. S., *Photogeological studies in the Surya project area* (Report Poona Univ. India, 1995) 35 pp.

POWAR, K. B. (1981), *Lineament Fabric and Dyke Pattern in the Western Part of the Deccan Volcanism and Related Basalts Provinces in Other Parts of the World*, Mem. GSI Ind. 3, 46–57.

RASTOGI, B. K., CHADHA, R. K., and RAJU, I. P. (1986), *Seismicity near Bhatsa Reservoir, Maharashtra, India*, Phys. Earth Planet. Inter. *44*, 177–199.

RASTOGI, B. K., *Correlation of filling history with seismicity near artificial water reservoirs*, NATO ASI Series, Partnership Sub-Series 2. Environment-Vol. 4, *Earthquakes Induced by Underground Nuclear Explosions* (eds. R. Console and A. Nikolaev) (Springer-Verlag 1995) pp. 343–352.

RASTOGI, B. K., CHADHA, R. K., and SARMA, C. S. P. (1995), *Investigations of June 7, 1988 Earthquake of Magnitude 4.5 near Idukki Dam in Southern India*, Pure appl. geophys. *145*(1), 109–122.

TALWANI, P., and ACREE, S. (1985), *Pore Pressure Diffusion and the Mechanism of Reservoir-induced Seismicity*, Pure appl. geophys. *122*, 947–965.

TALWANI, P., and RAJENDRAN, K. (1991), *Some Seismological and Geometric Features of Intraplate Earthquakes*, Tectonophys. *186*, 19–41.

(Received February 1, 1997, accepted July 14, 1997)

Pure appl. geophys. 150 (1997) 511–550
0033–4553/97/040511–40 $ 1.50 + 0.20/0

Pure and Applied Geophysics

Seismotectonics of the Koyna-Warna Area, India

Pradeep Talwani[1]

Abstract—Reservoir-induced seismicity has been observed near Koyna Dam, India since the early 1960s. In order to understand the seismotectonics of the region we analyzed available seismicity data from 1963 to 1995. Over 300 earthquakes with M ≥ 3.0 were relocated using revised location parameters (station locations, velocity model, station delays and V_p/V_s ratio). The spatial pattern of earthquakes was integrated with available geological, geophysical, geomorphological data and observations following the M 6.3 earthquake in December 1967, to delineate and identify the geometry of seismogenic structures. From this integration we conclude that the area lying between Koyna and Warna Rivers can be divided into several seismogenic crustal blocks, underlain by a fluid-filled fracture zone. This zone lies between ~6 and 13 km and is the location of the larger events (M ≥ 3.0). The seismicity is bounded to the west by the Koyna River fault zone (KRFZ) which dips steeply to the west. KRFZ lies along the N–S portion of the Koyna River and extends S10°W for at least 40 km. It was the location of the 1967 Koyna earthquake. The seismicity is bounded to the east by NE–SW trending Patan fault, which extends from Patan on the Koyna River, SW to near Ambole on the Warna River. Patan fault dips ~45° to the NW and was the location of the M 5.4 earthquake in February 1994. The bounding KRFZ and Patan fault are intersected by several NW–SE fractures which extend from near surface to hypocentral depths. They form steep boundaries of the crustal blocks and provide conduits for fluid pressure flow to hypocentral depths. Sharp bends in the Koyna and Warna rivers (6 km south of Koyna Dam and near Sonarli, respectively) are locations of stress build-up and the observed seismicity.

Key words: Reservoir-induced seismicity—case history, seismotectonics, Koyna-Warna earthquakes.

Introduction

The observed seismic activity in the Koyna-Warna area, Maharashtra State, India is unique. It is the only known location in the world where seismicity that began after the start of impoundment of a reservoir (Shivajisagar behind the Koyna Dam), has persisted for more than 30 years. The moderate seismicity has included at least six earthquakes with $M > 5.0$, including the largest, a M 6.3 event that occurred near Koyna at 4:21 a.m. (Indian Standard Time) on December 11, 1967 (22:51 UTC on December 10, 1967). It is also extraordinary because seismicity has been continuously monitored and also because of a number of studies that have been conducted. It is also interesting because until recently we lacked a clear

---

[1] University of South Carolina, Department of Geological Sciences, Columbia, South Carolina 29208, U.S.A.

understanding of the seismotectonics of this area. In order to understand the seismotectonics we reanalyzed the seismicity data for the period 1963–1995 and integrated it with available complementary lake level, geological, geophysical, geomorphic and neotectonic data. The results of the integration are the subject of this paper. The results suggest that the region between Koyna and Warna consists of several tectonic blocks and the earthquakes occur on the edges of these blocks.

1. Reanalyses of Seismicity Data

Seismicity was first observed in the Koyna area following the start of impoundment of Shivajisagar in 1961. The first seismological observatory to monitor seismic activity in the Koyna region was established in Koyna in 1963. Following the destructive Koyna earthquake of December 10, 1967, a seismological network consisting of seven stations was established during the years 1967 to 1972. A five-station Warna seismic network was established in 1990 to monitor the seismic activity in the neighboring Warna valley, about 35 km south of Koyna. This activity followed the impoundment of the Warna Reservoir. The quality of recording and analysis improved over the years. However, due to various factors the locations obtained by routine analyses (see e.g., Fig. 4.32b in GUPTA, 1992) were inaccurate and inadequate for meaningful tectonic interpretation of the recorded seismicity. The first step in the reanalyses of the seismicity data was to improve the accuracy of the epicentral and (later) hypocentral locations.

To improve the location accuracy the following location parameters, were addressed: (i) location of the seismic stations, (ii) velocity structure, (iii) delay time at a particular station and (iv) ratio of the P-wave to S-wave velocities (V_p/V_s).

In June 1994 all stations of the Koyna and Warna seismic networks were visited. The station coordinates obtained by using a hand held Global Positioning System (GPS) were compared with the ones in use before. The stations were found to be misplaced by an average of 2 km, with individual mislocations varying from a few meters to over 5 km. The new station coordinates are given in Table 1.

Most of the computer algorithms in use today for locating earthquakes assume a layered seismic velocity model. To choose among the various models that were available, we located a subset of well recorded events with the different models. Based on lower RMS values of travel-time residuals, KAILA's (1983) model (Table 2) was chosen for further analysis (see TALWANI et al. (1996) for details).

By iteratively calculating the RMS travel-time residuals for a selected set of earthquakes, the station delays were estimated for stations of the seismic network. These have been incorporated in Table 1. An average V_p/V_s ratio of 1.70 was obtained by plotting Wadati plots (plotting S-P times against P-wave arrival times), for a selected set of 40 well-timed earthquakes recorded in 1993–94.

Table 1

Revised coordinates and station delays for stations of Koyna-Warna seismic network

No.	Station	Station Code	Latitude °N		Longitude °E		Elevation (m)	Station Delay (sec)
			Degrees	Min	Degrees	Min		
1.	Alore	ALO	17°	28.59	73°	38.27	100	−0.07
2.	Chikhali	CKL	17°	14.83	73°	35.17	110	+0.07
3.	Chiplun	CPL	17°	30.96	73°	31.69	75	−0.03
4.	Govalkot	GOV	17°	32.94	73°	29.15	85	
5.	Kolhapur	KOL	16°	42.88	74°	14.53	565	+0.29
6.	Kokrud	KOK	17°	00.45	73°	58.87	555	+0.03
7.	Koyna (Quarry)	KNI	17°	24.50	73°	44.78	640	+0.06
8.	Mahabaleshwar	MAH	17°	55.39	73°	39.71	1360	
9.	Marathwadi	MRT	17°	13.28	73°	56.25	650	+0.09
10.	Pophali	PPL	17°	26.39	73°	40.98	200	
11.	Ratnagiri	RAT	16°	58.98	73°	18.65	40	+0.23
12.	Sakharpa	SKP	16°	59.75	73°	42.80	250	+0.04
13.	Satara	STA	17°	41.31	74°	00.89	650	+0.19
14.	Warnawati	WRN	17°	07.38	73°	53.09	580	+0.02

1A. Relocation of Koyna-Warna Earthquakes

The number of stations and their configuration changed over the years from a four-station network in 1963 to the current eleven stations (Table 1). (Of the 14 stations listed in Table 1, GOV, PPL and MAH are currently inoperative). The accuracy of the time imprinted on the records also improved after 1984–85. Over 90,000 earthquakes have been recorded of which over 1400 had $M \geq 3.0$. Consequently the relocation of the earthquakes, using revised location parameters, was divided into four time periods (Table 3).

Only 64 events were located for the period before the large event on December 10, 1967. These included 17 events with magnitudes between 2.0 and 3.0, Set A in Table 3. These events were located using data from KNI, GOV, STA and MAH stations (Table 1). There were over 54,000 earthquakes between December 10, 1967 and 1982. Of these there were 50 with $M \geq 4.0$. Due to poor absolute times at different stations for most events we used $(S\text{-}P)$ times to locate them. The S phase could be clearly distinguished on the distant stations and also on the horizontal component seismographs at various stations. Forty-eight events were located (Set B). Of the nearly 24,000 events recorded between 1983 and 1992, there were 177 with $M \geq 3.0$. Of these there were adequate data to accurately locate 109 events (Set C). Due to the addition of stations in the Warna network and faster recording speed on the MEQ-800 seismographs, earthquakes recorded after 1993 were placed in a different set. Of the 108 events with $M \geq 3.0$, 100 were located (Set D). A complete listing of the hypocentral parameters is given elsewhere (TALWANI et al.,

Table 2

P-wave velocity model

V_p km/s	Depth km
4.900	0.0
5.335	1.0
5.765	2.0
5.895	4.0
6.025	6.0
6.155	8.0
6.380	10.0
6.470	13.0
6.560	16.0
6.600	19.0
6.805	25.0
6.895	28.0
6.985	31.0
7.105	34.0
8.100	38.0

KAILA (Pers. Comm., 1983)

Table 3

Recorded and located earthquakes (1963–1995) (Koyna-Warna Region)

Set	Period	Earthquakes Recorded			Earthquakes Located Total	Quality				Remarks
		Total	$M \geq 3.0$	$M \geq 4.0$		A	B	C	D	
A	1963 to December 10, 1967	1,575	128	7	64	–	–	24	40	Data from three or more stations are available for 64 earthquakes (M 2.0 to 6.3)
B	December 10, 1967 to 1982	54,255	995	50	48	–	1	31	16	Earthquakes with $M \geq 4.0$ were selected for locations; 2 events could not be located due to insufficient data
C	1983 to 1992	23,984	177	9	109	–	6	72	31	Earthquakes of $M \geq 3.0$ were selected for locations; 68 events could not be located due to insufficient data
D	1993 to 1995 (April)	10,311	108	14	100	1	58	40	1	Earthquakes of $M \geq 3.0$ were selected for locations; 8 events could not be located due to insufficient data
	Total	90,125	1,408	80	321	1	65	167	88	

Table 4

List of Koyna-Warna earthquakes with $M \geq 5.0$ and other significant events

Date	HM	Magnitude MERI	Magnitude GUPTA (1992)	Magnitude USGS (m_b)	(M_s)
September 13, 1967		5.8*		Foreshock	
December 11, 1967		6.3	6.3		
December 24, 1967		5.0		5.2	
				Aftershock	
October 29, 1968		5.0			
October 17, 1973		5.1	5.1		
September 2, 1980		4.3		4.9	5.5
September 20, 1980	0728	4.7		4.9	4.3
September 20, 1980	1045	4.9		5.3	4.2
October 4, 1980	1637	4.1		4.5	
February 5, 1983	2253	4.4		4.2	
November 14, 1984	1158	4.4		4.6	
January 6, 1991	2213	4.8		4.4	
August 28, 1993	0426	4.9		4.9	4.5
September 3, 1993	2301	4.7		4.7	
December 8, 1993	0142	5.1		5.0	4.6
February 1, 1994	0930	5.4		5.0	

USGS magnitudes from Preliminary Determination of Epicenters.
* Magnitude of GUHA et al. (1970), revised to 5.2 in GUHA et al. (1974) and LANGSTON (1981) claims it is $M \leq 4.5$.

1996) and the larger events are listed in Table 4. The epicentral locations with quality D were not considered reliable, and as there were few with quality B or better, to study the spatial pattern of the seismicity we plotted the earthquakes with quality C or better for the various sets. The solution quality ratings, A–D, of the hypocenter indicate the general reliability of the solution. Quality rating A indicates excellent epicenter and a good focal depth; B, good epicenter and fair depth; C, a fair epicenter and poor depth, while D, indicates poor epicenter and depth.

1A.1 Spatial Distribution of Seismicity 1963–1995

Figure 1 shows the location of seismicity for the period between October 1964 and December 1967 (Set A, $M \geq 2.0$ QC). Most of the earthquakes are located north of the east-west segment of Koyna River and near the deepest part of the Shivajisagar (Koyna Reservoir). These earthquakes followed the initial impoundment of the reservoir and their spatial distribution suggests a possible causal association. Set B has been divided into two subsets: December 10, 1967 to 1973 and 1974 to 1982. (There was a M 5.1 event in October 1973). The distributions of $M \geq 4.0$ events occurring in these periods are shown in Figures 2 and 3. The earthquakes occurring between 1967 and 1973 (Fig. 2) are located to the south of the epicentral area of Set A and to the north of Warna River. The epicenters seem to be concentrated in a broad zone to the south of the right angle bend in the Koyna River. The epicenters lying between the Koyna and Warna rivers also suggest a NW-SE trend. This NW-SE trend is also seen in the locations of the aftershocks (solid dots) of the December 1967 event. The $M \geq 4.0$ events between 1974 and 1982 seem to define two areas of activity, an apparent E-W zone just to the south of Koyna River and a broad zone surrounding the Warna River (Fig. 3). The seismicity between 1983 and 1992 ($M \geq 3.0$, QC or better) is very widespread, covering a broad region from the Koyna to the Warna rivers (Fig. 4). The annual distribution shows generally decreasing seismicity from 1983 to 1987 in a broad area south of the Koyna River and extending to the Warna River. The seismicity in the vicinity of the Warna Reservoir appears to be spatially related to the annual filling (impoundment of Warna Reservoir began in 1985). This is particularly noticeable from 1988 onwards. The seismicity in 1992 ($M \geq 3.0$, QC or better), shown in solid dots, appears to be along a NNE-SSW segment which is collinear with the southern part of the Koyna Reservoir and Koyna River before the big bend (Fig. 4). This trend was better defined with subsequent earthquake activity.

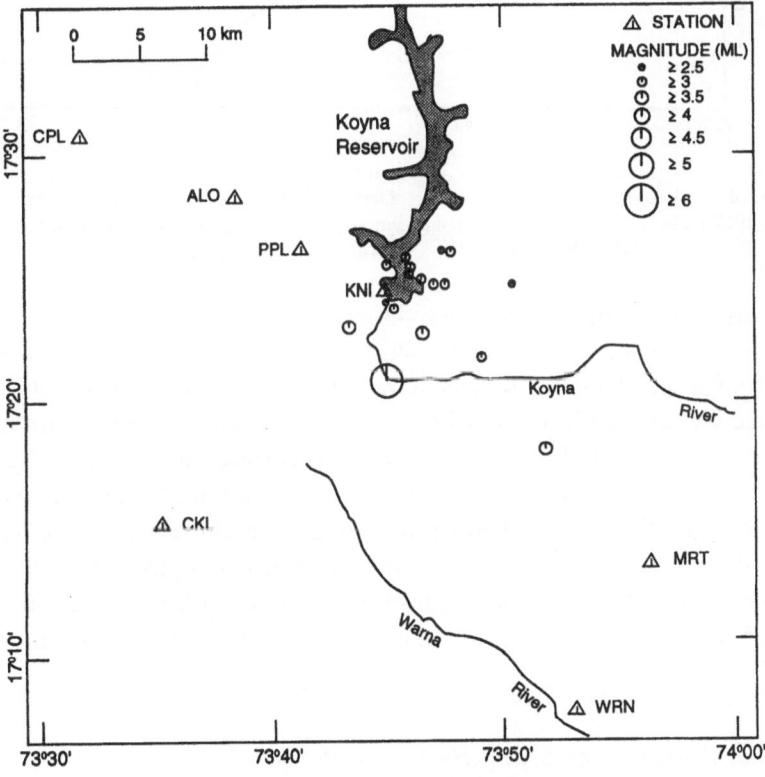

Figure 1
Relocation of earthquakes between 10/1964 and 12/1967.

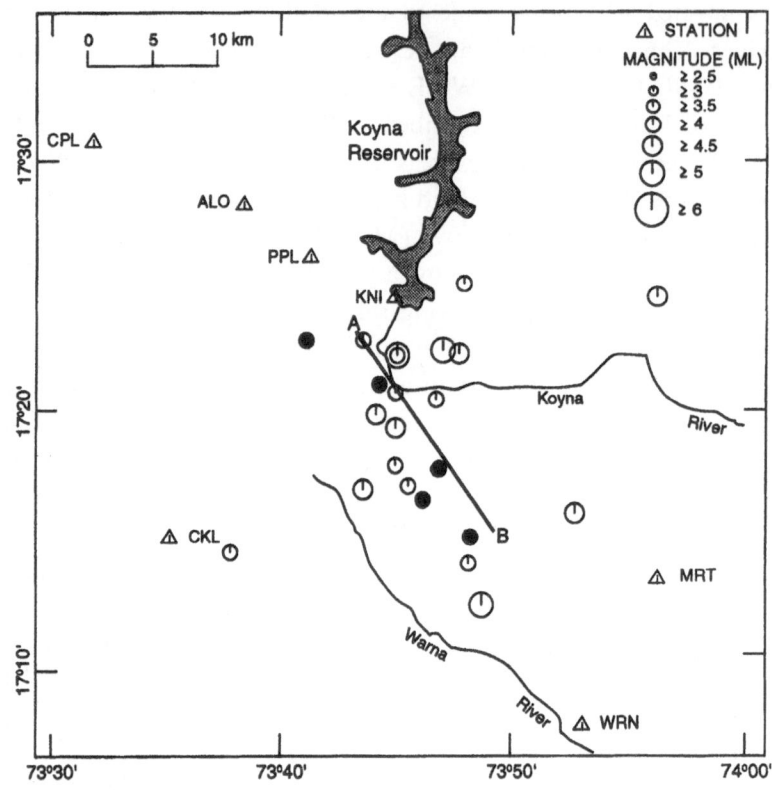

Figure 2

Relocation of earthquakes ($M \geq 4.0$) between December 10, 1967 and 1973. Aftershocks of the December, 1967 main shock are shown by solid symbols, they lie along a NW–SE trend AB, defined by the aftershocks of the February, 1994 event (Fig. 9).

The seismicity in the period 1993—April 1995 (Fig. 5) defines two clusters: A dense one near the upper reaches of Warna Reservoir and a smaller one south of the right angle bend in the Koyna River. The February 1, 1994, M 5.4 Koyna earthquake was located in the northern cluster. The annual distribution of seismicity shows that most of the seismicity was concentrated near Warna Reservoir in 1993. The pattern continued in 1994–95. In 1995 we notice that Warna seismicity had migrated to the west and northwest of intense clusters at the confluence of Bhogiv *nala* and Warna River. The outward migration of epicenters is characteristic of reservoir-induced earthquakes, suggesting that the observed seismicity in the vicinity of the Warna River is associated with the annual impoundment of Warna Reservoir.

1A2. Detailed Analysis of Seismicity for the Period 1993–1995

For the period starting with 1993 and through April 1995, 100 events with $M \geq 3.0$ were located; of these, 58 were located with a quality factor B and one with an A. It was decided to analyze this set of 59 events further to obtain the depth distribution. To check the reliability of the depths obtained for the 59 events, each event was further analyzed for the stability of the calculated depths. For each event the RMS error was calculated for the locations determined by using fixed depths (in the location program) varying from 1 to 20 km. The depths were also calculated for different starting depths. Figure 6 shows the plots for two events on January 03, 1994 at 08:57 and July 16, 1993 at 06:24. The calculated depths of these events, 11.2 km and 8.8 km agree well with the depth corresponding to the lowest RMS value (in the fixed depth vs. RMS plot). They also agree well with depth estimates obtained from various starting depths.

Of the 59 events that were chosen for the depth analysis 40 gave reliable depths. The locations of these earthquakes are shown in Figure 7. Similar to the location of the larger set of events for this period, their locations define two clusters, one near the bend of the Koyna River and the other near the confluence of Bhogiv *nala* and Warna River. The depth distribution of these events (Fig. 8) demonstrates that the earthquakes lies between the depths of 5 and 16 km with 80% of them lying between 7 and 13 km.

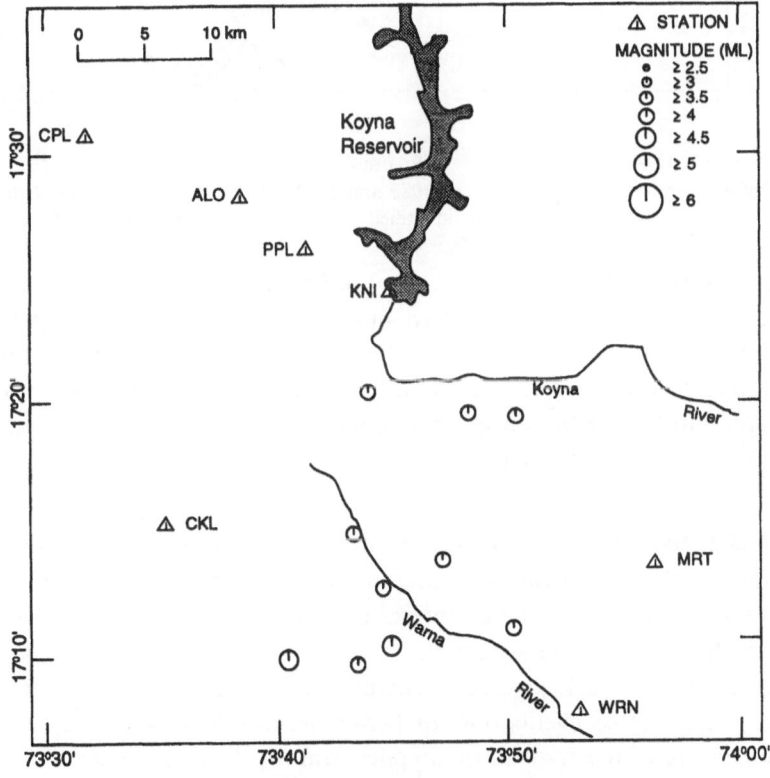

Figure 3
Relocation of earthquakes ($M \geq 4.0$) between 1974 and 1982.

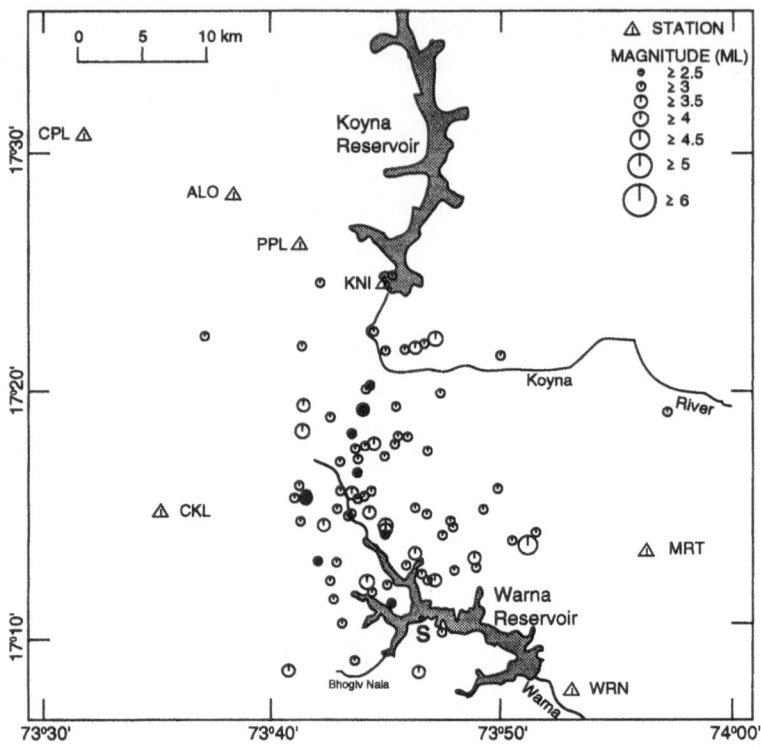

Figure 4
Relocation of earthquakes ($M \geq 3.0$) between 1983 and 1992. The earthquakes in 1992 (solid symbols)
define a NNE–SSW trend, identified as the Koyna River Fault Zone.

The results of the detailed analyses revealed that most of the earthquakes south
of Koyna River and the northwestern most of the Warna earthquakes lie in a
narrow NNE–SSW trending "channel" lying between about 6 and 14 km depths,
on a fault plane with a steep (possible SE) dip. No obvious fault plane was defined
by the hypocentral distribution of the second cluster located at the confluence of
the Bhogiv *nala* and Warna River.

1A.3 Detailed Analysis of Aftershocks of the February 1, 1994 Earthquake

The magnitude 5.4 earthquake that occurred on February 1, 1994 was the
largest event since the 1967 main shock. It was followed by 79 aftershocks with
$M > 1.0$ in the next 24 hours covering ~250 km area. The aftershock locations
varied in quality. To seek possible structures associated with the aftershocks we
chose the better located events (QC or better, RMS ≤ 0.25 sec and ERZ < 5.0).
ERZ is a measure of the error in the hypocentral depth calculations. This subset of
28 events defines three linear features (Fig. 9), two parallel features trending
NW–SE terminating in a NE–SW trend. Stereo plot view of these earthquakes

from the southeast and southwest show that the two sets of NW–SE trending hypocenters define the NE and SW boundaries of NW–SE trending blocks. The NE–SW trending hypocenters define the southeast boundary of the block. A NW–SE cross section through the main shock and adjacent NW–SE trending hypocenters (within 2 km of the line AB in Fig. 9) defines two parallel planes dipping to the NW at about 43° (Fig. 10). The two parallel planes (defined by the aftershocks) lie above and below the main shock. The fault plane solution for this event (Fig. 9) suggests normal faulting on a NE–SW plane dipping 45° to the northwest or southeast. The distribution of the hypocenters of the aftershocks suggests that the fault plane dips to the northwest.

Thus the hypocentral locations of (the better located) aftershocks of the February 1, 1994 earthquake define a block dipping northwest bounded to the northeast and southwest. The hypocenters lie in two roughly parallel northwest dipping planes, which (probably) define the top and bottom of the block. The hypocentral locations define the boundaries of the block which are faulted and fractured and provide conduits for fluid flow.

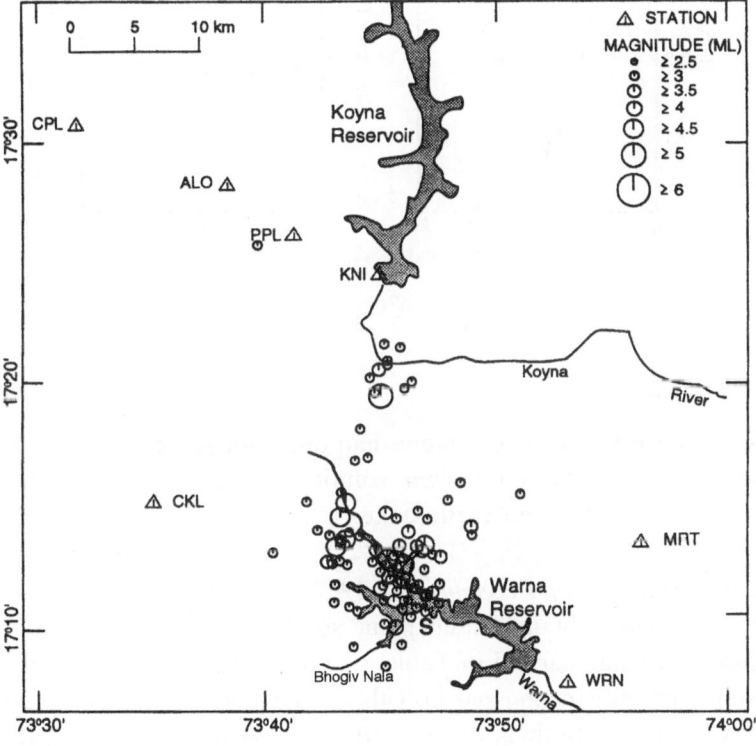

Figure 5
Relocation of earthquakes $M \geq 3.0$ earthquakes between 1993 and April, 1995. Increased seismicity near Warna River followed the impoundment of the Warna Dam. S shows the location of Sonarli.

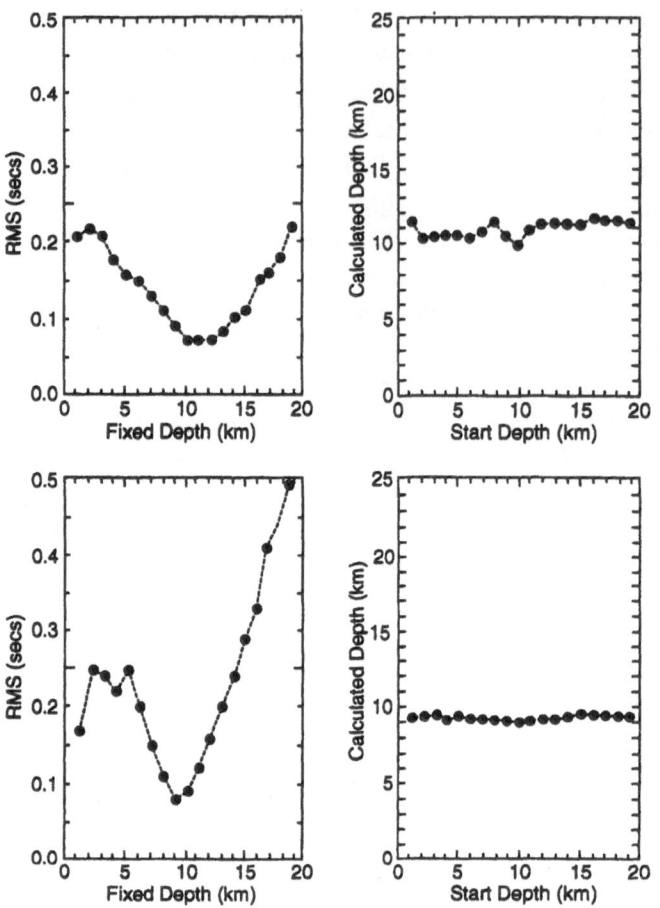

Figure 6
Depath analyses of two events on January 3, 1994 (top) and July 16, 1993 (bottom).

1B. Fault Plane Solutions

Various data suggest that there is more than one fault plane associated with the observed seismicity. Available fault plane solutions were reviewed and new ones obtained for well recorded recent earthquakes.

1B.1 Fault Plane Solutions of December 10, 1967 Main Shock

Several workers have obtained fault plane solutions for the 1967 main shock. The results have been summarized in Table 5, and the methods used to obtain the fault plane solutions are summarized in Table 6. Except for an early solution by GUPTA *et al.* (1969), all solutions suggest left lateral strike-slip faulting (Fig. 11). More recently GUPTA (1992) has suggested his preference for the left-lateral strike-slip solution. The strike of the fault plane obtained by various workers using different techniques is within about ±10° of N20°E–S20°W. (Khattri's solution is

Table 5

Source Mechanisms for the Main Shock
(December 10, 1967 at 22 h 51 m UTC)
Location 17° 20.8'N, 73° 44.88'E, 10.4 km deep

Fault Planes I			Fault Planes II			Fault Type	Author
Strike°	Dip°	Rake°*	Strike°	Dip°	Rake°		
206	66NW	165	110	74SW	24	Strike-slip	TANDON and CHAUDHURY, 1968
328	90	−90				Normal	GUPTA et al., 1969
217	72NW	176	126	84SW	18	Strike-slip	LEE and RALEIGH, 1969
170	80W	160	76	70SSE	11	Strike-slip	KHATTRI, 1970
201	75NW	0	111	90	15	Strike-slip	SYKES, 1970
203	70NW					Strike-slip	TSAI and AKI, 1971
202	80NW	177	112	88N	10	Strike-slip	BANGAR, 1972
190	78NW	−5°	111	85N	−12	Strike-slip with small normal component	SINGH et al., 1975
16	67E	−29	117	64	205	Strike-slip with small normal component	LANGSTON, 1976
a.206	78NW	0	116	90	12	Strike-slip	CHANDRA, 1977
b.206	51NW	−20	129	75NE	−40	Strike-slip	

* Rake angle measured (in the fault plane) counterclockwise from strike direction.

Table 6

Method used to obtain fault plane solution for the main shock

Method/Data	Remarks	Author
P-wave first motion data	From 89 stations in India and outside	TANDON and CHAUDHURY (1968)
P-wave first motion data	From WWSSN and Indian stations	GUPTA et al. (1969)
P-wave first motion data	From 26 WWSSN, 17 Indian, 24 ISC, 18 USCGS and 1 Canadian stations	LEE and RALEIGH (1969)
P-wave fist motion data	WWSSN stations	KHATTRI (1970)
P-wave first motion and S-wave polarization data		SYKES (1970)
Analysis of Rayleigh and Love wave data	6WWSSN stations. Also obtained $M_o = 1.8 \times 10^{26}$ dyne-cm	TSAI and AKI (1971)
P, PKP first motions and polarization or S-wave first motion data	WWSSN and Indian stations	BANGAR (1972)
Analysis of Rayleigh waves recorded on long-period vertical component seismograms	30WWSSN stations	SINGH et al. (1975)
Inversion of long-period body wave (P, pP, sP)	17WWSSN stations	LANGSTON (1976)
P-wave first motion data	WWSSN stations	CHANDRA (1977)

N10°W–S10°E). The fault was found to have a steep northwesterly dip (66°–80°) by most workers, however LANGSTON (1976) obtained an easterly dip. Most workers obtained pure strike-slip to a small reverse component along with the strike-slip motion, however SINGH et al. (1975) and LANGSTON (1976) obtained a small normal faulting component. LANGSTON's (1976) solution differed from other solutions, in that he suggested an easterly dipping fault. He used a generalized inverse method to analyze the first 25 s of the long-period P and SH wave forms recorded by the WWSSN. He inferred a shallow source, 4.5 ± 1.5 km, and suggested that such a shallow source would be associated with a complex source time function. As a result, the interference of P, pP and sP made apparent compressional P-wave polarities where the direct P wave was really dilatational. Thus taken together, the various fault plane solutions support left-lateral strike-slip faulting or a fault oriented from about N10°E–N20°E. This result is in general agreement with the conclusions of many recent studies (see e.g., GUPTA, 1992). LANGSTON's (1976) solution suggests a southeasterly dip, whereas other studies suggest a northwesterly dip. We will discuss the direction of the dip when comparing with other data.

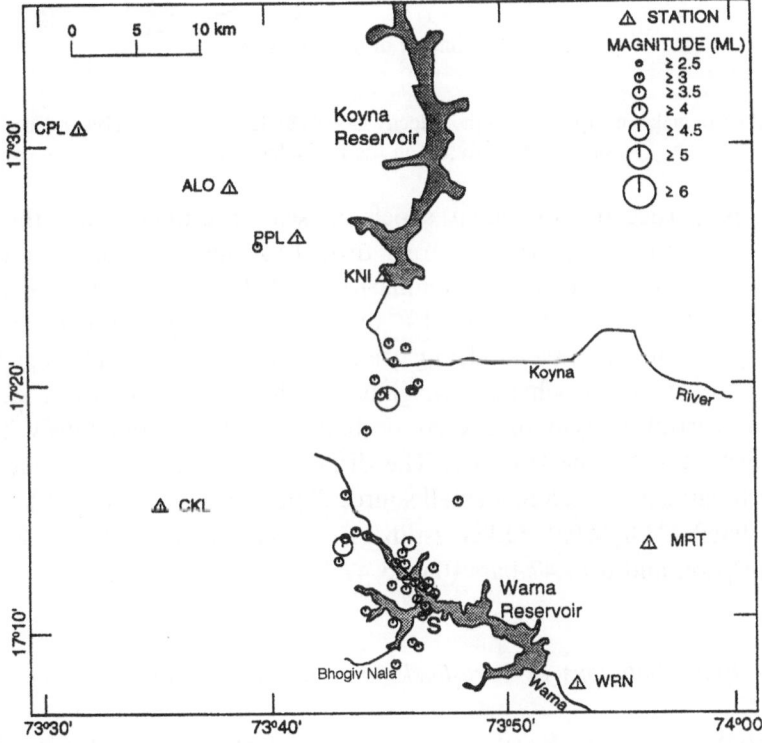

Figure 7
Earthquakes in 1993–1995 with best depth control. S shows location of Sonarli.

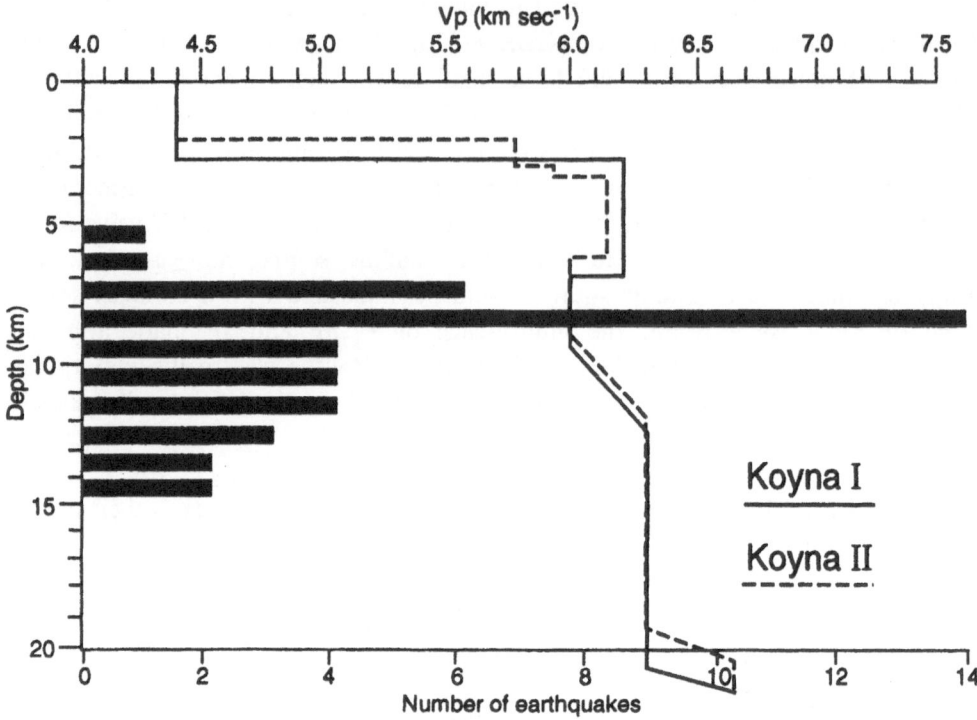

Figure 8
The depths of earthquakes compared with the velocity models by KRISHNA *et al.* (1983). Note how most of the seismicity lies in the low velocity zone.

Source properties for the main shock consist of estimates for the seismic moment (M_o), source dimensions, stress drop ($\Delta\sigma$) and displacement (u). The results of various studies have been summarized in Table 7. Estimates for M_o range between ~ 1 to 78×10^{25} dyne-cm (10^7 dyne-cm = 1 Nm). Estimates based on careful body and surface wave modeling are generally consistent and range between 3 and 18×10^{25} dyne-cm, whereas the value obtained by KHATTRI *et al.* (1977), based on a spectral analysis of teleseismic body waves, is anomalously high and probably overestimates the true M_o. The displacement spectra of strong ground motion seismograms resulted in a small source dimensions (2.5 km) and large stress drop (238 bars) (23.8 MPa). Other estimates ranged between 18 and 40 km for source dimension and 6 to 47 bars (0.6 to 47 MPa) stress drop.

1B.2 Fault Plane Solutions of Foreshocks and Aftershocks of December 10, 1967 Earthquake

LANGSTON (1981) and RAO *et al.* (1975) and LANSTON and FRANCO-SPERA (1985) obtained fault plane solutions for a foreshock and aftershocks of the December 10, 1967 Koyna earthquake. For the foreshocks, which occurred on

Table 7

Source properties of the main shock

$M_o \times 10^{25}$ dyne-cm	Source dimension km	$\Delta\sigma$ bars	u cm	Method	Author
18	23–40	6.2–19.8	108	Surface waves analysis Spectral analysis of Rayleigh waves	TSAI and AKI (1971) SINGH et al. (1975)
8.2					
77.6	17.9 ± 1.2 to 22.5 ± 5.8	47		Spectral analyses of teleseismic body waves	KHATTRI et al. (1977)
3.2 ± 1.4				Inversion modeling of body waves	LANGSTON (1976)
0.86	2.5	238	126.8	Displacement spectra from strong motion accelerograph records	GUPTA and RAM BABU (1993)

September 13, 1967, GUHA *et al.* (1970), had assigned magnitudes of 5.8, 4.5 and 4.5. LANGSTON (1981) examined the WWSSN records for the stations POO and NDI and argued that the magnitudes were closer to 4.0–4.5. A left-lateral strike-slip fault plane solution was obtained for the larger event (Table 8). A left-lateral strike-slip solution also was obtained for the M 5.0 aftershock on December 12, 1967 at 15h 48m by RAO *et al.* (1975). For a M 5.3 aftershock, which occurred at 0600h 18m on December 12, 1967, LANGSTON and FRANCO-SPERA (1985) obtained a normal faulting solution on a NW–SE trending fault (Table 8). A similar solution was obtained for a m_b 5.3 earthquake on September 20, 1980, using a moment tensor solution (DZIEWONSKI *et al.*, 1988). These fault plane solutions are shown in Figure 12.

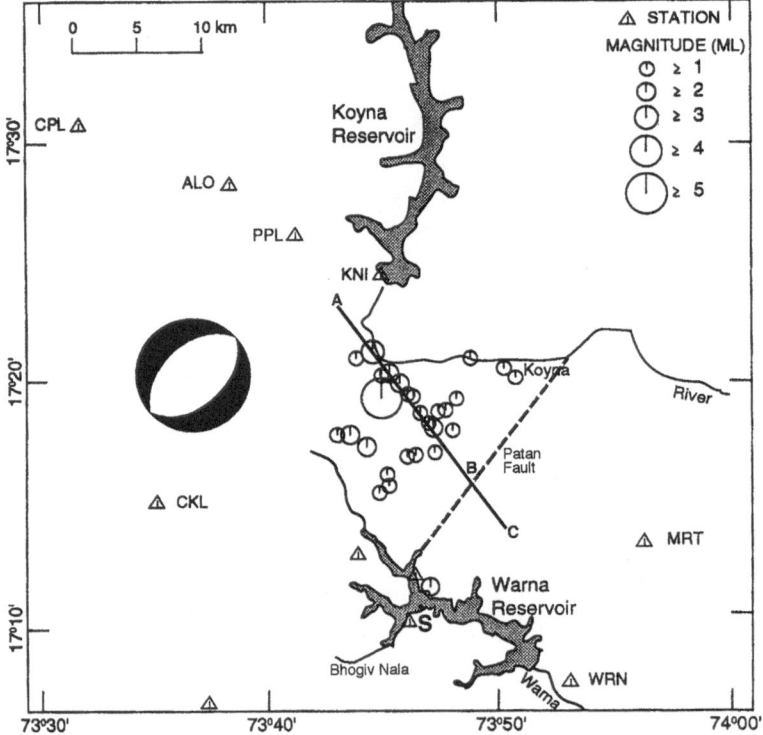

Figure 9
Location of subset of aftershocks of February 1, 1994 event that occurred within 24 hours. Only events with RMS ≤ 0.25 s or better have been plotted. The fault plane solution for main shock (TALWANI, 1994) shows normal faulting on a NW–SE trending fault. AB is inferred to be a NW–SE trending block boundary. B lies on the Patan fault inferred from trends of rivers (PATWARDHAN *et al.*, 1995). A vertical cross section along the line AC is shown in Figure 10.

Table 8

A. Fault plane solutions of a foreshock and aftershocks, B. Moment Tensor Solution of m_b 5.3 earthquake on September 20, 1980

Date			Origin Time			Magnitude	Fault Planes I			Fault Planes II			Fault Type	Author
Day	Month	Year	hour	min	sec		Strike°	Dip°	Rake°	Strike°	Dip°	Rake°		
A.														
13	9	1967	6	23	31	5.8 (GUHA)	20 ± 5	90 ± 15	0 ± 35				Strike-Slip	LANGSTON (1981)
						4.0–4.5	100 ± 20	40 ± 10	240 ± 20				Normal	LANGSTON and FRANCO-SPERA (1985)
12	12	1967	06	18	37	5.3	22	56	187	116	82	-35	Strike-slip	RAO et al. (1975)
12	12	1967	15	48	55	5.0								
B.														
20	9	1980	10	45	29	5.3	139	29	-111	342	63	-79	Normal	DZIEWONSKI et al. (1988)

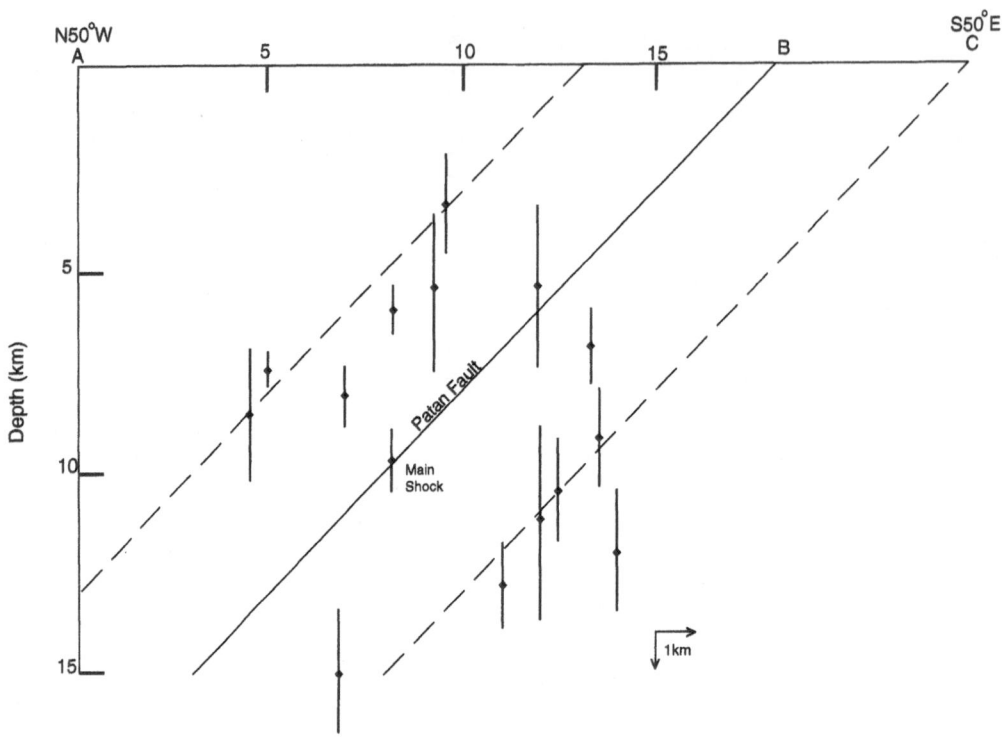

Figure 10
A NW–SE cross section showing earthquakes along AC in Figure 9. Only earthquakes lying within 2 km of AB have been plotted. The dip of the Patan fault, passing through the main shock, has been inferred from the fault plane solution. Its surface projection outcrops near the fault location inferred from changes in river course (B) by PATWARDHAN *et al.* (1995). Dashes lines have been drawn parallel to the dip of the Patan fault.

1B.3 Composite Fault Plane Solutions

RASTOGI and TALWANI (1980) relocated about 300 events occurring between 1967 and 1973. Based on the spatial distribution of the events, they identified three trends of epicenters and obtained composite fault plane solutions (CFPS) for them (Table 9). Two solutions suggested left-lateral strike-slip faulting on NE–SW planes with a dip of 80° to SE for one solution and 80° to NW for the other. The third solution suggested normal faulting on a NW–SE plane. GUPTA *et al.* (1980) obtained CFPS for eight events with $M \geq 4.0$. They used a combination of foreshocks and aftershock data to obtain the CFPS. For the eight events, they obtained strike-slip solutions for two events on roughly north-south striking planes. They obtained normal faulting for five events on generally NW–SE striking fault planes and reverse faulting for one event on a NE–SW striking fault plane (Table 9).

Table 9

Composite fault plane solutions

Date D	M	Y;	H:M	Fault Planes I Strike°	Dip°	Rake°	Fault Planes II Strike°	Dip°	Rake°	Fault type	Author
1967–1973											
NW trending zone				328	40	−98	318	50	−97	Normal	Rastogi and Talwani (1980)
NNE zone near dam				35	80	159	301	70	11	Strike-slip	Rastogi and Talwani (1980)
NNE zone W of dam				23	80	10	112	80	10	Strike-slip	Rastogi and Talwani (1980)
17	10	1973; (4FS + 8AS)*	15:24	5	76	−175	97	86	−15	Strike-slip	Gupta et al.(1980)
17	02	1974; (1FS + 2AS)	14:06	346	80	−90	166	10	−90	Normal	Gupta et al. (1980)
28	08	1974; (2AS)	20:20	358	80	−90	178	10	−90	Normal	Gupta et al. (1980)
11	11	1974; (1FS + 1AS)	15:11	316	40	−90	136	50	−90	Normal	Gupta et al. (1980)
20	12	1974; (1FS)	14:16	32	74	90	212	14	90	Reverse	Gupta et al. (1980)
2	12	1975; (1AS)	07:40	318	60	−90[1]	138	30	−90	Normal	Gupta et al. (1980)
				298	22	−90[2]	118	28	−90	Normal	Gupta et al. (1980)
24	12	1975; (2FS + 14AS)	13:25	258	30	−90[1]	78	60	−90	Normal	Gupta et al. (1980)
				293	70	−90[2]	112	20	−90	Normal	Gupta et al. (1980)
14	03	1976; (6AS)	05:16	350	62	−156[1]	91	70	−31	Strike-slip	Gupta et al. (1980)
				2	10	−170[2]	81	88	−100	Strike-slip w/normal comp.	Gupta et al. (1980)

* FS and AS are foreshocks and aftershocks.

[1,2] are two possible solutions for the earthquake.

1B.4 Fault Plane Solutions of Eight Events in 1993–1994

We obtained fault plane solutions for eight events with magnitudes between 3.7 and 5.4 that had been well recorded on stations of Koyna and Warna seismic networks (Fig. 13). Seven of the eight earthquakes are located near Warna and the eighth was the M 5.4 Koyna earthquake of February 1, 1994. Multiple solutions were obtained for some events because of a lack of constraints. The results are summarized in Table 10.

The M 5.4 event was associated with normal faulting. Of the seven Warna events, two each were associated with strike-slip and reverse faults and the remaining were associated with normal faults. The orientation of fault planes varied greatly.

1B.5 Discussion

The various fault plane solutions suggest that the prominent style of faulting is by left-lateral strike-slip motion on N–S to NNE–SSW striking faults. Normal faulting occurred on NW–SE to E–W striking faults and in some cases on NE–SW striking faults (especially for earthquakes occurring near Warna).

LANGSTON and FRANCO-SPERA (1985) obtained a normal fault mechanism with a strike of $100° \pm 20°$ and a dip of $40° \pm 10°$ and rake, $240° \pm 20°$ for a m_b 5.3

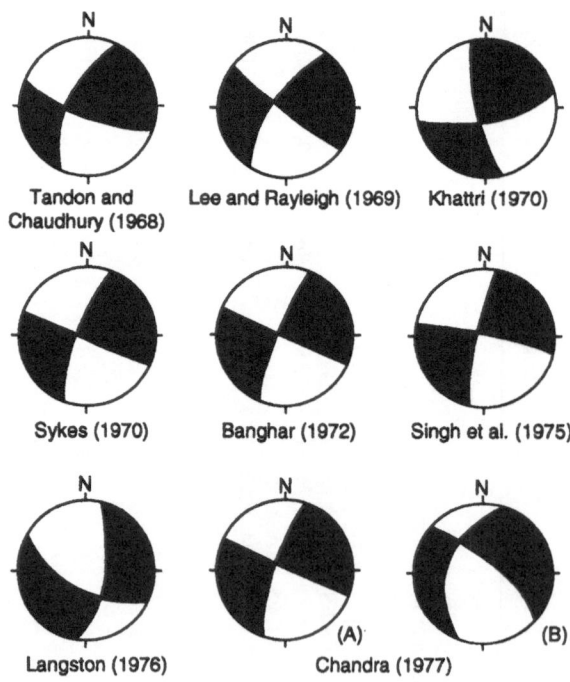

Figure 11
Various fault plane solutions for the Koyna December 10, 1967 main shock.

Table 10

Fault plane solutions for 1993–94 events

No.	Date D	M	Y	Origin Time H:M	Magnitude	Fault Planes I Strike°	Dip°	Rake°	Fault Planes II Strike°	Dip°	Rake°	Fault Type	Author
1	28	08	93	04:26	4.9	165	40	-180	75	90	50		This study
2	03	09	93	23:01	4.7	319	88	86	250	5	20	Strike-slip	This study
						296	90	-25	25	25	-180		
4	22	10	93	01:14	4.3	16	37	-25	135	70	-120	Strike-slip/normal	This study
5	08	12	93	01:42	5.1	54	32	-107	35	60	-100	Normal	This study
6	21	12	93	10:15	4.0	44	45	-49	173	58	-123	Normal	DZIEWONSKI *et al.* (1994)
						02	50	-160	260	75	-40	Normal/strike-slip Not well constrained	This study
7	22	01	94	11:12	3.8	319	56	135	75	65	50	Strike-slip w/reverse component	This study
8	01	02	94	09:30	5.4	45	35	-130	91	64	-105	Normal	This study
						50	45	-90	50	45	-90	Normal	TALWANI (1994)
9	01	03	94	16:07	3.7	325	65	160	63	72	27	Strike-slip w/reverse component	This study
						302	56	68	85	40	120	w/normal component	
						320	50	-130	12	50	-60		

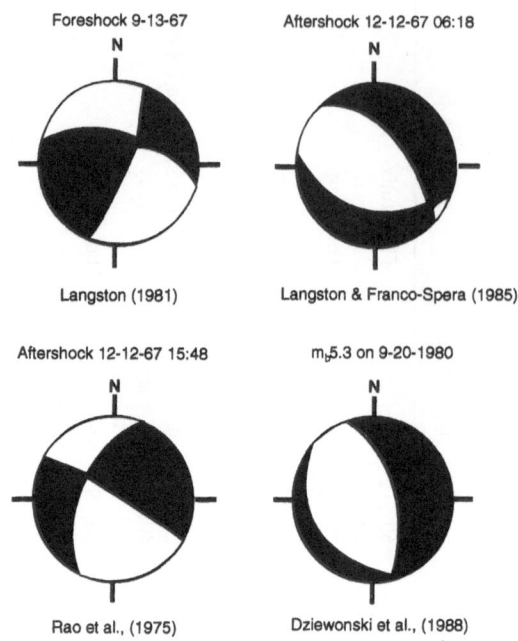

Figure 12
Fault plane solutions of the foreshock and aftershocks of the December 10, 1967 Koyna earthquake.
The fault plane solution of the September 1980 event is also shown.

aftershock of the December 10, 1967 event. The aftershock which occurred on December 12, 1967 at 06:18 UTC could not be located. However, the locations of six aftershocks with $M \geq 4.0$ that occurred from December 1967 to February 1968 (Fig. 2) also define a NW–SE fault zone. Although the locations of these events exhibit some scatter, they delineate a NW–SE zone. Interestingly this zone is almost exactly along the NW-SE trending boundary of a block defined by the aftershocks of the February 1, 1994 earthquake (Fig. 9).

The orientations of the fault plane solutions obtained by different authors will be used to infer the orientations and boundaries of various blocks delineated by the seismicity, geology and geomorphological features (next section).

2 Geology and Tectonics of the Area

To understand the seismotectonics of a region we compared the seismicity data with complementary data. These data consist of observations immediately following the 1967 earthquake, geomorphological and satellite imagery data and geological and geophysical observations. The various data were integrated to describe the tectonic framework.

2A. Observations Following the 1967 Koyna Earthquake

Very detailed investigations were carried out following the 1967 main shock by the officers of the Geological Survey of India (GIS Report, 1968). These observations described ground deformation, incidences of water level changes and general felt reports and perceived directions of felt ground motion. Widespread surface deformation was observed in the meizoseismal area (GSI Report, 1968; SATHE *et al.*, 1968). These are located in Figure 14. Slumping or rock slips (GSI Report, 1968) was observed in areas of high relief, hillsides and was parallel to the surface contours. The rock slips were rarely deep and were mainly due to weathered and open jointed columns of basalt coming down with the loose slope wash (GSI Report, 1968). At other locations the earthquake caused the development of ground fissures (mole cracks commonly observed in areas of strike-slip faulting). Although

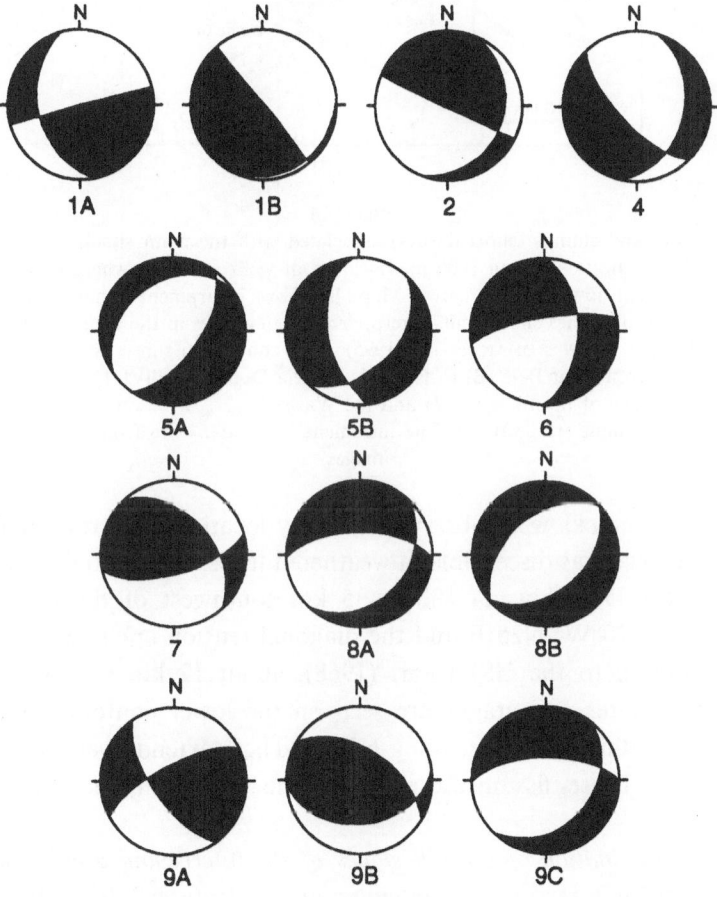

Figure 13
Fault plane solutions of selected events in 1993–94. See also Table 10.

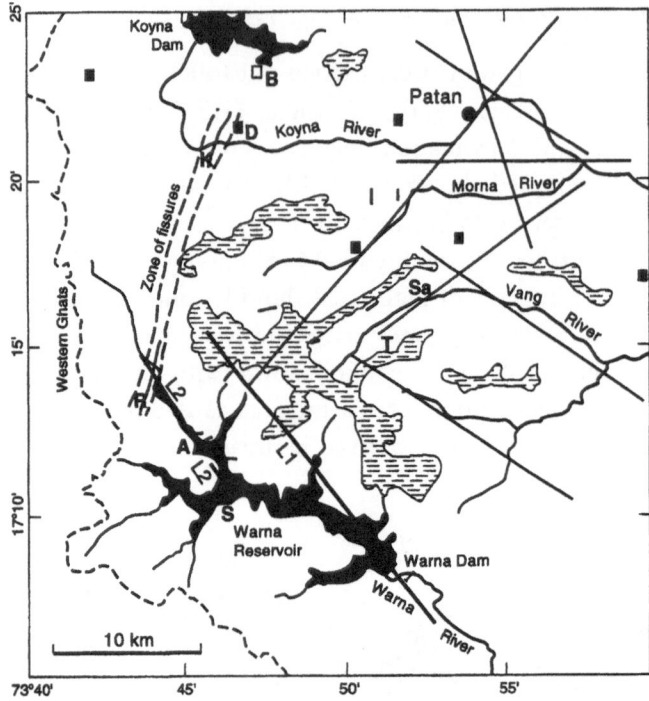

Figure 14

Location of fissures and slumps (short dashes) associated with the main shock. The hatched pattern
shows areas with elevation exceeding 1000 m. Location of wells is shown where water levels went up
[solid squares] and went down [open squares]. Map also shows escarpment trends along the continental
divide (dashed). The solid lines show faults interpreted from changes in the course of the Koyna River
and its tributaries by PATWARDHAN *et al.* (1995). The zone of fissures associated with the 1967
earthquakes extends from near Baje (B) to Randhiv (R) via Donichawadi (D) and Kadoli (K). Sonarli
(S) lies at the confluence of the Bhogiv *nala* and the Warna River. A, Sa and T show the locations of
Ambole, Salve and Tamine, respectively. The lineaments L_1 and L_2 are from LANDSAT in INSAT
images.

such fissures and cracks were observed at many locations, the regular pattern and
continuity of cracks was discernible in weathered traps and soil only between Nanel
and Kadoli near Donichiwada Village, 5 km southwest of the dam. The main
fissures trended N10°W–N25°E and the diagonal tension cracks trended N10°W–
N40°W. According to the GSI Reort (1968), about 12 km southwest of Kadoli
Village, small fissures and cracks are seen on the lower contours of the western
ridge face of Rundhiv Village trending N20°E. These extend over 50 m and have a
depth ≤1.8 m. These fissures were along the same alignment as those near
Donichiwada.

   "*Detailed examination of the hill slopes of the intervening area between Kadoli
and Rundhiv did not reveal the continuity of the features*" (GSI Report, 1968).
However, SATHE *et al.* (1968) suggest that the "*... zone of fissures near Donichiwada*

(*which includes Baje, Nanel, Kadoli*) ... *extend up to Rundhiv*". N–S trending *en-echelon* fissures were also observed near Morgiri, located about 10 km east of Kadoli, although they were not as well developed as at Kadoli (SATHE *et al.*, 1968). These authors further suggest that the NNE–SSW fissures observed near Kadoli were associated with the fault causing the December 1967 earthquake.

Changes in water levels were observed in streams and in wells over a widespread area. In most of the wells the rise in water levels varied between about 5 and 30 cm. In some streams up to 1.7 m rise in water level was reported. At three locations the water level decreased. Of the 23 locations for which we have reports of changes in water level, seven are located in the vicinity of the epicentral region of the observed seismicity. Two of these locations show an interesting pattern. The water level rose at Donichiwada and declined at Baje (Fig. 14). These two places are located on either side of the inferred fault based on ground fissures. In addition to changes in water levels, the water turned milky and turbid at several locations, providing evidence that the earthquake caused widespread changes in the ground water flow pattern.

The report by the GEOLOGICAL SURVEY of INDIA (1968) provided excellent details of the felt effects of the earthquake. Many of the reports described the perceived direction of motion at a particular location (see TALWANI *et al.*, 1996 for details). These observations are consistent with strike-slip faulting on a N10–15°E–S10–15°W trending fault.

2B. Geomorphological Observations

2B.1 Trend of the Koyna River

SAHASRABUDHE *et al.* (1971) mapped the geology of the area and also noted that the trends of the escarpment along the continental divide and the Koyna River 6 km to its east were similar. They noted that "... *the en-echelon displacement pattern of the hill trends and similar changes in the Koyna River courses can be made out in topographic sheets*". They noted that joints and shears were responsible for the topography and have, to a major extent, controlled the drainage pattern in the area. They concluded that shearing had been responsible for the *en-echelon* pattern of the ridges and the similar disposition of the Koyna valley.

Later authors, e.g., GUPTA *et al.* (1980), have suggested that the right angle bend in the Koyna River is fault controlled, whereas GUPTE (1968) attributes the bend to river capture.

2B.2 Evidence of Neotectonics

Reconnaissance geologic mapping and aerial reconnaissance near the Koyna Dam by HARPSTER *et al.* (1979) showed that the fissures that developed during the December 1967 event were the result of displacement on a pre-existing fault zone.

They named it the Donichiwada fault zone with an overall strike of N35°E. They noted that there were retaining walls that were offset as much as 1 m left-laterally and the presence of clay within the fault zone indicated that the Donichiwada fault was active.

2B.3 Interpretation of Aerial Photographs in the Koyna Region

DESHPANDE and JAGTAP (1971) described the main features in aerial photographs of the area taken from a height of 19,500 ft. The most prominent feature was the prominent steep, north-south fractures. The north-south fractures were found to be planar, systematically oriented and cross-cutting. Continuous individual or zones of fractures cut the cliffs, laterite caps and particularly, excavation flows (SNOW, 1982).

2B.4 Matching Erosional Surfaces

In order to seek geomorphic evidence of uplift, SNOW (1982) compared erosional surfaces on either side of the N–S section of the Koyna River using Survey of India topographic maps. On one profile, he noted displacement of the two oldest surfaces near Koyna Reservoir, indicating down faulting to the west. Thus the geomorphic evaluation of the E–W erosional surfaces indicates the presence of a N–S trending fault with the west side down thrown.

2B.5 Geomorphic Investigations Near Patan

PATWARDHAN et al. (1995) studied geomorphic features, especially the courses of the rivers Kera, Morna and Vang, the three major tributaries of Koyna in the Patan Taluka (a subdivision of a tax district, covering several villages). The town of Patan is located about 16 km SSE of Koyna Dam. They noted that "... the course of Koyna is almost linear, parallel to the western ghats right from the hills of Mahableshwar up to the present establishment of Koyna Nagar town and Helwak (Fig. 14). The river takes a sudden turn eastward and flows again in a linear valley up to Patan, where it is joined by the Kera which also takes a NNW–SSE course (roughly) parallel to the initial course of Koyna. **Just before Patan, the Koyna also takes a sudden northeasterly course** (emphasis added) and beyond Patan after flowing eastward for a short distance, it takes a linear southeasterly course right up to Karad where it joins the Krishna. Almost parallel to the course of the Koyna flow the tributaries Morna and Vang. The southeasterly course of the Koyna from Patan to Karad and a similar course of Varna (Warna) on which the Chandoli dam has been constructed are remarkably linear".

PATWARDHAN et al. (1995) further suggest that these geomorphological features (changing trends in the course of the rivers) "may be consequent to movements in deep seated faults with the pre-Trap formations ...". The reactivated basement faults are expected to follow major foliation directions and planes of weakness of the metamorphosed basement rocks. They further suggest that "the shear pattern observed in the thin Deccan basalt cover largely follows the movement directions of faults within the basement and has, in turn, controlled the linear courses of the rivers as well as facilitated the deep erosion by them. They infer several NE–SW, E–W and NW–SE trending faults or shear planes (Fig. 14). These directions agree well with those observed on a LANDSAT image by LANGSTON (1981).

One of the inferred faults trending N45°E–S45°W, based on the north-easterly courses of the Koyna River near Patan and Morna River near Morgiri, if extended further southwest is collinear with a NE–SW trending tributary of the Warna River located between the Ambole and Atoli. We infer this fault, herein named the Patan fault, to be a block boundary and will discuss its significance later. A fault associated with the NE–SW trend of the Vang River passes through Salve and Tamine and is roughly parallel to the Patan fault and separated from it by the ≥ 1000 m high ridge. The point C on Figures 9 and 10 lies near Tamine, suggesting that the fault (?) passing through Tamine also dips to the northwest.

2B.6 Fractures in the Warna Area

In order to discover the geologic features which could possibly be associated with the seismicity near Warna Reservoir, the available geologic literature was reviewed. Detailed and reconnaissance geologic mapping in the Warna Reservoir area was carried out by the GEOLOGICAL SURVEY of INDIA (GSI). Most of the reports deal with the mapping of various lava flows in the area. Fourteen flows were delineated and were found to be subhorizontal or with low gradients and there was an absence of vertical displacements. A large number of fractures had been identified on aerial photographs and REDDY and JERATH (1984) confirmed four major fractures by ground checks. All these fractures are along small *nalas* (streams), tributaries of Warna River. No evidence of displacement was found on any of the fractures.

In order to delineate the weak zones in the Warna Valley, PESHWA (1991) reviewed the available Remote Sensing data. His report is the most detailed review of the features found in Remote Sensing images, and documents their ground check. The area studied was the catchment area of the Warna River (Fig. 15).

He found there are two NE–SW trending fracture zones, one lying between Ambole and Aloti and the other along a line from Petlond towards Bhogiv. These trends appear to be younger than the NW–SE trend defining the course of the Warna River. The intersection of the two fracture systems seems to control the observed seismicity, and observations suggest the fractures themselves may be fault related.

The Warna events on the other hand appear to be associated with faults trending both NE–SW and NW–SE and are located near their intersection.

2C. Geological Observations

SAHASRABUDHE *et al.* (1971) carried out detailed geologic mapping in the
Koyna area following the December 1967 earthquake. They delineated 8 flows in
the Koyna River Valley in a 245-m thick column. They further noted that "... *the
flows have a very gentle (1° to 2°) westerly dip and are traversed by N–S to E–W
trending vertical joints and several N–S trending shear zones*". According to the
authors no conspicuous faults were noticed in the river valley.

Interestingly, SNOW (1982) noted that "... *crossing Koyna Dam is a vertical zone
1 to 3 m wide that had to be excavated and back-filled with concrete. In the tailrace
tunnel, three near-vertical, north–northeast to north–northwest zones pinch and swell,
with maximum widths as great as 10 m. Lavas on opposite sides of that zone are offset
3 m. Thus, it is consistent to regard all steep, continuous fractures and zones that
strike N15°W to N15°E strike as faults*".

Thus, field mapping does suggest the possible presence of steeply dipping faults.

2D. Geophysical Studies

A variety of geophysical studies have been carried out in the Koyna-Warna
area. These include deep seismic sounding on two east-west lines, detailed gravity
surveys in a narrow strip just south of the Koyna Dam and aeromagnetic surveys
over a broad area from the west coast of India and covering the study area.

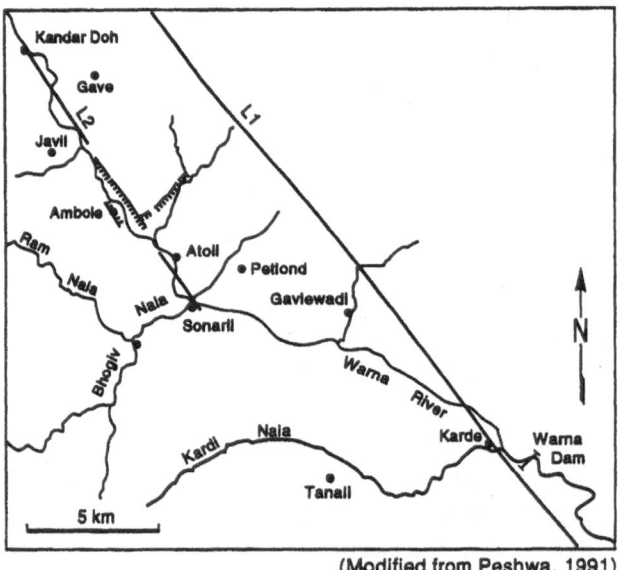

(Modified from Peshwa, 1991)

Figure 15
Locations of fractures and escarpments and streams (before impoundment) in the Warna Valley.
LANDSAT lineament L_2 lies along the course of Warna River NW of Sonarli. Lineament L_1 lies to its
NE.

2D.1 Deep Seismic Soundings

KAILA *et al.* (1981) described the results of Deep Seismic Soundings (DSS) on a 220-km long E–W profile from Guhagar on the west coast, passing through Koyna and east to Chorochi. The results revealed a number of reflection horizons below the Deccan Traps down to the Moho discontinuity. They found that below the Deccan Traps, the crustal section along this profile is cut into two blocks by a eastward dipping deep fault west of Koyna. The eastern block is further cut by another deep fault which affects only the deeper horizons including the Moho. They suggested that the Koyna earthquakes were due to movements on the faults east of Koyna.

2D.1.1 Reinterpretation of the Velocity Model

The multi-layer interpretation of the DSS data yielded a 14 layer model (Table 2) (KAILA, 1983). Reinterpretation of the DSS data along two east-west profiles using synthetic seismogram modeling revealed the presence of two low velocity zones (KRISHNA *et al.*, 1989). These two low velocity zones were found to occur in the upper crust and in the lower crust. On the Koyna II profile (located 60–70 km to the north and parallel to the Koyna I profile through Koyna) the velocity reduction occurs at a 6.0-km depth and there is a 3-km thick transition zone from 8.5 km to 11.5 km depth. The upper crustal low velocity zone was found to be one km deeper in the Koyna I profile. The second low velocity zone occurred at a depth of 26.0 km in the two profiles.

Interestingly the depth of most of the well located earthquakes (Fig. 8) lie in the low velocity zone interpreted by KRISHNA *et al.* (1989).

2D.2 Detailed Gravity Profiles

KAILASAM and MURTHY (1971) observed gravity along five short (~ 400 m) profiles across the Koyna River, south of the Koyna Dam, before the big bend near Helwak. The data suggested a possible shear zone (fault?) in a N–S direction parallel to the river course and to its west. The authors further suggested that the inferred fault from gravity data "... *is a buried one within the trap and extends deeper into the sub-trap formations and possibly into the basement rocks*".

2D.3 Aeromagnetic Data

In an effort to determine the thickness of the Deccan Trap rocks in the Koyna area, aeromagnetic data were acquired on 13 short profiles, each about 100 km long and spaced at 4 km (NEGI *et al.*, 1983). Data were recorded at elevations of 1220, 1524 and 2134 m, above mean sea level. The flight elevation 1220 m AMSL is only about 200–600 m about the ground surface and the flight lines were spaced 4 km apart. This led to the appearance of several high frequency (pseudo) anomalies (Fig. 6 in NEGI *et al.*, 1983). The pseudo-anomalies are absent in the residual anomaly map for the flight at 2134 m AMSL (Fig. 8 in NEGI *et al.*, 1983). Two

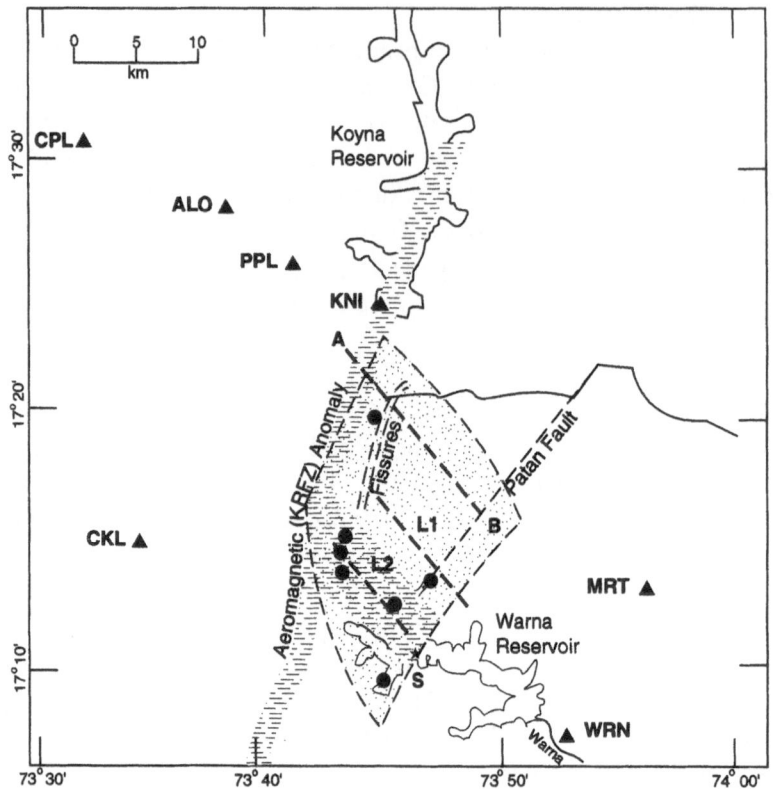

Figure 16

Interpreted structural boundaries from the aeromagnetic data. The inferred seismotectonic features are shown. The Koyna-River-Fault-Zone (KRFZ) is based on the pattern of aeromagnetic anomalies. The NW–SE pattern of aeromagnetic anomalies includes a portion of the Warna River and lies along lineament L_2. Other NW–SE block boundaries are indicated by lineaments L_1 and L_2 and line AB. The better located seismicity (1993–95) is roughly located within a seismogenic zone (stippled pattern) enclosed by the KRFZ to the west and Patan fault to the east (and a parallel fault to its southeast). The larger events in 1993–95 are shown by solid dots. The zone of fissures associated with the 1967 earthquake lies to the east of KRFZ.

major features were interpreted from the aeromagnetic maps (Fig. 16). The first a broad NNE–SSW zone, extends from north of the Koyna Dam, 40 km to its SSW. This feature abuts against a strong NW–SE anomaly along the upper reaches of the Warna River. This NW–SE anomaly is coincident with the lineament L_2 in Figure 15.

Thus the aeromagnetic data support the two major trends (NNE–SSW) along the (N–S) portion of the Koyna River and inferred as the main fault from various data discussed and the NW–SE trend of the Warna River upstream of Sonarli (Fig. 16).

This paper is concentrated towards Koyna as most of the data and studies pertain to the Koyna area. We realize that most of the current seismicity is in the Warna area. It is the Warna area that now bears careful watching and study.

3. Seismotectonics of the Koyna-Warna Area

In this section we combine various geological, geophysical, geomorphic data with seismicity data to delineate and describe the geometry of seismogenic structures. The results of this exercise show that the Koyna-Warna area is criss-crossed with several steeply-dipping fractures and faults. Some of these faults extend to great depths (>10 km). The geophysical and seismicity data also suggest the presence of a fluid-filled fracture zone where most of the larger events occur.

3A. A Fluid-filled Fracture Zone

Due to poor control on various parameters needed to accurately locate hypocenters, earlier estimates of the depth of seismicity in the Koyna-Warna region ranged from near surface to over 40 km. Those estimates were wrong. By carefully analyzing the hypocenters for accuracy in their depth estimates, we found that most of the seismicity is shallower than 15 km. Further, we found that most of the seismicity occurs between about 7 and 13 km depth. The detailed and careful inversion of deep sounding seismic data by KRISHNA et al. (1989) demonstrated that there is a low-velocity zone between the depths of about 6 and 13 km. A low velocity zone in the upper crust can be due to hotter rocks in which case we would expect plastic deformation and few earthquakes, or it could be due to fluid-filled fractured rocks. The observation that the low velocity zone is coincident with seismicity is interpreted to mean that it is associated with fluid-filled fractures. (There is abundant literature which demonstrates that fluid-filled fractures occur in mid-crust and are associated with a decrease in their seismic velocities see e.g., GAJEWSKI and PRODEHL 1987; WENZEL and SANDMEIER, 1992).

Consequently our first conclusion is that in the Koyna Warna area there is a fluid-filled upper crustal layer associated with a low velocity zone and most of the larger events occur in it.

3B. Koyna-River Fault Zone

In this section we present our interpretation of the 40–50 km long N10–15°E–S10–15°W trending Koyna-River fault zone (KRFZ).

AUDEN (1954), SAHASRABUDHE et al. (1971) and PATWARDHAN et al. (1995) among others have suggested that many subvertical fractures/faults encountered widely in the basalt flows in the Deccan Traps are related to throughgoing fractures in the underlying crystalline rocks. Movements on these faults in the crystalline basement control the geomorphic features in the overlying basalt e.g., the course of rivers, escarpments, cliffs, etc. Hence by studying the geomorphic features it is possible to infer subsurface faults underlying the basalt layers.

Evidence for the existence of the KRFZ comes from many sources. A study of aerial photographs (DESHPANDE and JAGTAP, 1971) revealed the presence of a predominance of N–S trending fractures. Surficial evidence of N–S faulting comes from a study of ancient geomorphic surfaces (terraces). SNOW (1982) demonstrated that they indicate faulting along the N–S stretch of the Koyna River with the west side downthrown. Evidence of shallow faulting was also obtained from the short E–W detailed gravity and magnetic profiles running just south of the Koyna Dam and crossing the Koyna River (KAILASAM and MURTY, 1971). They indicated a shallow N–S fault parallel and to the west of the Koyna River. The lateral extent of this fault is between the Koyna Dam and the right angle turn to the east in the Koyna River. Both the aeromagnetic data (NEGI et al., 1983) and seismic refraction data (KAILA et al., 1981) show that on a E–W profile through Koyna, the basalt is thicker under Koyna than under Alore to its west. The aeromagnetic data also show that there is a N–S to N10°E–S10°W trending fault along the Koyna River that extends at least 40 km to the south. Major aeromagnetic anomalies abut and terminate against this trend.

The 1967 main shock has also been interpreted to have occurred along the KRFZ. Various fault plane solutions (Fig. 11) suggest left-lateral strike-slip fault on a N10°E–S10°W steeply dipping fault. The seismicity data also support the KRFZ. The $M \geq 3.0$ earthquakes in 1992 define a N10°E–S10°W trend (Fig. 4), that is parallel and to the east of the magnetic anomaly defining the KRFZ. The ground fissures that were associated with the December 1967 earthquake are parallel and 1–2 km to the east of the magnetic anomaly and the 1992, $M \geq 3.0$ epicenters (Fig. 16). A 3-D stereo view from SSW of the best located earthquakes and lying on the postulated KRFZ reveals that the KRFZ defines a zone of activity, about 3–4 km wide and lying between depths of ~ 8 and 14 km. As most of these earthquakes are deeper than ~ 8 km and the ground fissures are parallel and ~ 1–2 km to the east, the inferred dip of the fault is 70–80° to the northwest, in general agreement with the focal mechanism solutions.

Another intriguing observation is that the surface projection of the KRFZ seems to lie between two wells that showed opposite coseismic response to the December 1967 earthquake. The water level in the well at Donichiwada rose whereas it went down in the well at Baje (Fig. 14). The response of the public to the 1967 main shock was also varied. Villages lying in a broad NNE-SSW zone parallel to the fault, but extending to large distances on either side, all reported sensing the ground motion in a N–S or NE–SW direction. This motion is consistent with what would be expected for a left-lateral strike-slip fault.

All these observations confirm the conclusions based on the fault plane solutions and seismicity that the 1967 Koyna earthquake occurred on a N10°E–S10°W fault and was associated with left-lateral strike-slip motion. The newer seismicity data also show that along the KRFZ the seismicity is generally deeper than about 6–7 km. We also note the KRFZ forms a western boundary of the seismicity (Fig.

16). We interpret this observation to mean that the KRFZ is the fractured edge of a crustal block. The regression relation between surface rupture length and magnitude (WELLS and COPPERSMITH, 1994) is log SRL = 3.22 + 0.69 M where SRL is the surface rupture length and M is Moment magnitude. For M = 6.3, the surface rupture length is 13.4 km. This length is consistent with the observed extent of ground fissures.

3C. Patan Fault

A NE–SW trending fault was postulated by PATWARDHAN et al. (1995), based on anomalous trends in the Koyna and Morna rivers near Patan (Fig. 14). If we extend this fault to its SW it lies along fault and escarpments near Ambole on the Warna River mapped by PESHWA (1991) and seen on LANDSAT data. We name this fault, the Patan fault.

The aftershocks of the M 5.4 earthquake on February 1, 1994 were well located and those in the first twenty-four hours occupy an area of ~250 sq km (Fig. 9). The epicenters of these aftershocks delineate a sharp SE boundary which is to the NW and parallel to the Patan fault. Some of the aftershocks and the main event (February 1, 1994 shock) were associated with normal faulting on a 45°NW dipping fault plane. If we project this fault plane to the surface (B on Figs. 9 and 10), it coincides with the Patan fault as described on the surface from geomorphic data. The absence of any aftershocks to the SE of the surface projection of the Patan fault suggest that it acts as a dipping boundary of a block.

Fault plane solutions of some events near Sonarli on the Warna River are also associated with normal faulting on NE–SW trending faults. PATWARDHAN et al. (1995) also suggested the presence of another NE–SW trending fault to the SE of the Patan fault. This fault passes through Salve and Tamine (Fig. 14). Interestingly this fault and the Patan fault lies to the SE and NW of a sharp, elongate NE–SW trending ridge with elevation exceeding 1000 m. We do not know the dip of the southern fault and speculate that it too dips to the NW, as is suggested by the deeper earthquakes in Figure 10.

3D. NW Trending Boundary Faults

A variety of data suggest the presence of steep NW trending faults that lie between the Koyna and Warna rivers and divide the area into distinct blocks.

The aftershocks of the February 1, 1994 earthquakes define one such block boundary (Fig. 9). A 3-D stereo view of this feature along NW–SE indicates that the earthquakes define a sharp NW–SE zone ~1–2 km wide and 10 km long extending from near surface to ~8 km depth (see also Fig. 10).

Another edge of a NW–SE trending block is defined by the aftershocks lying to the SW and parallel to the first set. This second set of aftershocks (Fig. 9) lies along a major NW–SE lineament seen on the LANDSAT and INSAT images (L1 in Figs. 14 and 15).

There were at least 5 aftershocks with $M \geq 4.0$ of the December 1967 M 6.3 event (Fig. 2). Although we could not obtain very accurate locations of these events, they are probably good to 2–3 km. These $M \geq 4.0$ events define a broad NW–SE zone. On Figure 9, AB shows the NW–SE trend of aftershocks of the February 1, 1994 earthquake. Note how the aftershocks of the 1967 main shock also lie along AB. This observation suggests that both the 1967 aftershocks and the February 1, 1994 aftershocks occurred along the same NW–SE fracture zone defining the NE edge of a tectonic block.

LANGSTON and FRANCO-SPERA (1985) obtained a fault plane solution of the aftershock on 12-12-67. It had a NW-SE strike and was associated with normal faulting. Composite fault plane solutions of several events lying in a NW–SE direction by RASTOGI and TALWANI (1980) also yielded normal faulting on NW–SE striking planes.

The Warna River upstream of Sonarli (Fig. 16) is also parallel to the NW trends described above and lies between a well developed NW–SE anomaly on the aeromagnetic map. Interestingly the NW trend and the seismicity continues only up to the KRFZ and does not extend beyond it. The seismicity along this stretch of the Warna River extends from near surface to ~ 10–12 km depth.

The seismicity that followed the impoundment of Warna Dam (after 1992) also shows epicentral growth along this trend. Faults along this trend were also inferred by PESHWA (1991) on LANDSAT data and by field checking.

Thus various geological and seismicity data delineate several NW–SE trending faults.

3E. E–W Trending Faults

The right angle turn in the Koyna River near Helwak has been noted by many workers and interpreted to suggest the presence of an E–W fault. Further support for this view comes from the mapping of two prominent E–W aeromagnetic anomalies (see Fig. 8 in NEGI et al. (1983)). On the E–W trend of the Koyna River (after the bend near Helwak) we note a broad 8–10 km long aeromagnetic high to the north of the river and a parallel low to its south. We interpret this observation to be a manifestation of an E–W fault.

The presence of topographic highlands (> 1000 m) to the north and south of the river (Fig. 14) further suggests that the river has carved its path along the weak zone that faulting represents.

3F. Major Kinks and Intersections

Several authors have shown that intersections of faults or bends in the trends of faults serve as location for stress build-up (see e.g., KING and NABELEK, 1985; TALWANI, 1988; ANDREWS, 1989). We note a major concentration of seismicity near Sonarli on the Warna River. There the major NW–SE trend of the Warna River changes by 40° at the intersection with the Bhogiv *nala* (Figs. 5, 7, 14). This intersection of trends is the location of the most intense seismicity following the filling of the Warna Reservoir. The other major bend/intersection is near and to the south of the right angle bend in the Koyna River. This area has been the focus of intense past and ongoing seismicity.

4. Conclusions

Relocation of seismicity between 1963 and 1995 shows that it is widespread and extends from near the Koyna River to south and southwest of the Warna River. From an integration of an assortment of data with the seismicity pattern we conclude that the area lying between Koyna and Warna rivers can be divided into several seismogenic crustal blocks underlain by a fluid-filled fracture zone. This fluid-filled fracture zone lies at depths between ~ 6 and 13 km and is the location of most of the larger events ($M \geq 3.0$). The seismicity (as indicated by the better located earthquakes in 1993–95 and also the larger events) is bounded to the west by the Koyna River fault zone (KRFZ) (Fig. 16). KRFZ lies along the N–S portion of the Koyna River and extends S10°W for at least 40 km and dips steeply to the west. The current seismicity along the KRFZ lies at depths between ~ 6 and 14 km. We infer that the 1967 Koyna earthquake occurred along the KRFZ and was associated with a parallel set of ground fissures located 1–2 km to its east. The seismicity is bounded on the east by the NE–SW Patan fault (and possibly the parallel fault (?) through Tamine). This fault extends from Patan on the Koyna River SW to near Ambole on the Warna River. Patan fault dips $\sim 45°$ to the NW possibly terminating in the KRFZ. The February 1994 M 5.4 occurred on the Patan fault at its intersection with KRFZ. The bounding KRFZ and Patan fault are intersected by several NW–SE fractures which extend from near surface to hypocentral depths. They form steep boundaries of the blocks and have surface manifestations. They are discernable on satellite images. These NW–SE fractures provide conduits for fluid pressure flow to hypocentral depths. Geomorphic and aeromagnetic data support the view that the E–W leg of the Koyna River is fault controlled. Sharp bends in the Koyna and Warna rivers (6 km S of Koyna Dam and near Sonarli, respectively) are locations of stress build-up and the observed seismicity.

Acknowledgements

I wish to thank Mr. Sawalwade and Mr. K. D. Kumara Swamy of the Maharashtra Engineering Research Institute (MERI). They were largely responsible for all the data processing and relocation of earthquakes. My thanks also to Mr. Pendse, Secretary of Irrigation, Government of Maharashtra and Mr. Lagwankar, former Director of MERI for sending Mr. Sawalwade and Mr. Kumara Swamy to the University of South Carolina, and for their interest in this study. I also want to thank Chuck Langston and Evelyn Roeloffs for their insightful and helpful reviews. Finally, I want to thank Mrs. Lynn Hubbard for word processing and Ron Marple and Bob Trenkamp for their assistance with the figures.

REFERENCES

ANDREWS, D. J. (1989), *Mechanics of Faults Junctions*, J. Geophys. Res. *94*, 9389–9397.
AUDEN, J. B. (1954), *Erosional Patterns and Fracture Zones in Peninsular India*, Geol. Mag. *91*(2), 89–101.
BANGHAR, A. R. (1972), *Focal Mechanism of Indian Earthquakes*, Bull. Seismol. Soc. Am. *62*, 603–608.
CHANDRA, U. (1977), *Earthquakes of Peninsular India—A Seismotectonic Study*, Bull. Seismol. Soc. Am. *67*, 1387–1413.
DESHPANDE, B. G., and JAGTAP, P. N. (1971), *Interpretation of aerial photographs of Koyna Region*. In *Symposium on Koyna Earthquake: Indian Journal of Power and River Valley Development*, pp. 25–26.
DZIEWONSKI, A. M., EKSTRÖM, G., FRANZEN, J. E., and WOODHOUSE, J. H. (1988), *Global Seismicity of 1980: Centroid-Moment Tensor Solutions for 515 Earthquakes*, Phys. Earth and Planet. Inter. *50*, 127–154.
GAJEWSKI, D., and PRODEHL, C. (1987), *Seismic Refraction Investigation of the Black Forest*, Tectonophysics *142*, 27–48.
GEOLOGICAL SURVEY OF INDIA (Officers of the) (1968), *A Geological Report on the Koyna Earthquake of 11th December, 1967, Satara District, Maharashtra State*, Unpublished Report (GSI) 242 pp. [Referred to in the text as GSI Report, 1968].
GUHA, S. K., GOSAVI, P. D., VARNA, M. M., AGARWAL, S. P., PADALE, J. G., and MARIWADI S. C. (1970), *Recent Seismic Disturbances in the Shivajisagar Lake Area of the Koyna Hydroelectric Project, Maharashtra, India, Central Water and Power Research Station, Poona, India*.
GUPTA, H. K., *Reservoir-Induced Earthquakes* (Elsevier Publishers, Amsterdam 1992) 364 pp.
GUPTA, H. K., NARAIN, H., RASTOGI, B. K., and MOHAN, I. (1969), *A Study of the Koyna Earthquake of December 10, 1967*, Bull. Seismol. Soc. Am. *59*, 1149–1162.
GUPTA, H. K., RAM KRISHNA RAO, C. V., RASTOGI, B. K., and BATIA, S. C. (1980), *An Investigation of Earthquakes in Koyna Region, Maharashtra, for the Period October 1973 Through December 1976*, Bull. Seismol. Soc. Am. *70*, 1833–1847.
GUPTA, I. D., and RAM BABU, V. (1993), *Source Parameters of Some Significant Earthquakes Near Koyna Dam, India*, Pure appl. geophys. *140*, 403–413.
GUPTE, R. B. (1968), *The Koyna Earthquake*, Geol. Soc. India Bull. *5*, 37–41.
HARPSTER, R. E., CLUFF, L. C., and LOVEGREEN, J. R. (1979), *Active Faulting in the Deccan Plateau Near Koynanagar, India*, Geol. Soc. Am. Abstr. Progr. *11*(7), 438–439.
JAIN, M. S., GHODKE, S. S., and GAJBHIYE, N. G. (1971), *The Koyna earthquake of 11.12.1967 and the damage caused by it*. In *Symposium on Koyna Earthquake, Indian Journal of Power and River Valley Development*, pp. 55–60.
KAILA, K. L. (1983), *Personal Communication*.

KAILA, K. L., REDDY, P. R., DIXIT, M. M., and LAZARENKO, M. A. (1981), *Deep Crustal Structure at Koyna, Maharashtra, Indicated by Deep Seismic Soundings*, J. Geol. Soc. India *22*, 1–16.

KAILASAM, L. N., and MURTHY, B. G. K. (1971), *A short note on gravity and seismic investigations in the Koyna area*. In *Symposium on Koyna Earthquake, Indian Journal of Power and River Valley Development*, pp. 27–30.

KHATTRI, K. N. (1970), *The Koyna Earthquake—Seismic Studies*, Proc. Symposium on Earthquake Engg., School of Research and Training in Earthquake Engg., University of Roorkee, Roorkee, India.

KHATTRI, K. N., SAXENA, A. K., and SINVHAL, H. (1977), *Determination of Seismic Source Parameters for the 1967 Earthquake in Koyna Dam Region, India, Using Body Wave Spectra*, Proc. Sixth World Conference on Earthquake Engineering, New Delhi, India *V.2*, 308–316.

KING, G., and NABELEK, J. (1985), *The Role of Bends in Faults in the Initiation and Termination of Earthquake Rupture*, Science *228*, 984–987.

KRISHNA, V. G., KAILA, K. L., and REDDY, P. R. (1989), *Synthetic seismogram modeling of crustal seismic record sections from the Koyna DSS profiles in western India*. In *Properties and Processes of Earth's Lower Crust*, Am. Geophys. Union Geophys. Monogr. *51*, IUGG 6, 143–157.

LANGSTON, C. A. (1976), *A Body Wave Inversion of the Koyna, India, Earthquake of December 10, 1967 and Some Implications for Body Wave Focal Mechanisms*, J. Geophys. Res. *81*, 2517–2529.

LANGSTON, C. A. (1981), *Source Inversion of Seismic Wave Form: The Koyna India Earthquake of 13, September 1967*, Bull. Seismol. Soc. Am. *71*, 1–24.

LANGSTON, C. A., and FRANCO-SPERA, M. (1985), *Modeling of the Koyna, India, Aftershock of 12 December 1967*, Bull. Seismol. Soc. Am. *75*, 651–660.

LEE, W. H. K., and RALEIGH, C. B. (1969), *Fault Plane Solution of the Koyna (India) Earthquake*, Nature *223*, 172–173.

NEGI, J. G., AGRAWAL, P. K., and RAO, K. N. N. (1983), *Three-dimensional Model of the Koyna Area of Maharashtra State (India) Based on the Spectral Analysis of Aeromagnetic Data*, Geophysics *48*, 964–974.

PATWARDHAN, A. M., KARMARKAR, N. R., PANASKAR, D. B., and MASHRAM, D. C. (1995), *Seismicity Impact in Patan Taluka, District Satare, Maharashtra*, J. Geol. Soc. India *46*, 275–285.

PESHWA, V. V. (1991), *Geological Studies of Chandoli Dam Site Area Warna Valley, Sangli Dist., Maharashtra State*. Studies Based on Remote Sensing Techniques. Unpublished Report to Maharashtra Engineering Research Institute, Nashik, Department of Geology, University of Pune, 45 pp.

RAO, B. S. R., PRAKASA RAO, T. K. S., and RAO, V. S. (1975), *Focal Mechanism Study of an Aftershock in the Koyna Region of Maharashtra State, India*, Pure appl. geophys. *113*, 483–488.

RASTOGI, B. K., and TALWANI, P. (1980), *Relocation of Koyna Earthquakes*, Bull. Seismol. Soc. Am. *70*, 1843–1868.

REDDY, K. N., and JERATH, O. (1984), *Consolidated Report on the Geological Setting of Parts of Warna Valley, Sangli, Kolhapur and Satara Districts, Maharashtra*, (Geological Survey of India, Progress Report for the Field Season 1982–83).

SAHASRABUDHE, Y. S., RANE, V. V., and DESHMUKH, S. S. (1971), *Geology of the Koyna Valley*. In *Symposium on Koyna Earthquake, Indian Journal of Power and River Valley Development*, pp. 47–54.

SATHE, R. V., PADKE, A. V., PESHWA, V. V., and SUKHATANKAR, R. K. (1968), *On the Development of Fissures and Cracks in the Region Around the Koyna Nagar Earthquake Affected Area*, J. of Univ. of Poona, Science and Technology Section, *34*, 15–19.

SINGH, D. D., RASTOGI, B. K., and GUPTA, H. K. (1975), *Surface Wave Data and Source Parameters of Koyna Earthquake of December 10, 1967*, Bull. Seismol. Soc. Am. *65*, 711–731.

SNOW, D. T. (1982), *Hydrology of Induced Seismicity and Tectonism: Case Histories of Kariba and Koyna*, Geol. Soc. Am. *Special Paper 189*, 317–360.

SYKES, L. R. (1970), *Seismicity of the Indian Ocean and a Possible Nascent Island are Between Ceylon and Australia*, J. Geophys. Res. *75*, 5041–5055.

TALWANI, P. (1988), The Intersection Model for Intraplate Earthquakes, Seis. Res. Lett. *59*, 305–310.

TALWANI, P. (1994), *Ongoing Seismicity in the Vicinity of the Koyna and Warna Reservoirs, A Report to the United Nations Development Programme and Department of Science and Technology*, Government of India, August, 85 pp.

TALWANI, P., KUMARA SWAMY, S. V., and SAWALWADE, C. B. (1996), *Koyna Revisited: The Reevaluation of Seismicity Data in the Koyna-Warna Area, 1963–1995*, Univ. South Carolina Tech. Report (Columbia, South Carolina) 343 pp.

TANDON, A. N., and CHAUDHURY, H. M. (1968), *Koyna Earthquake of December 10, 1967*, India Meteorol. Dept. Seismol. Rept. *59*, 12 pp.

TSAI, Y-BEN, and AKI, K. (1971), *The Koyna, India, Earthquake of December 10, 1967 (Abstract Only)*, Trans. Am. Geophys. Union *52*, 277.

WELLS, D. L., and COPPERSMITH, K. J. (1994), *New Empirical Relationships Among Magnitude, Rupture Length, Rupture Width, Rupture Area and Surface Displacement*, Bull. Seismol. Soc. Am. *84*, 974–1002.

WENZEL, F., and SANDMEIER, K.-J. (1992), *Geophysical Evidence for Fluids in the Crust Beneath the Black Forest, SW Germany*, Earth-Science Reviews *32*, 61–75.

(Received October 30, 1996, accepted June 25, 1997)

Pure appl. geophys. 150 (1997) 551–562
0033–4553/97/040551–12 $ 1.50 + 0.20/0

❚ Pure and Applied Geophysics

Delineation of Active Faults, Nucleation Process and Pore Pressure Measurements at Koyna (India)

R. K. Chadha,[1] H. K. Gupta,[1] H. J. Kumpel,[2] P. Mandal,[1]
A. Nageswara Rao,[1] Narendra Kumar,[1], I. Radhakrishna,[1]
B. K. Rastogi,[1] I. P. Raju,[1] C. S. P. Sarma,[1] C. Satyamurthy[1]
and H. V. S. Satyanarayana[1]

Abstract—Earthquakes continue to occur in the vicinity of Shivaji Sagar Lake since its creation by the Koyna Dam in 1962. The seismicity peaked in 1967 with a M 6.3 earthquake which claimed over 200 human lives and destroyed the Koyna township. Earthquakes of $M \geq 4$ occur every year following an increase of water level in the reservoir. During 1973, 1980 and 1993–94 earthquakes exceeding magnitude 5 occurred. Most earthquakes of $M \geq 4$ are associated with pronounced foreshocks and aftershocks. Starting Sepember 1993, seismic monitoring was vastly improved with the deployment of additional close-by stations (analog and digital). The focal parameters now available have enabled delineation of the active faults and deciphering of the earthquake nucleation process. During 1995–96, 13 boreholes were drilled to depths of 130 to 250 m and measurement of water levels in these wells was initiated. A preliminary analysis of one year's data from a borehole 1 km south of Koyna reveals tidal signatures, indicating connection of the well to a confined aquifer which is favorable for detection of pore pressure anomalies induced by crustal strain. We hope to improve our understanding of the genesis of reservoir-induced earthquakes at Koyna with these new measurements.

Key words: Reservoir-induced seismicity, nucleation, pore pressure.

Introduction

Earthquakes started soon after the initial impounding of the Shivaji Sagar Reservoir behind the Koyna Dam (height 103 m, capacity 2.78 km³) in 1962. The seismicity continues with earthquakes of $M \geq 4$ occurring every year and $M \geq 5$ which occurred during 1967–68, 1973, 1980 and 1993–94. To date, about 100,000 shocks down to zero magnitude have been recorded, with over 100 tremors of $M > 4.0$. Since 1993 there has been an increase in seismic activity with most of the earthquakes occurring around the newly impounded Warna reservoir of height 80 m, capacity 1 km³, 25 km south of Koyna. The dam was filled to over 60 m height during 1993.

---

[1] National Geophysical Research Institute, Hyderabad 500 007, India.
[2] Geologisches Institut, University of Bonn, Nussallee 8, D-53115 Bonn, Germany.

Our earlier studies (GUPTA *et al.*, 1972a,b; GUPTA and RASTOGI, 1976; GUPTA, 1992) indicate that the seismicity is reservoir-induced. Recently, RAJENDRAN *et al.* (1996) have discussed the correlation of space-time patterns of seismicity with the Koyna reservoir level. TALWANI *et al.* (1996) have related this seismicity with the geomorphology and possible faults in the area. From the fault plane solutions and hypocenters distribution they have suggested existence of NE trending fault(s), dipping to the NW. Since 1993, we have considerably improved monitoring with the deployment of several analog and digital seismographs (Fig. 1) close to the epicentral area (with 5 or more stations at 6 to 15 km distance from the epicenters). With the improved event locations, it is now possible to delineate active faults in the region and study the nucleation process of earthquakes.

Several theoretical studies have been carried out to understand the role of increased pore pressure in inducing earthquakes around reservoirs (GUPTA and RASTOGI, 1976; TALWANI and ACREE, 1985; SIMPSON *et al.*, 1988). These studies have clearly brought out the role of pore pressure diffusion in controlling the disposition and migration of earthquakes. To understand the continued seismicity of the Koyna-Warna region, measurement of pore pressure in 130 to 250 m deep boreholes, began in 1995. These measurements should provide additional information in understanding the mechanism responsible for the continued seismicity in this region.

This paper presents the results of our ongoing work in the Koyna region which includes delineation of seismically active faults, the nucleation/migration process of foreshocks, and pore pressure measurement.

Hypocentral Locations and Delineation of Active Faults

Figure 2 shows the epicenters of shocks of M 2.0–5.4 which occurred during September 1993 through December 1995 in the Koyna-Warna region. These earthquakes were located using Hypo71pc program (LEE and VALDES, 1985) and phase data from 19 stations operated by the National Geophysical Research Institute, Hyderabad and the Maharashtra Engineering Research Institute, Nasik. Most of these hypocenters are of B and C quality. As seen from the standard errors in the HYPO71 program, the accuracy in epicentral locations is <1 km and that of hypocenter is <2 km for a majority of the earthquakes. The accuracy in depth determination has been checked in two other ways: i) the depth determined from waveform analysis (local earthquake moment tensor analysis, being reported elsewhere) for 80% of the earthquakes is within 1 km of the value obtained from HYPO71 and ii) the depth determined for many quarry blasts for Bauxite mining, 8 km west of Warna Dam is obtained to be 0.6 km or less (S. S. Rai, personal communication). However, the actual error in depth determination may be slightly more for some events. The root-mean-squares error of the P residual times is less than 0.2 s. The trend of epicenters defines two NNE-SSW trending faults, one passing through Koyna and extending southward for a length of 30 km and the other through Warna reservoir and extending for a length of about 20 km. These trends are seen predominantly in Figure 3 for magnitudes 2.0 to 2.9 and less clearly in Figure 4 for magnitudes 3.0–3.9 ranges. The Warna reservoir which is newly impounded seems to be more active, presently and perhaps because of this the two trends are overlapping in this area and giving an impression of a NW trend. Studies of earlier data during 1967–76 did indicate a conjugate set of trends in NNE-SSW

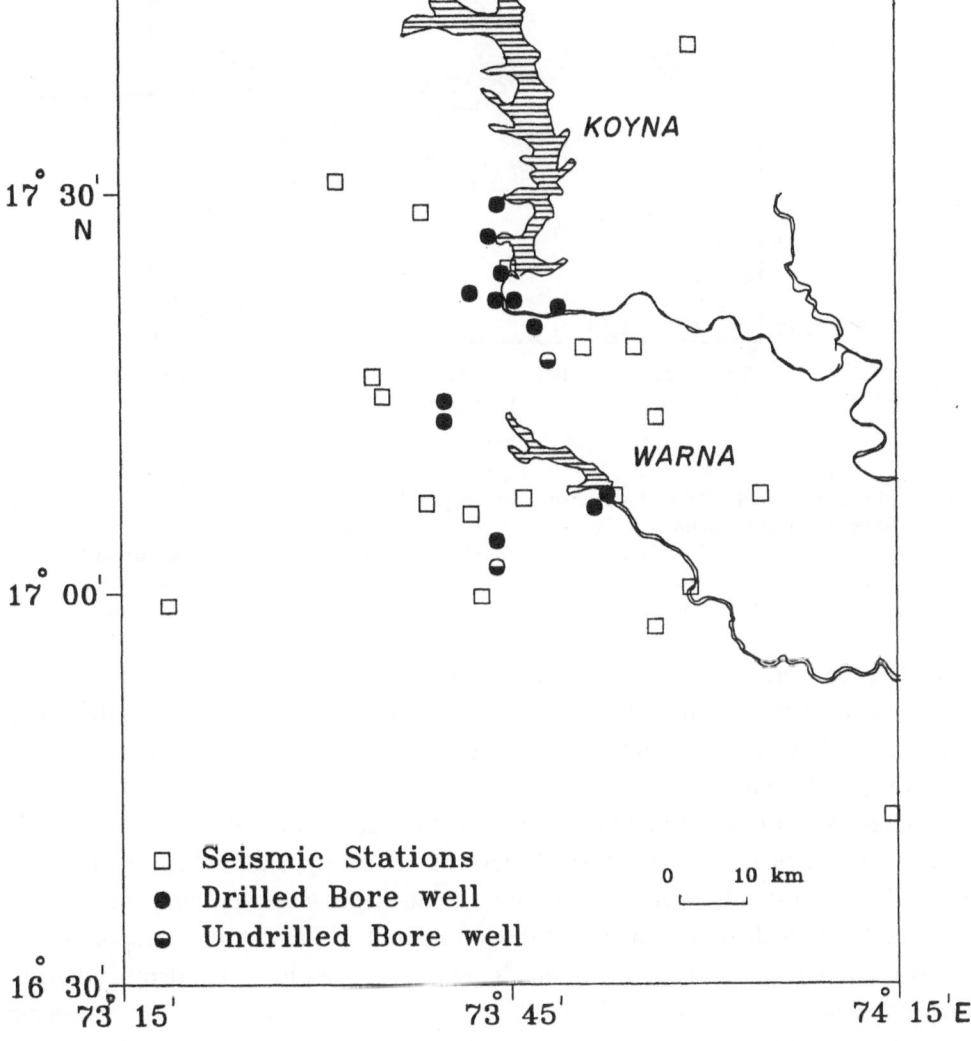

Figure 1
The location of seismic stations and 15 bore wells in the Koyna-Warna area. Locations of additional six bore wells are yet to be decided.

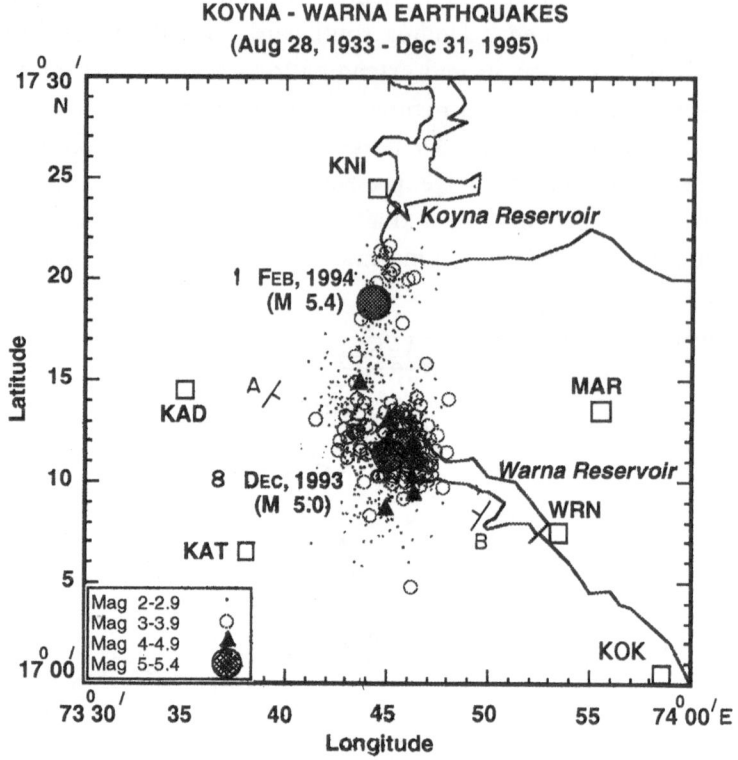

KOYNA - WARNA EARTHQUAKES
(Aug 28, 1933 - Dec 31, 1995)

Figure 2

Earthquakes of magnitudes 2.0 to 5.4 located during September 1993–December 1995. Locations of six seismic stations in close proximity are also shown. Epicenters of earthquakes with ERH < 1 km and ERZ < 2 km only are plotted in Figures 2, 3 and 4. Depth section along line A-B is shown in Figure 5. All the earthquakes have been considered for the depth section. The station symbols are: KNI- Koyna; MAR- Marathwadi; KAD- Kadvai; KAT- Katwali; WRN- Warna; KOK- Kokrud.

and NW-SE directions (RASTOGI and TALWANI, 1980; GUPTA *et al.*, 1980). LANGSTON (1976) has also shown predominance of lineament trends in these two directions. However, our studies indicate that the present activity is mostly along the NNE-SSW trend.

Figure 5 illustrates the hypocentral depth distribution along a WNW-ESE section perpendicular to the two clusters of seismicity and indicates two near vertical faults extending down to about 14 km depth corresponding to the two clusters. Most of the hypocenters are down to about 14 km depth corresponding to the two clusters. Most of the hypocenters are down to about 10 km depth, whereas, only a few are deeper. Hence, the fault may be considered to be active for a depth of at least 10 km. A clear deficiency of hypocenters at 1–4 km depth range is observed.

Figure 6 shows the composite fault plane solution for the earthquakes located during the study period and indicates two nodal planes striking NNE and NW. The plane striking NNE (N20°E) with a dip of 56° towards NW is considered as the fault plane which exhibits left-lateral strike-slip movement. The NNE trend agrees with the trend of seismicity as well as the direction of ground rupture caused by the main earthquake of December 10, 1967 (SAHASRABUDHE et al., 1969; SEEBER et al., 1996). Incidently, the same direction of fault was found from the focal mechanism solution for the main earthquake (TANDON and CHAUDHURY, 1968). The discrepancy between 56° dip of the fault plane obtained from the fault plane solution and near vertical dip from the depth section could be due to uncertainty in any of the two estimates.

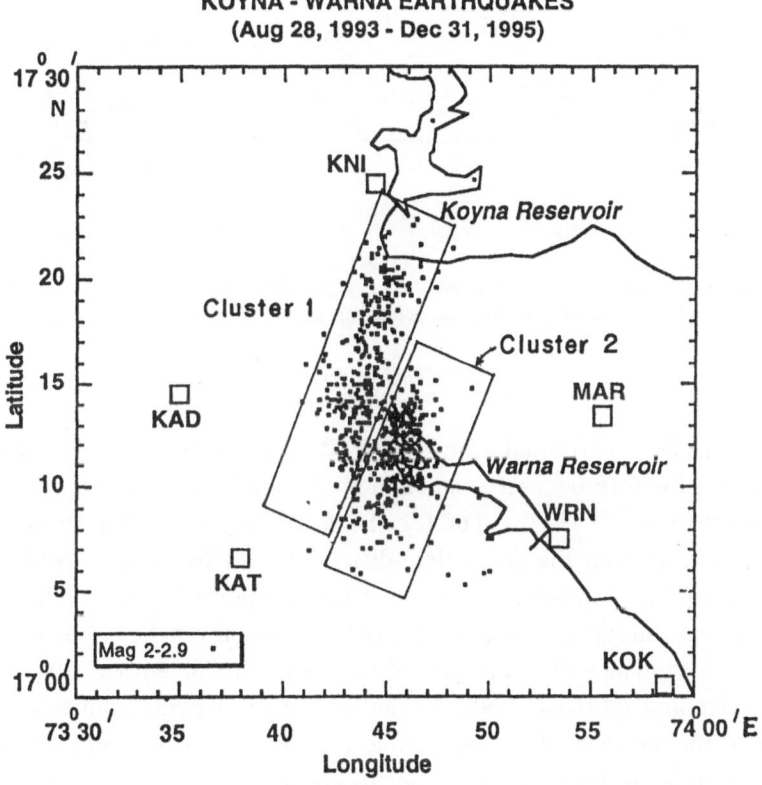

Figure 3

Earthquakes of magnitude 2.0–2.9 located during September 1993–December 1995. Two NNE-SSW trends of epicenters are shown by rectangles. Cluster 1 is for epicenters located south of Koyna reservoir and Cluster 2 is across the Warna reservoir.

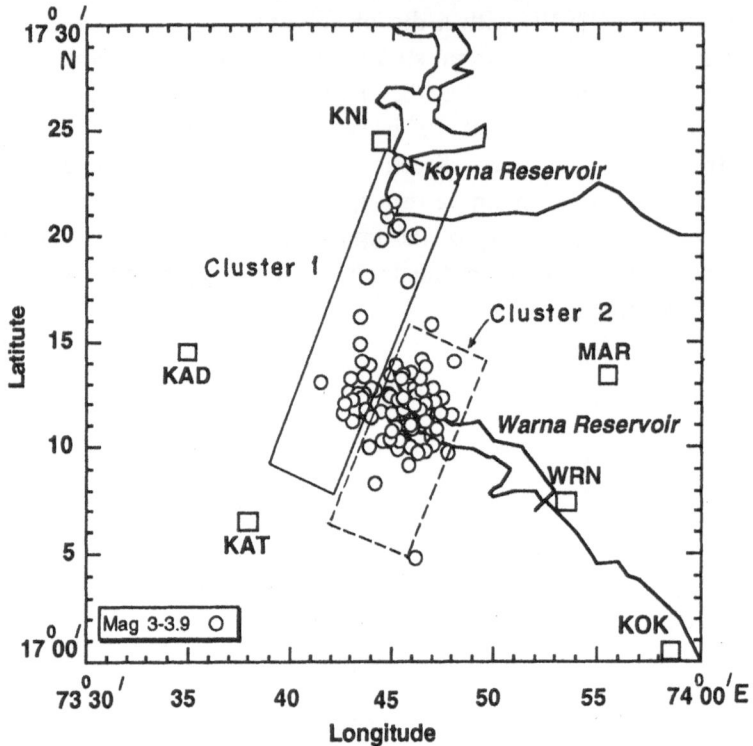

KOYNA - WARNA EARTHQUAKES
(Aug 28, 1993 - Dec 31, 1995)

Figure 4

Earthquakes of magnitude 3.0–3.9 located during September 1993–December 1995. The trend for cluster 2 is not well defined in this magnitude range.

Nucleation Process

The main 1967 earthquake and other Koyna shocks of $M > 4.0$ have been accompanied by foreshock-aftershock activity corresponding to Mogi's type-II pattern (GUPTA *et al.*, 1972b). The foreshock activity occurs for about 20 days. It has been observed that the depth-time distribution of foreshocks displays a nucleation process in the quasi-dynamic state of rupture preceding the main rupture, and the foreshocks during the nucleation process can play a key role in forecasting damaging earthquakes (ISHIDA and KANAMORI, 1978; MOGI, 1985; OHNAKA, 1992; DODGE and BEROZA, 1995). In the case of the Koyna main shocks, we tried to examine the shape of the seismic zone for foreshocks, to detect the nucleation process. With accurate hypocentral locations, we could decipher the precursory nucleation process for some larger shocks during 1993–95. This is illustrated through one example of the December 8, 1993 earthquake as shown in Figure 7. The upper part of this figure shows the epicenters of the main shock and foreshocks

500 hours before the main shock. The shocks which have occurred within a distance of 5 km from the epicenter of the December 8, 1993 main shock and within a period of 500 hours of the main shock were examined for the nucleation process. Time vs. depth plots for these foreshocks are shown in the lower part of Figure 7. The fracture is inferred to have nucleated at shallow depth in the brittle layer and then propagated in a cone shape to cause the main shock at a depth of 8.4 km, i.e., near the base of the seismogenic layer as modeled by OHNAKA (1992).

Pore-fluid Measurements

Under a collaborative program initiated during 1995 between the University of Bonn, Germany, and the National Geophysical Research Institute, Hyderabad, India, a network of 21 deep observation wells are planned to be drilled close to the seismicity areas around the Koyna-Warna reservoirs to monitor and study water-

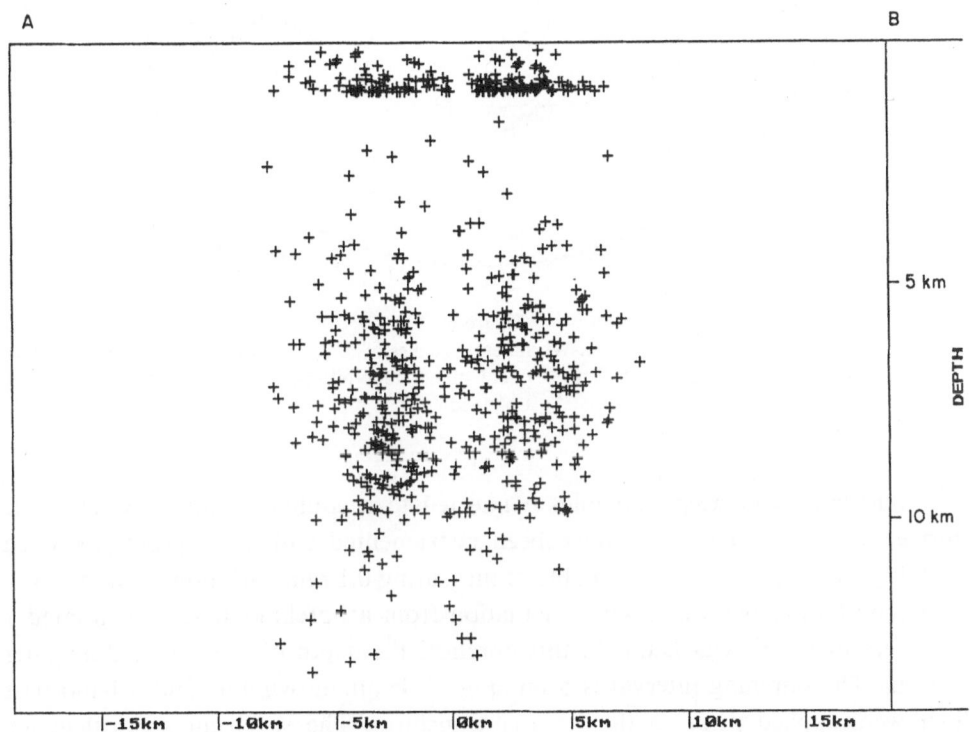

Figure 5
Depth distribution of hypocenters along a WNW-ESE trending line (shown as A-B in Fig. 2) which is perpendicular to the NNE-SSW trending two seismicity clusters. Separation of two clusters at 0 km position on line AB can be inferred. Hypocenters of earthquakes of $M \geq 2.0$ and with RMS < 1 km and ERH < 2 km which occurred during September 1993 to December 1995 have been included. Vertical exaggeration is two times.

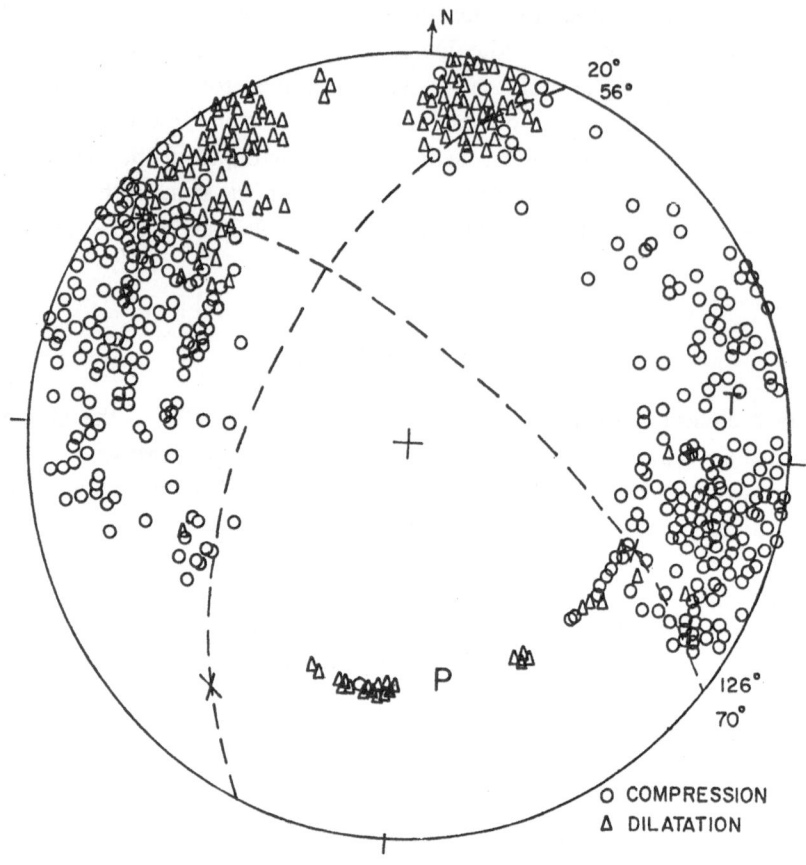

Figure 6

Composite fault plane solution for Koyna-Warna earthquakes of $M \geqq 2.0$ during 1993–95. Direction of P' arrivals, wherever clear are plotted. Two nodal planes in NNE-SSW and NW-SE directions are obtained.

level fluctuations in deep boreholes connected to a confined aquifer system. The thirteen wells drilled to date, have been instrumented with microprocessor-based sensitive pressure transducers capable of measuring 0.1 mm variation in water level.

Figure 8(a) shows water-level fluctuation from a borehole at Rasati situated 1 km south of the Koyna Dam. In this borehole the depth of the confined acquifer is 70 m. The sampling interval is 5 minutes. A Hanning window and a band-pass filter were applied prior to the Fourier transform. The spectrum was calculated using the program ETERNA 3.0 of WENZEL (1993) and a band-pass filter of 145 coefficients, with high-pass characteristics for the low frequency part (damping −37 db at 7.5 deg/h, −153 db below 1.5 deg/h). Figure 8(b) shows the spectrum of the raw data for the period 24 July–27 December 1995. The spectral analysis reveals a clear existence of tides in the piezometric water-level hydrograph. The presence of the semi-diurnal Lunar tide $M2$ proves that the aquifer system connected to the well is confined, indicating that the recorded fluctuation reflects pore pressure variations in a water-saturated formation. The dominance of the solar constituent $S2$ over $M2$ and probably of $S1$ (as part of the unresolved harmonics $P1$, $S1$ and

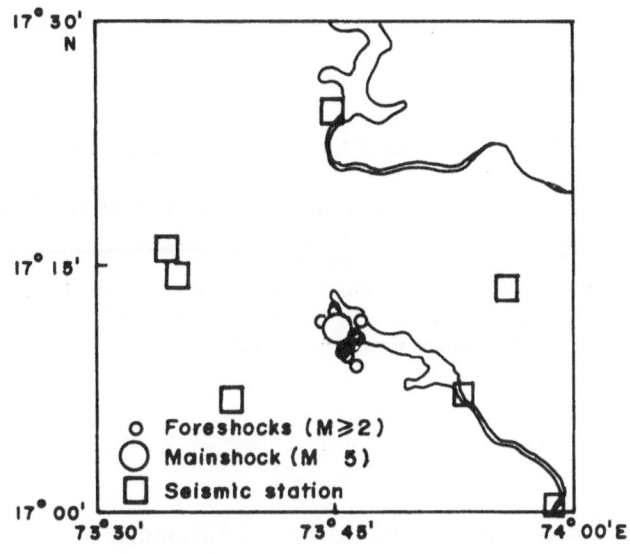

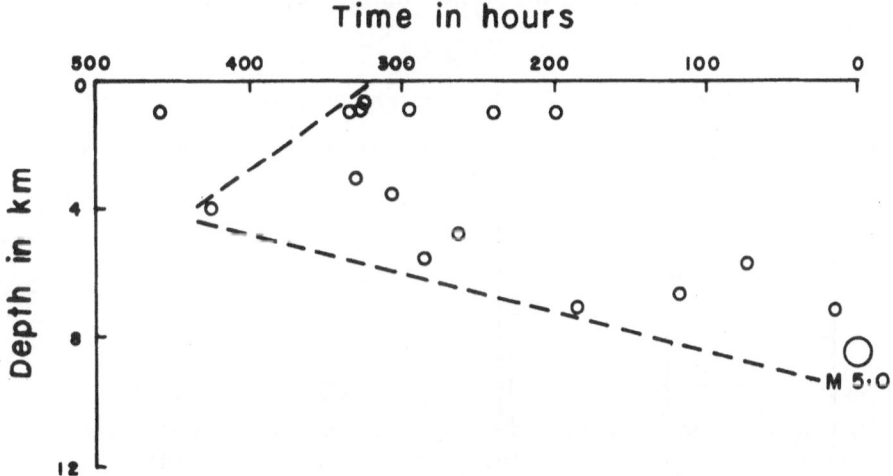

Figure 7

Epicenters of the main shock of magnitude 5.0 on December 8, 1993 and its foreshocks which have occurred within an area of 5 km radius and 500 hours before the main shock are shown in the upper part of the figure (the other shocks which have occurred outside the target area are not shown). Temporal distribution of the foreshocks with depth is shown in the lower part of the figure. The foreshocks have migrated from shallow depths (about 1 km) in a cone shape, resulting in the occurrence of the main shock at 8.4 km depth. Earthquakes of $M \geq 2.0$ have been considered for this figure.

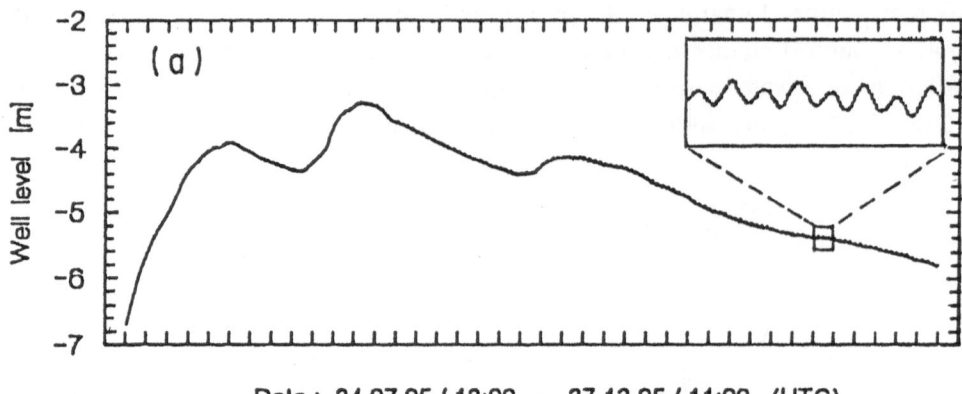

Dala : 24.07.95 / 12:00 - 27.12.95 / 11:00 (UTC)

Figure 8(a)

Water-level variations in meters from the ground level at Rasati well during July 24, 1995 through December 27, 1995. Each division on time axis represents 4 days. A blow-up of a portion showing tidal fluctuations for four days is also shown (inset).

$K1$) over $O1$, as well as the presence of higher harmonics $S3$ and $S4$ is interpreted as an inverse barometric effect, signifying thermomechanical stresses from the heating and cooling of rocks at the surface during day and night.

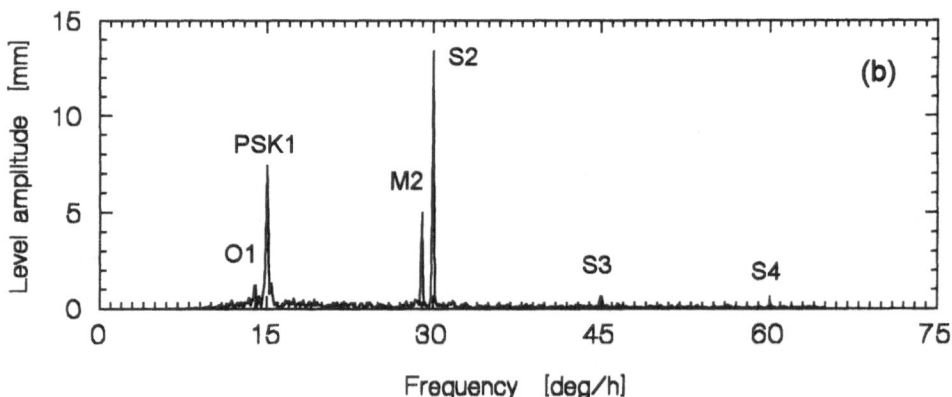

Figure 8(b)

Amplitude spectrum of Rasati well level data with resolved tidal peaks showing diurnal bands around 15 degree/hr, and semi, ter, and quarter diurnal bands around 30, 45 and 60 degree/hr, respectively. $O1$ = Principal lunar diurnal; $P1$ = Principal solar diurnal; $K1$ = Luni-solar diurnal; $M2$ = Principal lunar semi-diurnal and $S2$ = Principal solar semi-diurnal.

Discussion and Conclusions

The continued seismicity exceeding 34 years near the Koyna Dam indicates the presence of high stresses in the area. The seismicity has been correlated with the filling of the Shivaji Sagar Lake created by the Koyna Dam. This correlation is based on the start of seismicity soon after the beginning of filling in 1962, increase in seismicity after almost every annual filling and shallow depth (<10 km) of hypocenters. The seismic zone extended to about 30 km south of Koyna, where the newly built Warna Dam was impounded to 60 m in 1993. The present seismic activity which began in August 1993 is mostly confined to the northwestern part of the Warna reservoir. It was observed that $M \geqq 5.0$ earthquakes had occurred in 1967, 1973, 1980 when the rate of filling had exceeded 12 m/week in the Koyna Dam (GUPTA, 1985). In 1993 this rate of filling was exceeded at the Warna Dam.

The hypocenters located during August 1993 to December 1995 have delineated two NNE-SSW faults, one passing through Koyna and the other through Warna reservoirs. The foreshocks of some earthquakes of magnitude >4 have shown temporal migration, possibly indicating a nucleation process where the fracture initiates at shallow depths and propagates to the hypocentral depth of the main shock. However, further studies are required to understand this process.

It has long been recognized that water-level fluctuations in confined aquifers, which behave like a porous elastic material, reflect changes in crustal strain induced by earth tides and atmospheric loading (BREDEHOEFT, 1967). The sensitivity of well levels to crustal strain indicates that they are also useful in detecting tectonic strain. There are several well documented cases in the world where water-level fluctuations were observed in deep bore wells during the pre, co- and post-seismic stages. These fluctuations are believed to reflect pore pressure changes associated with redistribution of stresses in near and far fields of the dislocation source. Hence, water-level changes in a well, in response to earthquakes or aseismic fault creep can be detected if a calibration based on the response to earth tides is known in a well. The preliminary results from an experiment mounted for the first time in the seismically active Koyna region on the west coast of India indicate that the wells being monitored are responsive to crustal strain. However, these studies are based on observations over a very short duration. The full monitoring program, after the completion of 21 boreholes, is scheduled to continue for a considerable time so that the data acquired over various monsoon and inter-monsoon periods will become vital input parameter in understanding the relationship between pressure changes (well level signals) and stress perturbations due to Koyna reservoir and local tectonics.

Acknowledgement

Maharashtra Engineering Research Institute (MERI) provided us with the seismic phase data of the stations operated by the Koyna and Warna projects. The assistance rendered by site engineers is gratefully acknowledged. The project was supported by the Department of Science and Technology, New Delhi.

REFERENCES

BREDEHOEFT. J. D. (1967), *Response of Well Aquifer Systems to Earth Tides*, J. Geophys. Res. *72*, 3075–3087.

DODGE, D. A., and BEROZA, G. C. (1995), *Foreshock Sequence of the 1992 Landers, California, Earthquake and its Implication for Earthquake Nucleation*, J. Geophys. Res. *100*, 9865–9880.

GUPTA, H. K., RASTOGI, B. K., and NARAIN, H. (1972a), *Common Features of the Reservoir Associated Seismic Activities*, Bull Seismol. Soc. Am. *62*, 481–492.

GUPTA, H. K., RASTOGI, B. K., and NARAIN, H. (1972b), *Some Discriminatory Characteristics of Earthquakes near the Kariba, Kremasta and Koyna Artificial Lakes*, Bull. Seismol. Soc. Am. *62*, 493–507.

GUPTA, H. K., and RASTOGI, B. K., *Dams and Earthquakes* (Elsevier, Netherlands 1976) 227 pp.

GUPTA, H. K., RAO, C. V. R. K., RASTOGI, B. K., and BHATIA, S. C. (1980), *An Investigation of Earthquakes in Koyna Region, Maharashtra for the Period October 1973 through December 1976*, Bull. Seismol. Soc. Am. *70*, 1833–1847.

GUPTA, H. K. (1985), *The Present Status of Reservoir-induced Seismicity Investigations with Special Emphasis on Koyna Earthquakes*, Tectonophys. *118*, 257–279.

GUPTA, H. K., *Reservoir-induced Eartquakes* (Elsevier, Amsterdam, 1992) 355 pp.

ISHIDA, M., and KANAMORI, H. (1978), *The Foreshock Activity of the 1971 San Fernando Earthquake, California*, Bull. Seismol. Soc. Am. *68*, 1265–1279.

LANGSTON, C. A. (1976), *A body-wave Inversion of the Koyna, India, Earthquake of December 10, 1967, and Some Implications for Body-wave Focal Mechanisms*, J. Geophys. Res. *81*, 2517–1529.

LEE, W. H. K., and VALDES, C. M. (1985), *HYPO71PC: A Personal Computer Version of the HYPO71 Earthquake Location Program*, U.S. Geol. Surv. Open File Report 85–749, 43 pp.

MOGI, K., *Earthquake Prediction* (Academic Press, San Diego, Calif. 1985) 355 pp.

OHNAKA, M. (1992), *Earthquake Source Nucleation: A Physical Model for Short-term Precursors*, Tectonophys. *211*, 249–178.

RAJENDRAN, K., HARISH, C. M., and KUMARASWAMY, S. V. (1996), *Re-evaluation of Earthquake Data from Koyna-Warna Region: Phase-I*, Report Center Earth Science Studies, Trivandrum.

RASTOGI, B. K., and TALWANI, P. (1980), *Relocation of Koyna Earthquakes*, Bull. Seismol. Soc. Am. *70*, 1849–1868.

SAHASRABUDHE, Y. S., RANE, V. V., and DESHMUKH, S. S. (1969), *Geology of the Koyna Valley*, Proc. Symp. Koyna Earthquake, Ind. J. Power River Valley Development, 47–54.

SEEBER, L., EKSTRÖM, G., JAIN, S. K., MURTY, C. V. R., CHANDAK, N., and ARMBRUSTER, J. G. (1996), *The 1993 Killari Earthquake in Central India: A New Fault in Mesozoic Basalt Flows?*. J. Geophys. Res. *101*, 8543–8560.

SIMPSON, D. W., LEITH, W. S., and SCHOLZ, C. H. (1988), *Two Types of Reservoir-induced Seismicity*, Bull. Seismol. Soc. Am. *78*, 2025–2040.

TALWANI, P., and ACREE, S. (1985), *Pore Pressure Diffusion and the Mechanism of Reservoir-induced Seismicity*, Pure appl. geoph. *122*, 947–965.

TALWANI, P., KUMARSWAMY S. V., and SAWALWADE, C. B. (1996), *The Re-evaluation of Seismicity Data in the Koyna-Warna Area*, Report Univ. South Carolina, Columbia, South Carolina, U.S.A., 109 pp.

TANDON, A. N., and CHAUDHURY, H. M. (1968), *Koyna Earthquake of December 10*, 1967, India Meteorol. Dept., Sci. Rep. No. *59*, 12 pp.

WENZEL, H. G. (1993), *Earth Tides Analysis Program System ETERNA* 3.0, Geodetic Institute, University of Karlsruhe, Germany.

(Received March 13, 1997, accepted July 30, 1997)

Pure appl. geophys. 150 (1997) 563–583
0033–4553/97/040563–21 $ 1.50 + 0.20/0

| Pure and Applied Geophysics

Seismic and Aseismic Slips Induced by Large-scale Fluid Injections

F. H. CORNET,[1] J. HELM,[2] H. POITRENAUD[3] and A. ETCHECOPAR[4]

Abstract—A 3600 m deep well has been used to conduct large water injection tests in the Rhine Graben. The total volume injected during the fall 1993 reconnaissance program reached 44000 m³. Induced seismicity was monitored with both a downhole and a surface seismic network. About 20000 events have been recorded by the downhole tools and 165 events with the surface network. The largest observed magnitude reached 1.9, as determined from signal duration observed on the surface network. Borehole televiewer observations show that some slip events were larger than 4 cm at the borehole wall, a value much larger than the slip motion associated with microseismic events, as evaluated from events' magnitude. It is concluded that these observed slip events were aseismic. This implies that induced seismicity is not a good marker for the efficiency of this hydraulic stimulation. It only helps to identify zones of high pore pressure during injection.

Key words: Fluid injection, induced seismicity, aseismic slip, downhole monitoring network, surface monitoring network.

1. Introduction

An experimental research program on the possibility of developing geothermal energy in the central upper Rhine Graben has been ongoing since 1988 (KAPPELMEYER *et al.*, 1991). For this purpose a site is being developed at Soultz, about 40 km to the northeast of Strasbourg and about 35 km to the southwest of Karlsruhe (Fig. 1). The site is located within a geothermal anomaly identified by temperature measurements in the superficial sediment deposits. At this location, the sedimentary cover exhibits a thickness ranging from 1400 m to 1500 m. Below 1500 m, a granite formation is encountered and is the target for developing a geothermal reservoir. In 1993 the site included one 3600 m deep borehole (GPK1) and four observation wells reaching the granite (EPS1, 2200 m; 4616, 1414 m; 4550, 1500 m; 4601, 1604 m). Temperature at 3600 m is close to 160°C.

A large-scale water injection was undertaken in September and October 1993 in GPK1 in order to obtain a characterization of the hydraulic behavior of the rock

---

[1] Institut de Physique du Globe de Paris, France.
[2] Formerly with Ecole et Observatoire de Physique du Globe de Strasbourg, France, now at Stag Geological Consultants, Aldermaston, United Kingdom.
[3] Formerly with IPGP, now at Dowell Schlumberger, Clamart, France.
[4] Schlumberger, Clamart, France.

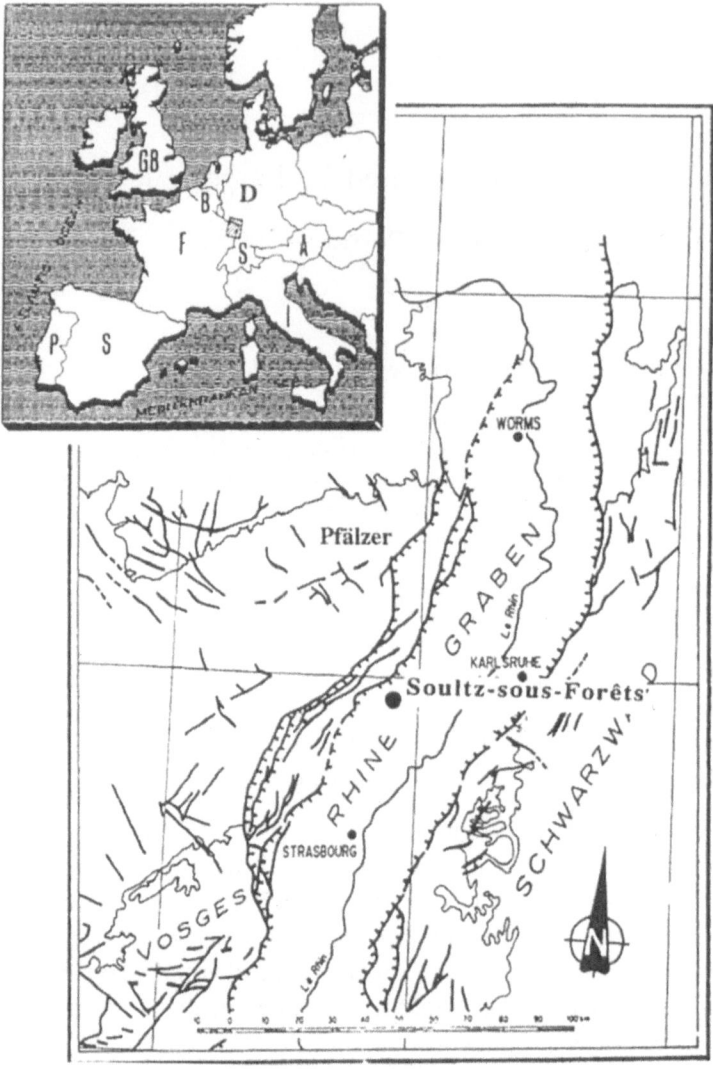

Figure 1
Location of Soultz-sous-fôrets, in the Rhine Graben, where the European Hot Dry Rock Experimental
site is located. The main faults on both sides of the Graben are indicated. The eastern fault system is
dipping more steeply than the western.

mass between 2800 m and 3500 m, together with information regarding the regional
stress field. During this injection, induced seismicity was monitored with both a
downhole and a surface seismic network. The former helped to accurately locate the
events, while the latter helped constrain focal mechanisms and events magnitude.
At the end of this injection test, an ultrasonic imaging log was conducted in order
to identify effects of this injection on the fracture network.

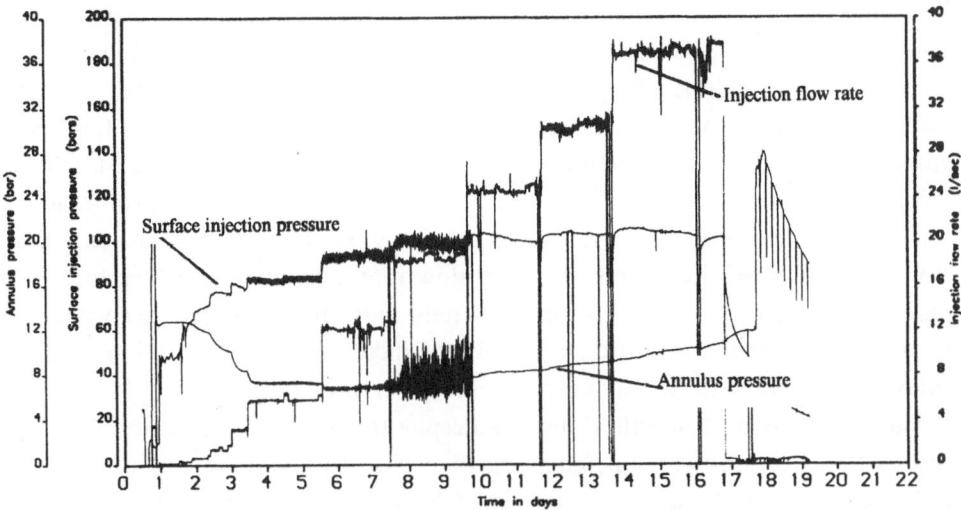

Figure 2
Surface well-head pressure, injection flow rate and annulus pressure for the September 1993 injection test in well GPK1.

After a brief presentation of the injection program, of the seismic monitoring system and of features of the seismic cloud induced by the injection, this paper presents the results from the slip motions as observed directly in the well. Comparison between source parameters of microseismic events as deduced from simple source models and observed slip motions helps to demonstrate the existence of nonseismic slip motions induced by this water injection.

2. The 1993 Large Water Injection at the Soultz Hot Dry Rock Experimental Site

The well GPK1 is cased down to the granite. In addition, a liner was cemented between 2850 and 2800 m but was hanging free from surface to 2800 m, in order to limit difficulties with thermal expansions or contractions caused by the alternate injection-production tests.

Injection proceeded in two periods. First (September injection), the well was sanded in, up to 3400 m, in order to avoid stimulating a large fault intersected around 3490 m. Then the sand was washed out and injection took place along the complete open hole section (October injection). During the first phase, injection proceeded for 15 days with a stepped incremental flow rate ranging from 1 l/sec to 36 l/sec. Once injection rate reached 6 l/sec, each step lasted 48 hours and the rate increments were equal to 6 l/sec (Fig. 2). The total injected volume reached 25300 m³. During the second phase, injection started at 41 l/sec and was later increased to 50 l/sec; a total volume of 19300 m³ was injected.

During the first phase, the well-head injection pressure rose progressively with each flow rate increment. But once the flow rate reached 24 l/sec, it stabilized (Fig. 2) so that the downhole overpressure at 2850 m was 9.1 MPa (about 10 MPa well-head pressure). Meanwhile the pressure in the annular, i.e., in the space between the liner and the borehole wall, increased progressively up to 1.2 MPa.

Spinner logs were run every 24 h in order to monitor the flow rate profiles in the well (CORNET and JONES, 1994; EVANS *et al.*, 1996). They outline the fact that, below a 10 MPa well-head pressure, each flow rate increment, and therefore each pressure increment, resulted in proportionately more flow reaching deep parts of the well. But once the pressure stabilized, the increment in flow rates resulted in proportionately more flow leaving the upper portion of the well (Fig. 3).

The stabilization of injection pressure despite the increase in injection flow rate suggests that fractures were opening, i.e., the walls of the opened fractures had no mechanical contact. However, as discussed later in this paper, no real hydraulic fracture was subsequently observed with the ultrasonic borehole imaging log.

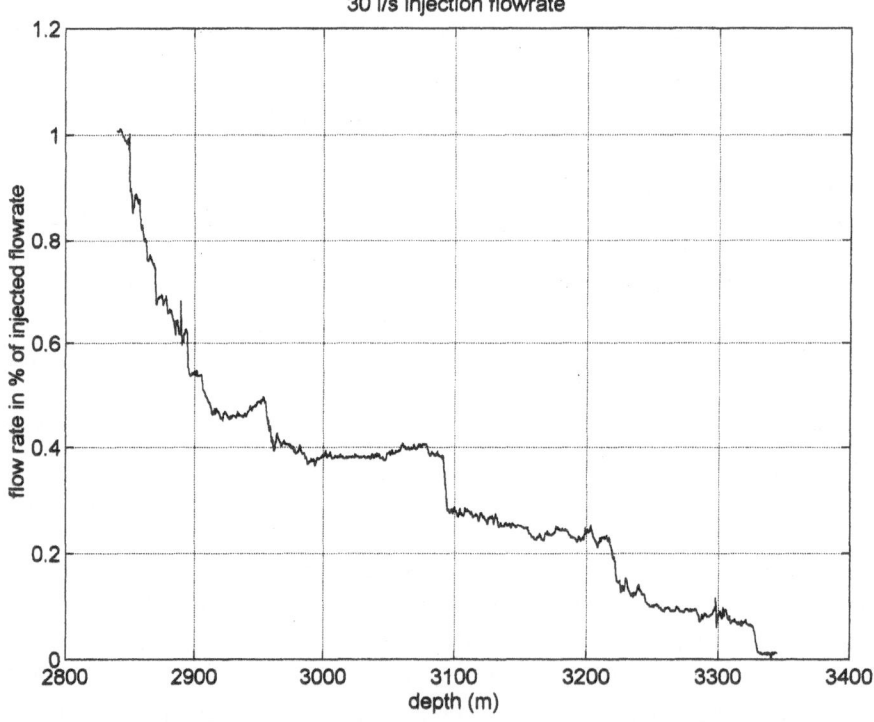

Figure 3

Variation of flow rate with depth (Spinner log) in GPK1 during the October 1993 injection, when injection flow rate was 30 l/sec. Casing shoe is at 2850 m. 60% of the flow is lost before 3000 m. The data have been corrected for local variations in cross section and uncertainties on depth (after K. EVANS, 1996).

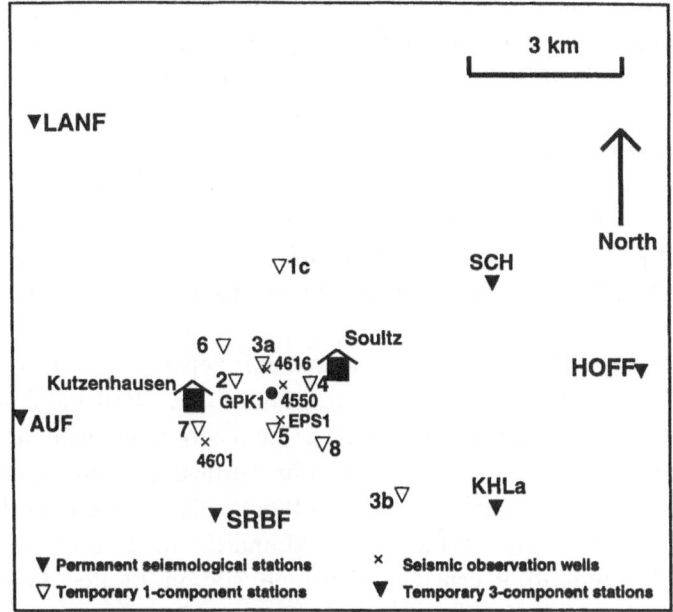

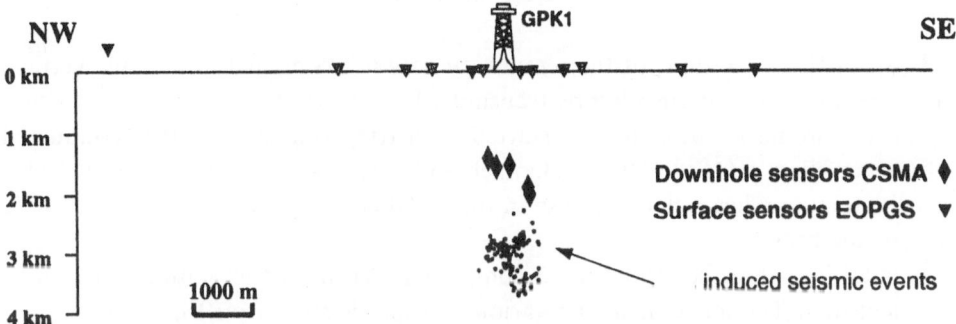

Figure 4
Location of the seismic stations for both the surface and the downhole network. Vertical cross section shows the relative opening of both networks.

3. Monitoring Induced Seismicity

3.1 Recording Network

The microseismicity induced by these injections has been monitored with two networks. The first one (Fig. 4) involves downhole sensors cemented at the bottom of the wells 4616, 4550, 4601 and hydrophones in the well EPS1 (CORNET and JONES, 1994; JONES et al., 1995). Each sensor contains 4 accelerometers. One is vertical while the other three are 109.5° away from each other and from the vertical direction; they exhibit a flat response up to 1 KHz and are anti-alias filtered at 1500 Hz. The sampling rate is 5000 sps/channel and the total duration of the recorded signal is 1.6 sec. Each sonde output, after gain, is 1000 V/g. Recording is triggered by an automatic detection routine.

The surface network (HELM, 1995) includes two different types of stations: a permanent network of 3 stations designed to continuously monitor the local natural seismicity and a temporary network of 11 stations, which is deployed only for the duration of the injections so as to obtain satisfactory azimuthal coverage for focal mechanism determination.

The three stations of the permanent network are respectively at 6.5 km (Langenberg site, LANF), 2.5 km (Surbourg site, SRBF) and 7.0 km (Hoffen site, HOFF) from the GPK1 well head. The Langenberg site includes three orthogonally oriented 0.5 Hz seismometers and one vertical accelerometer. The Hoffen and Surbourg stations consist of three orthogonally oriented accelerometers and one vertical seismometer. Data are digitized and multiplexed at each site and then sent by UHF radio to a central acquisition station located next to the downhole network acquisition station. Recording is automatically triggered; it is performed at a rate of 150 samples per second per channel and it lasts 64 seconds. Example of an artificial signal (blast in a quarry located about 15 km to the south of the site) recorded on the surface network and on the downhole network is shown on Figure 5.

Eight of the 11 stations of the temporary network consisted of a single vertical 1 Hz seismometer. All signals are transmitted by UHF to a central acquisition system. Recording is automatically started by a triggering system. It is conducted with a sampling rate of 180 sps/channel and lasts 64 seconds per event. This acquisition system has a flat response in the 1–50 Hz frequency domain, with a 72 dB dynamic range.

In addition, three autonomous 3-component seismometric stations have also been deployed. The location of the various sensors is shown on Figure 4.

3.2 Induced Microseismic Activity

Results from the downhole network for the September–October 1993 injection experiment have been presented (CORNET and JONES, 1994; JONES *et al.*, 1995). About 20000 events have been recorded. They outline a zone extending grossly about 400 m on both sides of GPK1 in the N 150°E direction. Figure 6 shows the location of events for three different depth intervals. It clearly outlines different cloud orientation, depending on depth: The cloud is north-south between 2700 m and 2900 m but N 145°–160°E between 3200 m and 3600 m.

It has been argued (CORNET and JONES, 1994) that these differences are linked to the relative value of pore pressure with respect to the *in situ* stress field. Between

2700 and 2900 m, where most of the flow is leaving the well, the pore pressure is larger than the normal stress supported by preexisting fractures and fractures are opened. These fractures are not perpendicular to the minimum principal stress, while the global flow direction, as derived from microseismicity, is. Indeed a 40 m high hydraulic fracture conducted at 2000 m and a thermal fracture observed at 3000 m are both oriented approximately north-south.

Below 3000 m, the pore pressure is smaller than the normal stress supported by preexisting fractures and failure occurs by shear, according to a Coulomb type failure criterion. Hence induced seismicity occurs in a direction about 30° off the maximum principal stress direction. This point will not be discussed further here and attention will focus on results obtained with the surface network.

Altogether, the surface network has recorded a total of 165 events: 127 events during the September injection and 38 events during the October injection. The magnitude of these events has been evaluated from the signal duration. This magnitude scale has been calibrated then with respect to the regional natural seismicity recorded on the permanent network. Magnitudes range from −0.5 to 1.9. Given the data recorded on the downhole network, this implies that the sedimentary cover filters out all events with a magnitude smaller than −0.5.

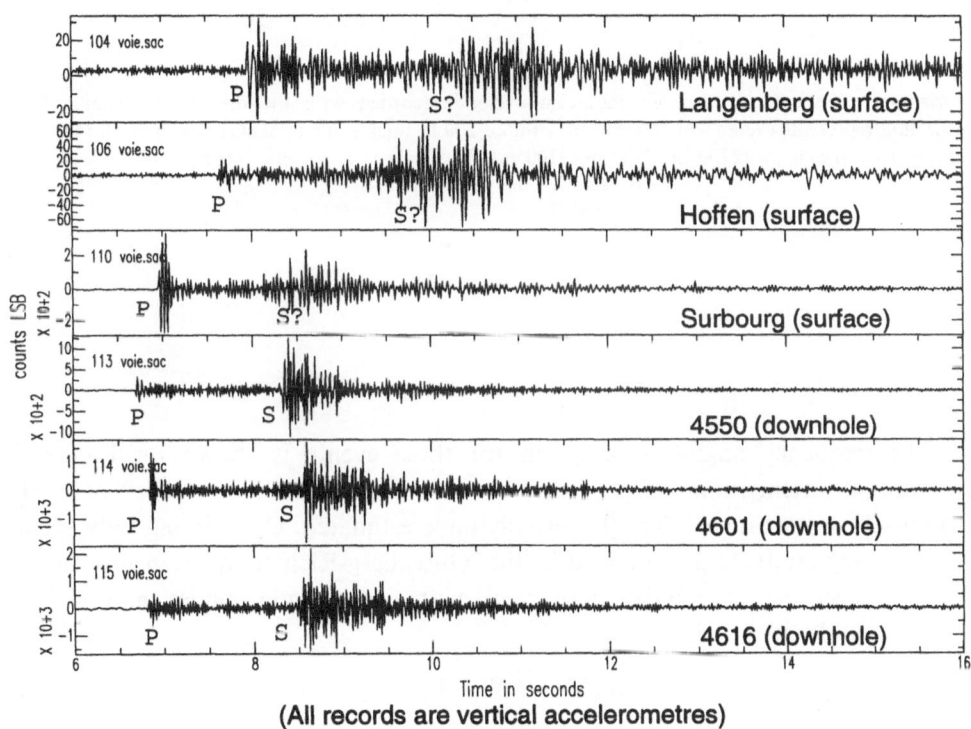

(All records are vertical accelerometres)

Figure 5

Example of signal recorded on both the downhole and the surface network. Source was a blast in a nearby quarry.

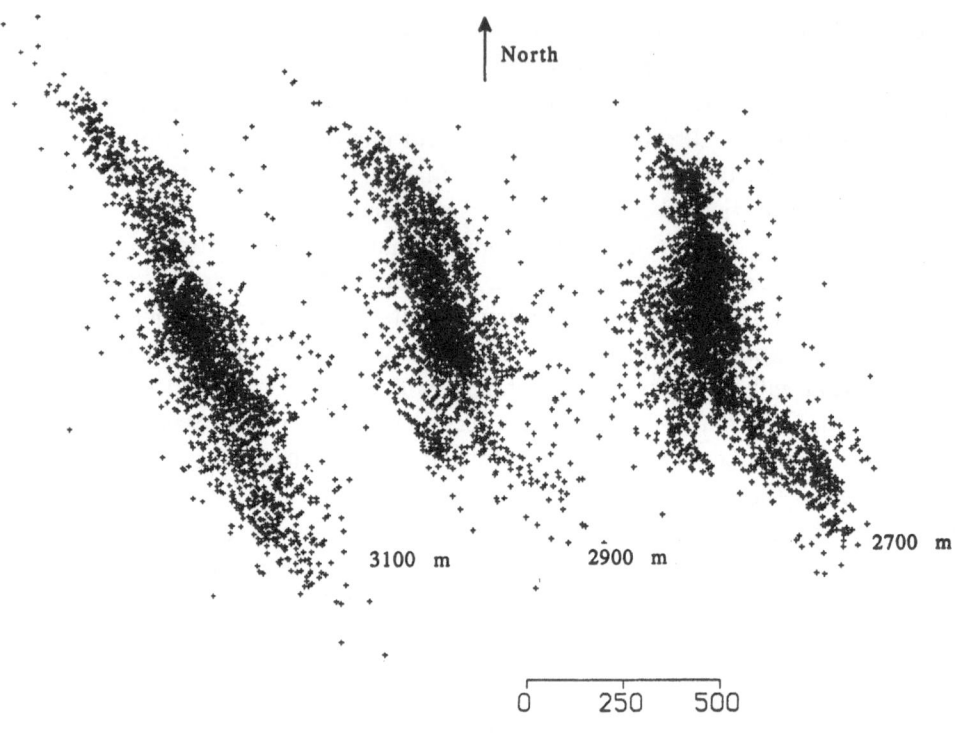

Figure 6

Horizontal cross sections through the seismic cloud recorded with the downhole stations. Each projection represents events which occurred within a 200 m thick horizontal slice centered on the depth indicated on the figure (2700 m, 2900 m, 3100 m). The scale is in meters. Nearly 20000 events were recorded with the downhole network.

The frequency-magnitude diagram for these events is shown on Figure 7. Because of attenuation effects, it is certain that the number of events with magnitude in the range -0.5–0 is grossly underestimated. Considering only events with a magnitude larger than 0.5, the Gutenberg-Richter relationship (linear regression between the number N of events with a magnitude smaller than M_L and the magnitude M_L) yields:

$$\log(N) = 3.59 - 1.26 \, M_L. \tag{1}$$

Some estimates of the source radii and of the stress drops associated with the largest events have been obtained. They will be discussed after the shear motion observations, as derived from the ultrasonic imaging log, have been presented.

4. Direct Observation of Slip Motions in the Injection Well

4.1 Ultrasonic Imaging Logs

Two ultrasonic imaging logs have been conducted in GPK1, one before and one after the injections. They were run between 3170 m and 2853 m for the first log and between 3104 m and 2853 m for the second one. The logging speed was 255 m/hr (850 ft/hr) so that vertical resolution is 1 cm.

The principle of this log has been discussed by ZEMANEC et al. (1970). An ultrasonic pulse is sent in a given azimuth, which reflects at the borehole wall and travels back to the sensor where it is recorded. Two data are retrieved from the reflected signal: peak amplitude and arrival time. The peak amplitude is used for obtaining an image of the borehole wall. The arrival time, by difference with emission time, yields the distance between the sensor and the borehole wall in the sensor azimuth. The sensor rotates continuously during logging (rotation velocity was 7.5 rotations per sec) so that a complete image of the well is obtained together with the distance between the well and the tool. 180 signals are emitted during one complete rotation so that angular resolution is 2°. This type of tool is commonly used for identifying borehole breakouts together with the dip and azimuth of fractures interesecting the well (e.g., ZOBACK et al., 1985).

At Soultz, comparison of the logs run before and after the injection experiment shows no long vertical fracture in the upper 75 m of the well, where more than half the injected flow was lost when the flow rate became greater than 24 l/sec (Fig. 3). Instead the logs derived from signal amplitude analysis reveal that inclined preexisting fractures have become considerably more pronounced. However the amplitude processing does not provide means to evaluate the amplitude of motion along these fractures. Between 3025 m and 3090 m, some axial fracturing (Fig. 8) is noticed but

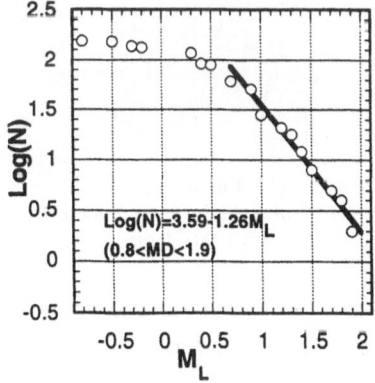

Figure 7

Gutenberg-Richter frequency-magnitude diagram for induced seismic events recorded with the surface network. A total of 164 events were recorded and the largest magnitude was 1.9.

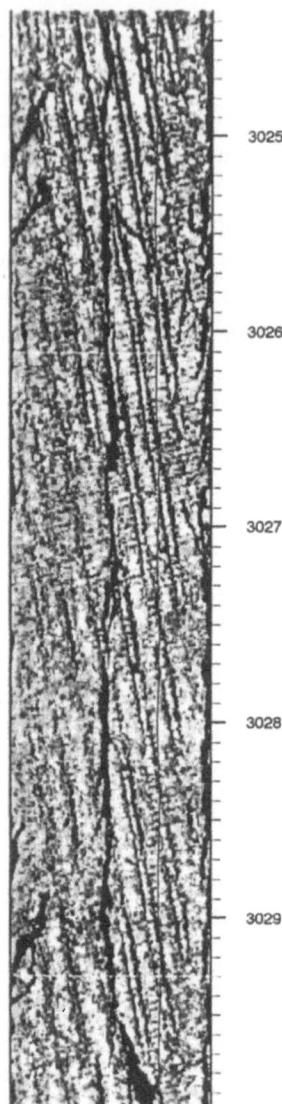

Figure 8
Trace of the north-south vertical fracture observed around 3030 m. Notice on Figure 3 that no flow is
lost at this depth.

no fluid was lost at this depth according to the spinner logs. Hence it has been concluded that the water injection resulted in the opening of preexisting fractures in the upper part of the well, and thermal cracking at some other locations (initial borehole downhole temperature was 155°C while downhole temperature was less than 40°C at the end of injection).

The tool which has been used for this investigation (Schlumberger's UBI) provides the possibility of determining, at all depths, the P-wave velocity of the fluid filling the well. Accordingly, the time readings can be converted with good accuracy in terms of distances between the tool and the borehole wall. Precision on the distance determination has been evaluated at ± 3 mm; this should not be confused with the resolution of the tool. Indeed, the precision concerns the absolute value of the reading while considerably smaller variations in distance may be identified. This measurement capacity has been advantageous for determining the amount of slip which affected some of the fractures.

4.2 Slip Motion Amplitude Determination

When slip occurs along a fracture, the borehole geometry is modified in a characteristic manner (Fig. 9). Azimuthal distance measurements identify zones of discontinuity. When the discontinuity magnitude is measured at the various locations where the fracture intersects the well, the slip vector magnitude and direction can be evaluated. The ultrasonic borehole imaging tool povides means to measure such azimuthal distance discontinuities.

One of the main difficulties of taking advantage of distance measurements is that the location of the tool within the well is never known. For this reason, a method has been developed (HAYMAN et al., 1994) for determining an accurate cross-section geometry without knowledge of the exact cross-sectional position of the tool. It is based on the determination of the radius of curvature of the well within a chosen sector.

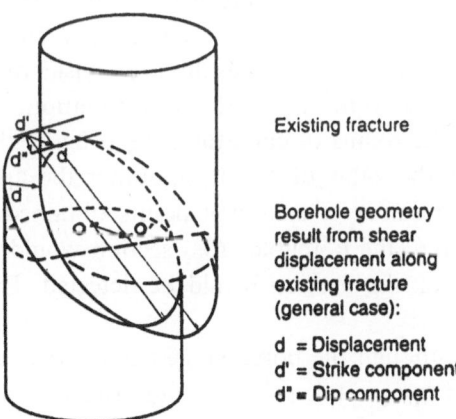

Existing fracture

Borehole geometry result from shear displacement along existing fracture (general case):

d = Displacement
d' = Strike component
d" = Dip component

Figure 9
Shear motion along a preexisting fracture. The radius of curvature technique described in text measures the projection of the slip vector in the borehole cross section (vector OO'). Note that if the fracture is subparallel to the borehole axis, a large shear motion does not result in a significant decrease in borehole diameter.

The mean borehole cross section is assumed to be circular before slip has occurred. Since the transit time yields the distance between the tool and the borehole wall in a given azimuth, the angular coordinates of P consecutive points may be used to determine the mean radius of curvature of the borehole for the corresponding angle of investigation. The change in position of the tool within the borehole cross section is assumed to be negligible during these P measurements (recall that the rotation velocity is 7.5 rotations per second with 180 signals per rotation, so that in most cases the P measurements occur within 0.02 sec). In practice P varies between 10 and 25, so that the sector for which the radius of curvature R is determined ranges from 20 to 50°. R is evaluated for each sampling azimuth as the radius of curvature computed for a $2P°$ sector centered on the corresponding sampling azimuth. The center of the well is determined from this radius of curvature determination. When the well is regular, all centers of curvature coincide with the center of the borehole cross section. If slippage has occurred, the center of the well as computed for one sector of the borehole wall, does not coincide with that determined from the other sector (Fig. 9).

The vector defined by these two centers is the projection, in the cross section of the well, of the slip motion which occurred along the fracture plane. Since the dip and azimuth of the fracture plane are known from the borehole image, and since the orientation of the logging tool is also known, the relative displacement of both sides of the fracture can be evaluated.

The difficulty in implementing this analysis is that the borehole wall often exhibits positive or negative asperities which must be eliminated from the calculation of the mean radius of curvature. Asperities generate a scatter with respect to the mean radius of curvature and this scatter is kept smaller than a given error value S, which ranged between 3 mm (precision of the tool) and 2.5 cm for this analysis. If the scatter is larger than the chosen value S, the determination of the radius of curvature is not undertaken. Results are considered as significant when small variations in P and S do not affect the determination. The smaller the angle $2P$ used for computing the radius of curvature, the rougher the apparent borehole cross section; the larger the value of P, the smoother the apparent cross section. Ultimately, if the total number of sampling points obtained during one rotation (180) was to be considered, the borehole cross section would be fitted to a perfect circle and no asperities or slip motion would be detected. Examples are given in Figure 10.

In a few instances, although the borehole geometry was not circular before the injection, comparison of images taken before and after injection clearly outlined a slip motion. For these cases (fracture at 2976 m in Table 1) the initial borehole geometry has been parameterized and this reference geometry has been used then to evaluate the slip vector, in a manner identical to that applied when the borehole is circular. Thus, the method can be extended to noncircular cross sections. However the time spent for the analysis becomes very significant.

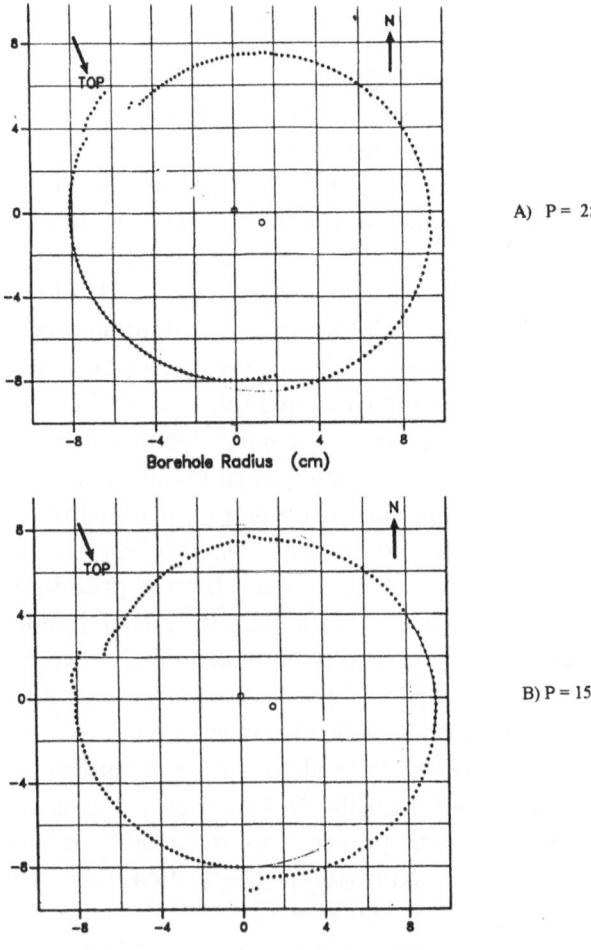

Figure 10

Determination of the amplitude S of the slip vector projection in the borehole cross section. A) the radius of curvature is determined with 25 consecutive points ($S = 1.3$ cm); B) the radius of curvature is determined with 15 points ($S = 1.3$ cm). The difference in the reading of the amplitude of the slip vector projection as obtained in both cases is smaller than 1 mm.

For fractures that are subparallel to the borehole axis, measurements can be conducted in numerous cross sections and resolution on the slip vector magnitude reaches a few millimeters. For fractures which are subperpendicular to the borehole axis, only one or two cross sections can be investigated and resolution is significantly decreased. In practice, because of difficulties with the roughness of the borehole surface, the resolution of the method has been found to be equal to about 0.4 cm. Any slippage smaller than 0.4 cm cannot be determined with certainty and has not been considered here.

4.3 Results

Analysis of the ultrasonic imaging log has been conducted only for the upper 250 m of the open hole section of the well. Within this depth interval, slip motion for six different fractures has been larger than or equal to 0.4 cm and could be measured properly (POITRENAUD, 1995). Except for the slip at 2976 m which occurred in a noncylindrical section, for the 5 other measurements it was verified that the borehole cross section in intact zones above and below the corresponding fracture zone was within 0.5 mm of the values measured before injection. It was also observed that the borehole radii of curvature within the sections in the slipping zones were generally within 2 mm of that of the intact borehole section above and below the fracture. Results are presented in Table 1. For all fractures but one (2887 m), no slippage has been detected on the initial log. Therefore the measured slip vector is considered to have been entirely caused by the increment in fluid pressure. For the fracture at 2887 m, the initial slip motion detected on the log run before injection has been determined to be 0.24 cm. The slip vector evaluated after injection has been found to be parallel to that observed prior to injection. In Table 1, only the modulus of the slip vector as observed after injection is given.

The mean direction of the thermal fracture identified between 3025 m and 3098 m is N 173°E $\pm$ 1°. This is taken as a good constraint on the maximum horizontal principal stress direction within this depth range. It has been noted however, that within short depth intervals this vertical fracturing comprises a set of *en echelon* smaller fractures, well aligned with the N-S direction. This is interpreted as local perturbations of the maximum principal stress orientation around the vertical direction. Such local stress perturbations have often been described in recent articles (CORNET, 1992; SHAMIR and ZOBACK, 1992; BARTON and ZOBACK, 1994; CORNET and YIN, 1995). At Soultz, they are considered to be associated with the

Table 1

Determination of slip motions induced in the well GPK1 during the September–October 1993 injections. Z is distance from well head (in m) according to the logging depth meter, β is the dip direction of fracture plane (positive eastward), α is the dip of fracture plane, λ is the strike of the slip vector, A is amplitude of the slip motion (in cm). Uncertainties are noted ε with the respective subscript. SX is the mean amplitude of the slip vector measured within the cross section of the borehole. Frame of reference is north, east, vertical positive downward. Within the fracture plane, rotation is positive from strike direction toward dip (downward positive).

Z (m)	β	α	λ	A (cm)	ε_λ	ε_A (cm)	SX (cm)	ε_{SX} (cm)
2966	105	84	110	4.7	5	0.7	0.5	0.1
2867	259	62	304	2.2	3	0.1	1.45	0.07
2976	269	61	218	0.8	15	0.2	0.5	0.05
2887	298	75	271	0.85	8	0.3	0.28	0.1
2973	273	78	198	0.4	10	0.06	0.22	0.04
2925	48	86	99	4.3	13	1.3	0.5	0.14

existence of slip along natural fractures which have occurred prior to or during the drilling operation.

5. Discussion

One of the major difficulties encountered by the development of Hot Dry Rock Geothermal reservoirs is to determine the direction of major fluid flow away from the wells. One of the most utilized methods for identifying these directions is to locate the induced seismicity and to consider that this mapping provides a good representation of the main flow directions.

CORNET and YIN (1995) have argued that, in fact, induced seismicity does not help locate zones of high-fluid flow but rather zones of high-fluid pressure. By mapping pore pressure from induced seismicity analysis they showed that the further away the induced microseismic events, the lower the flow rate at the event location.

Results presented here have demonstrated that some very significant fault motions have been generated at the borehole wall. The question which is to be discussed now concerns the possibility of deriving the magnitude of these slip motions from induced microseismic observations.

It is well known that the frequency content of seismic signals is influenced by source characteristics such as the size of the rupture area, the mean dislocation amplitude, but also by the dynamics of the rupture process (e.g., BRUNE, 1970; COCHARD and MADARIAGA, 1994). Unfortunately, the frequency characteristics of the recording systems for the surface microseismic monitoring networks do not allow a significant frequency content analysis. For the largest events the frequency analysis cannot be conducted either with the downhole network because of its dynamics: although this network provides satisfactory recording for most events, it is saturated on the largest events discussed hereafter. However some orders of magnitude for the source sizes for these largest events, and the corresponding dislocation amplitude, may still be retrieved from the event magnitudes.

As shown by KANAMORI and ANDERSON (1975), some empirical relationship exists between magnitude and seismic moments for major earthquakes. PEARSON (1982) and MAJER and McEVILLY (1979) have developed the same kind of relationship for microseismic events induced by forced fluid injections. Their observations come from the Geysers geothermal field in Northern California and the Fenton Hill Hot Dry Rock site in New Mexico:

$$\log(Mo) = 17.27 + 0.77\ M, \qquad (2)$$

where Mo is the seismic moment of a seismic event while M is its magnitude. Accordingly, given the fact that the largest observed magnitude observed at Soultz was 1.9, we conclude that the largest seismic moment was about 5.4×10^{18} dyne.cm. For the smaller events which did not saturate the downhole recording system, the seismic moments have been retrieved from signal frequency analysis (R. Jones personal communication). Interestingly, the results obtained with equation (2) yield systematically a larger moment estimate than that derived from signal spectrum analysis. Thus the values derived subsequently for the source characteristics correspond to overestimates.

For a pure uniform shear motion over a circular crack embedded in an elastic isotropic rock, the mean stress drop is a function of the source dimension (e.g., KANAMORI and ANDERSON, 1975):

$$\Delta\sigma = 7\pi/16 * G * D/a \tag{3}$$

where $\Delta\sigma$ is the mean stress drop associated with slippage, G is the shear modulus of the rock, D is the dislocation amplitude and a is the radius of the circular crack. If this shear motion is dynamic, the corresponding seismic moment is:

$$Mo = G * S * D = 16/7 * \Delta\sigma * a^3 \tag{4}$$

where S is the area of the source. Equation (4) provides means to evaluate the dislocation amplitude which is to be anticipated from a magnitude 1.9 event. It also allows an evaluation of the seismic moment which would be generated by dynamic circular sources with dislocation amplitude equal to the shear motions described in Table 1. This requires that the stress drop $\Delta\sigma$ be evaluated.

Both hydraulic fractures and thermal fractures are essentially coaxial with the borehole axis. Hence it is considered that the vertical stress is a principal stress direction. (Hydraulic fractures and thermal fractures are theoretically perpendicular to the minimum principal stress direction so that when the far-field principal stress directions are off by more than 20° from the borehole axis, *en echelon* fractures are usually observed at the wellbore wall.) Near the 2900 m depth, where most slippages have been measured in the well, the vertical stress is equal to about 77 MPa, for a rock density of 2.65 g/cm^3. An underestimate of the minimum principal stress is given by the maximum injection pressure. This value is an underestimate because it neglects thermal stresses which are known to be significant, given the observation of thermal fractures at the wellbore wall and given the low temperature in the well at the end of injection. The overpressure at 2900 m was about 9.1 MPa so that the fluid pressure was 38.1 MPa.

Focal mechanisms of induced seismicity are mostly normal faulting events with some events showing a strike slip characteristic (Fig. 11). Hence it is concluded that the vertical stress is the maximum principal stress and that the maximum horizontal principal stress is subequal to the vertical stress. Now, if the friction coefficient of the rock is taken somewhere between 0.7 and 0.85, it is found that the minimum principal stress must be somewhere between 39.5 and 42 MPa for the rock mass to be in equilibrium, on the assumption that favorably oriented fractures exist in the

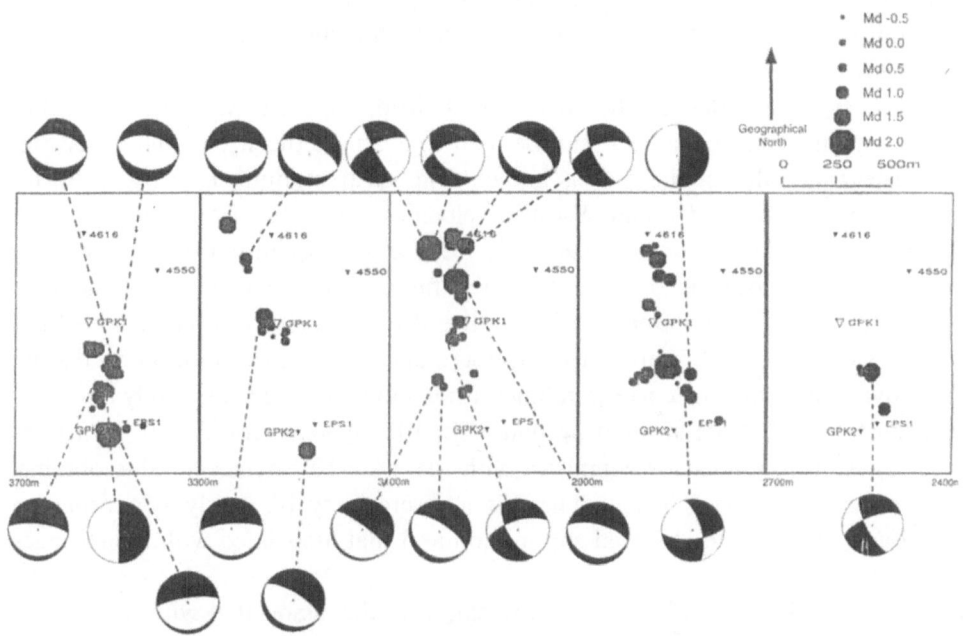

Figure 11
Location and focal mechanisms of events recorded with the surface network. Location is based on downhole network arrival times. Note that no large event is located within 100 m of the well in the 2900 m depth range. Focal mechanisms are predominantly normal faulting events with some strike slip events.

rock and that the local pore pressure is about equal to hydrostatic (29 MPa). Since the maximum horizontal principal stress is close to north-south, favorably oriented fractures are oriented north-south with a dip angle controlled by the friction coefficient (i.e., here dip angles around 60°). The analysis of fractures intersecting the well (GENTER et al., 1991) has shown numerous fractures oriented north-south with a dip ranging from 50 to 80°, hence potential planes of rupture do exist. This estimation of the minimum principal stress is slightly larger than the fluid pressure observed at the end of injection (between 1.6 and 4 MPa). This difference has been attributed to thermal stresses. In fact, given that it takes an overpressure equal to about 5 MPa for microseismicity to be observed, the minimum principal stress at 2900 m is likely to be closer to 45 MPa than to 40 MPa.

For the sake of this discussion, only the shear stress drop associated with fault motion is to be estimated. A value equal to 19 MPa (i.e. $(\sigma_{max} - \sigma_{min})/2$) is a sure overestimate of this stress drop, while a value equal to 9 MPa seems to be a more realistic overestimate. Indeed, most of the fractures which have sheared are subvertical (i.e., nearly perpendicular to the minimum principal stress direction), especially those with the largest shear displacement. In addition, as mentioned before, the fact that the injection pressure stabilized once the injection flow rate reached 24 l/sec is taken as proof that, for higher flow rates, the fractures were mechanically opened, i.e., the shear stress on the flowing fractures was completely released.

These two estimates of the shear stress drop, together with the dislocation measurements given in Table 1, can be used to evaluate the size of the dislocation if the shear modulus of the rock is known. The shear modulus (2×10^4 MPa) has been estimated from *P*- and *S*-wave velocities and also from measurements conducted in the laboratory (BARIA *et al.*, 1996). From equation (3), it is concluded that a 4.7 cm dislocation requires a circular fault with a 68 m radius if the stress drop is equal to 19 MPa, and a 143 m radius if the stress drop is equal to 9 MPa. Smaller stress drop will require larger fracture diameters. If this motion occurs in a single dynamic event, the corresponding seismic moments are respectively 3.6×10^{20} dyne.cm for the 9 MPa stress drop and 8.9×10^{19} dyne.cm for the 19 MPa stress drop. These values must be compared with the 5.4×10^{18} dyne.cm value obtained for the largest observed event: There is a discrepancy of nearly two orders of magnitude between the observed slip motion and that associated with the strongest observed microseismic event.

Following similar reasoning, the dislocation radius associated with a 5.4×10^{18} dyne.cm seismic moment may be evaluated. It is found to be about 25 m for a 9 MPa stress drop. Hence, it would take about 25 events with magnitude 1.9 to completely release the shear stress on a circular fracture with a 125 m radius, when no event with magnitude larger than 1.5 has been observed near the well in the 2800–2900 m depth interval while at least two fractures exhibited slip motions larger than 4.0 cm within the same depth interval.

The conclusion that shear motions observed at the well are nonseismic does not imply that source parameters derived from frequency analysis are erroneous. It only implies that some quasi-static motion may affect faults, before or after unstable slip motion occurs. Furthermore, it is very likely that the quasi-static shear motion which has been observed near the well is linked to the injection procedure and that quasi-static motion is much smaller when fractures are not mechanically opened. Nevertheless, it is well known that some creep occurs along natural faults (e.g., EVANS *et al.*, 1981), so a similar process is to be expected with induced seismicity.

It may be argued that slip may have occurred by an infinite set of dynamic but very small sources such that the events went undetected. It should be noted however that IHMLE and JORDAN (1994) and McGUIRE *et al.* (1996) have described shear motions along long fault sections that occur at low velocities before major earthquakes. Our results suggest that a similar behavior may occur during large water injection experiments. The significant conclusion is that large shear displacements occurred which cannot be retrieved from analysis of signals as recorded with the present recording system. Given the important results obtained in classical seismology with broadband sensors, it may prove worthwhile to adopt likewise broadband systems for induced seismicity monitoring.

We have endeavored to investigate the correlation between these shear motions and the zones of significant fluid flow. This raised a difficulty because absolute depths are not known precisely either for the imaging log or for the spinner logs used for characterizing flow characteristics. The depth discrepancy between the two borehole imaging logs run before and after injections is equal to 4 m at the casing shoe (2850 m), even though the depth measurement for the imaging logs was derived from accelerometer data recorded on the tool. The depth accuracy on the spinner log is slightly less since depth was determined from cable length measurements conducted at the well head, while it is well known that for long cable lengths, the downhole tool moves with variable velocity. This implies that a discrepancy of 5 to 10 m is to be expected between the spinner logs and the imaging logs depth measurements. For example, a 6 m discrepancy between flow log and imaging log is identified for the unambiguous zone at 2966 m (imaging log depth). When the well was tested for production, the significant flow zones observed between the casing shoe and 3000 m, are observed at 2860 m, 2870 m, 2894 m and 2960 m (EVANS et al., 1996). They amount to approximately 65% of the total production flow. Thus, three out of four of the significantly flowing fractures which intersect the well within the depth interval in which imaging analysis was conducted, correspond to fractures which underwent measurable shear motion. It must be mentioned however that, during injection, other very significant flow zones have been identified at depths where fractures exhibit no measurable shear motion. These results illustrate that the most significant long-term effect of the hydraulic stimulation, which is the quasi-static shearing of preexisting fractures, is not easily retrieved from the seismic activity recorded on site.

6. Conclusion

The large flow rate water injection which was conducted at Soultz, in the well GPK1 during fall 1993 has induced large shear motions along preexisting fractures. The slip magnitude has been found to be larger than 4.0 cm for two fractures and reached values in the centimeter range for another 4 fractures.

During this large volume injection (44500 m³) about 20000 microseismic events have been recorded with downhole stations anchored within the granite formation where injection took place, i.e., below the 1500 m thick sedimentary cover. Only 165 events, with a maximum magnitude equal to 1.9, were recorded with the surface network. This surface network provided good constraints for some of the focal mechanisms of the largest events and was used to evaluate their magnitude.

Given the observed amplitude of the shear motions in the wellbore and the estimated shear stress drop associated with them, it is concluded that the radius of these shearing zones is larger than 120 to 150 m, while the radius of the largest seismic events has been estimated to be about 30 m, if the same stress drop is assumed.

It is concluded that the most significant shear motions induced by these water injections occurred aseismically so that source parameters determined from the frequency analysis of induced seismic signals cannot be used to evaluate the efficiency of the stimulation.

Acknowledgments

John Helm benefited from a European Community scholarship during this work and H. Poitrenaud from a Schlumberger grant. This work was conducted under financing from Europe Community (DG XII), French Agence pour l'Environnement et la Maitrise de l'Energie, Schlumberger and French CNRS. The injection experiment was designed by one of us but engineered by Socomine which conducted a nearly perfect work despite some initial strong reservations about the concept. The portable surface seismic network is that of the Ecole et Observatoire de Physique du Globe de Strasbourg. Camborne School of Mines Associates operated the downhole network. K. Evans provided the corrected flow logs. While writing this paper, F. H. Cornet was on sabbatical leave at Lawrence Berkeley National Laboratory—Earth Sciences Division, with E. Majer. A. McGarr, R. Jones, K. Evans, M. Carra, L. Riverra, T. Reuchle and an anonymous reviewer assisted in improving the original manuscript. We wish to express our sincere thanks to all.

REFERENCES

BARIA, R., BAUMGARTNER, J., and GERARD, A. (1996), *European Hot Dry Rock Programme 1992–1995*; 327 p., Extended Summary for contract JOU2-CT92-0115 to European Community (DG XII), Socomine, Soultz/forets, France.

BARTON, C. A., and ZOBACK, M. D. (1994), *Stress Perturbation Associated with Active Faults Penetrated by Boreholes: Possible Evidence of Near-complete Stress Drop and New Technique for Stress Magnitude Measurements*, J. Geophys. Res. *99* (B5), 9373–9390.

BRUNE, J. N. (1970), *Tectonic Stresses and the Spectra of Seismic Shear Waves from Earthquakes*, J. Geophys. Res *75*, 4997–5009.

COCHARD, A., and MADARIAGA, R. (1994), *Dynamic Faulting under Rate-dependent Friction*, Pure appl. geophys. *142* (3/4), 419–445.

CORNET, F. H., *In situ stress heterogeneity identification with the HTPF tool*. In *Proceedings 33rd U.S. Rock Mech. Symp.* (Tillerson and Wawersik, eds.) pp. 39–48 (Balkema, Rotterdam 1992).

CORNET, F. H., and JONES, R., *Field evidences on the orientation of forced water flow with respect to the regional principal stress directions*. In *Proc. 1st North Amer. Rock Mech. Symp.* (Nelson and Laubach, eds.) pp. 61–69 (Balkema, Rotterdam 1994).

CORNET, F. H., and YIN, J. (1995), *Analysis of Induced Seismicity for Stress Field Determination and Pore Pressure Mapping*, Pure appl. geophys. *45* (3/4), 677–700.

EVANS, K. F., BURFORD, R. O., and KING, G. C. P. (1981), *Propagating Episodic Creep and the Aseismic Behavior of the Calaveras Fault North of Holister, California*, J. Geophys. Res. *86* (B5), 3721–3735.

EVANS, K. F., KOHL, T., HOPKIRK, R. J., and RYBACH, L. (1996), *Studies of the Nature of Nonlinear Impedance to Flow within the Fractured Granitic Reservoir at the European Hot Dry Rock Project Site* at *Soultz-sous-Fôrets*, *France*, Bundesamt für Bildung und Wissenschaft, Projekt 93.0010—Final Report; 144 pp; Swiss Federal Inst. of Tech, Institute of Geophysics Zürich.

GENTER, A., MARTIN, P., and MONTAGGIONI, P. (1991), *Application of FMS and BHTV Tools for Evaluation of Natural Fractures in the Soultz Geothermal Borehole GPK1*, Geoth. Sci. and Tech. *4* (3), 189–214.

IHMLE, P. F., and JORDAN, T. H. (1994), *Teleseismic Search for Slow Precursors to Large Earthquakes*, Science *266*, 1547–1551.

HAYMAN, A., PARENT, P., CHEUNG, P., and VERGES, P. (1994), *Improved Borehole Imaging by Ultrasonics*, Soc. Petr. Eng. Paper 28 440.

HELM, J. A. (1995), *The Natural Seismic Hazard and Induced Seismicity of the European Hot Dry Rock Geothermal Energy Project at Soultz-sous-Fôrets (Bas-Rhin, France)*, These de Doctorat, 197 pp; Ecole et Observatoire de Physique du Globe de Strasbourg, Strasbourg, France.

JONES, R. H. BEAUCE, A., JUPE, A., FABRIOL, H., and DYER, B. C., *Imaging induced microseismicity during the 1993 injection tests at Soultz-sous-Fôrets*. In *Proc. World Geothermal Congress*, Florence, Italy, vol. 4, pp. 2665–2669 (Int. Geotherm. Assoc. Pub., San Diego 1995).

KANAMORI, H., and ANDERSON, D. L. (1975), *Theoretical Basis of some Empirical Relations in Seismology*, Bull. Seismol. Soc. Am. *65* (5), 1073–1095.

KAPPELMEYER, O., GERARD, A., SCHLOEMER, W., FERRANDES, R., RUMMEL, F., and BENDERITTER, Y. (1991), *European HDR Project at Soultz-sous-Fôrets, General Presentation*, Geotherm. Sci. and Tech. *2* (4), 263–289.

MAJER, E. L., and MCEVILLY, T. V. (1979), *Seismological Investigations at the Geysers Geothermal Field*, Geophysics *44*, 246–269.

MCGUIRE, J. J., IHMLE, P. F., and JORDAN, T. H. (1996), *Time Domain Observations of a Slow Precursor to the 1994 Romanche Transform Earthquake*, Science *274*, 82–85.

PEARSON, C. (1982), *Parameters and a Magnitude Moment Relationship from Small Earthquakes Observed during Hydraulic Fracturing Experiments in Crystalline Rocks*, Geophys. Res. Lett. *9* (4), 404–407.

POITRENAUD, H. (1995), *Application des mesures par « Ultrasonic Borehole Imager » a la determination de glissements sur fractures preexistantes*, Rapport de stage DESS de l'Instit. de Phys. du Globe de Paris; Schlumberger-Riboud Product Center; 106 pp.

SHAMIR, G., and ZOBACK, M. D. (1992), *Stress Orientation Profile to 3.5 km Depth near the San Andreas Fault at Cajon Pass, California*, J. Geophys. Res. *97* (B4), 5059–5080.

ZEMANEK, J., GLENN, E., NORTON, C. J., and CARDWELL, R. L. (1970), *Formation Evaluation by Inspection with a Borehole Televiewer*, Geophys. *35*, 254–269.

ZOBACK, M. D., MOOS, D., MASTIN, L., and ANDERSON, R. N. (1985), *Wellbore Breakouts and in situ Stress*, J. Geophys. Res. *90* (B7), 5523–5530.

(Received January 30, 1997, accepted August 6, 1997)

Pure appl. geophys. 150 (1997) 585–603
0033–4553/97/040585–19 $ 1.50 + 0.20/0

Pure and Applied Geophysics

Induced Microseismicity and Procedure for Closure of Brine Production Caverns

CHRISTOPHE MAISONS,[1] ERIC FORTIER[1] and MARC VALETTE[2]

Abstract—Elf-Atochem has been solution mining a salt formation in southern France for the last twenty years and the problem arises of the disposition of deep caverns once brine production ceases. The regulatory authorities have required procedures for sealing off the caverns to make them environmentally safe and to monitor the post sealing behaviour of the caverns.

This paper describes the deep downhole seismic monitoring carried out during the experimental studies conducted to determine the appropriate means of monitoring and to validate closure safety criteria. The results confirm the accuracy of downhole permanent seismo-acoustic technology to monitor the sealing processing, even in strong conditions (2000 meter depth and 100°C). The microseismicity has been characterised and located using triaxial hodogram analysis, however the microseismicity shows many clusters of seismic "doublet" events allowing the use of relative analysis.

The system has been installed before sealing and displays a high sensitivity. Under these conditions, the microseismicity has yielded useful data on the geomechanical behaviour of the site with respect to the final objectives. The seismicity is associated with the brine flow. The seismicity induced by the salt leaching process is a means to emphasize weak zones (fractures) that could again become active after closure. A closure test reveals that a direct relationship exists between the seismicity and the bleeding off brine from the caverns. Microseismic monitoring results combined with field observations advance the development of the closure procedure.

Key words: Salt leaching, induced seismicity, seismic monitoring, closure of brine production cavern.

Introduction

For 20 years, ELF-ATOCHEM has been exploiting, by solution mining, a salt formation ranging from a depth of 1900 m to 2800 m (Figure 1), in Vauvert (South of France). Solution mining consists of dissolving the salt by circulating fresh water through cracks produced by hydraulic fracturing between two (or sometimes three) wells (Figure 2). The produced brine is piped to chlorine and soda plants at Lavera and Fos (Marseille). In the western part of the site, the salt has been leached between pairs or triplets of wells for twenty years and the problem of the disposition of the deep caverns arises once brine production ceases.

At the end of mining, the caverns are filled with brine and are roughly cylindrical in shape with diameters of around 30 m and a height of 500 m. From

---

[1] Géostock, 92563 Rueil Malmaison, France.
[2] Elf-Atochem, 30600 Vauvert, France.

experience acquired during the construction and monitoring of many solution-mined caverns in salt, it is known that sealing off production wells pairs causes the pressure of the brine in the cavern to rise due to creep and thermal effects.

Creep phenomenon attempts to equalise the pressure differential between the fluid inside the cavern and the geostatic pressure. The creep process attempts to reduce the cavern volume to the extent that the fluid pressure inside the cavern is less than the geostatic pressure. Reverse creep occurs when the cavern pressure exceeds geostatic pressure. Creep has a balancing effect which tends to limit fluid pressure from exceeding geostatic pressure. For a given volume, warming effect acts to increase the fluid pressure as long as the temperature of the brine is less than the temperature of the salt formation. Post closure heating of the cavern fluid may cause a considerable pressure rise (around 10 bar per 1°C), this is the factor that could lead to the cavern pressure rising above the geostatic pressure. The "creep" and "warming" processes interact over time, and from field tests and computer simulations a conservative behaviour model (Figure 3) has been proposed (YOU *et al.*, 1994). In the case of Vauvert this conservative model shows that abandonment by plugging could lead to overpressures exceeding geostatic pressure and fracturing

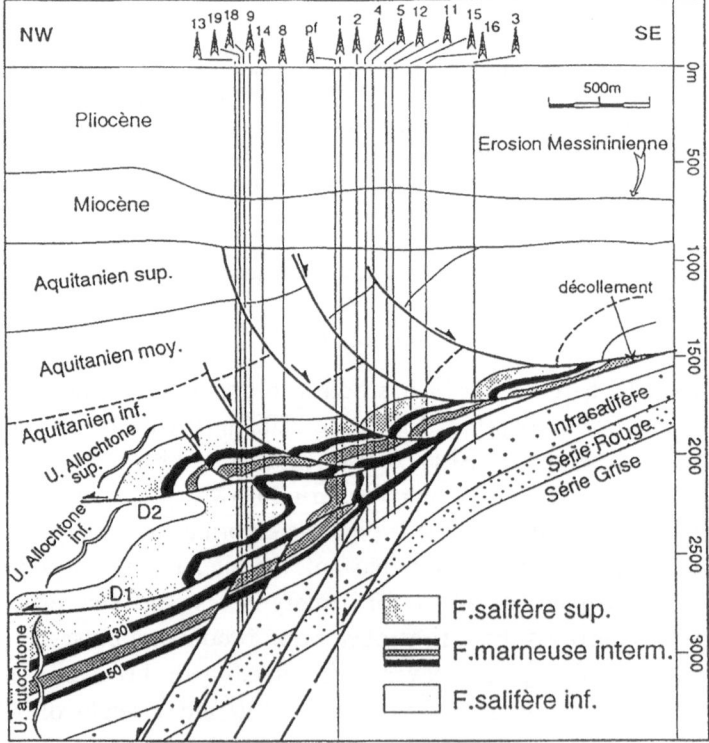

Figure 1
Geological cross section of the ELF-ATOCHEM Vauvert field (from VALETTE, 1991).

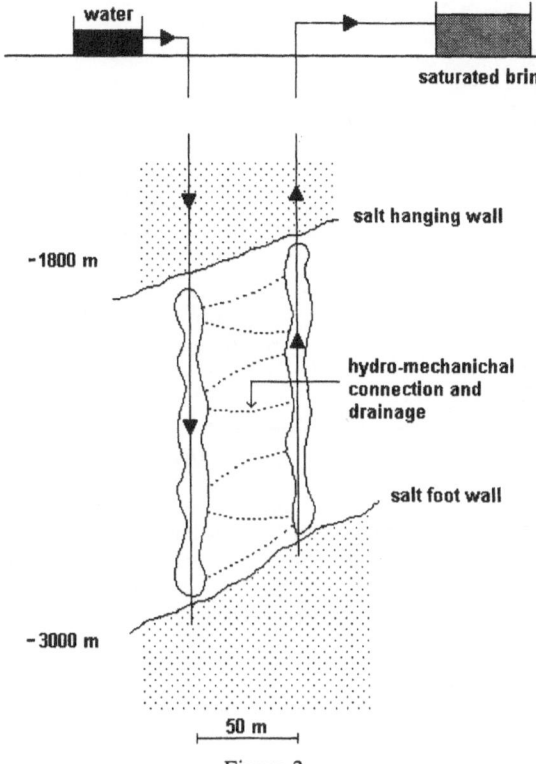

Figure 2
Schematic diagram of the salt leaching exploitation procedure.

pressure. In that case, the fluid overpressure inside the cavity might have three types of consequences:

— producing fractures between wells,
— producing fractures induction up the cemented casing annulus,
— producing fractures propagating along existing natural weak zones.

To abandon the caverns, the authorities are requesting assurances that there will be no significant fracturing after closure, the principal reason being the prevention of any salt contamination of the burdigalian rocks, the first water-bearing strata 900 m below ground level. In addition to criteria and sealing procedures which must limit the risks of fracturing, the authorities have required the installation of appropriate systems to monitor the postclosure behaviour of the caverns. In this context, seismic monitoring associated with thermal monitoring has been proposed. The temperature monitoring carried out in each well aims at detecting any annulus flow of hot brine up the well toward the upper levels. The seismic monitoring aims to provide prevention by early detection of potential fracturing around the caverns once they have been sealed off. In the first experimental sealing procedure proposed to the authorities, the wells are closed but not sealed off. The objectives are to more precisely calibrate thermal and rheological behaviour. Additionally the objectives are to evaluate the accuracy and efficiency of the monitoring systems recommended.

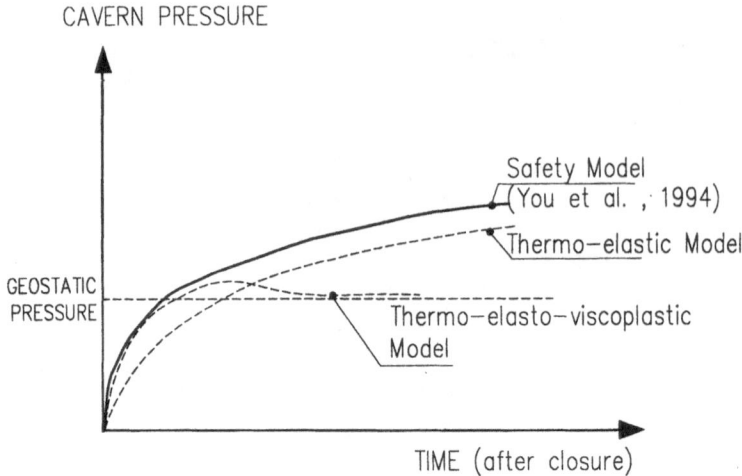

Figure 3
Definition of the safety model for salt cavern closure. In this model, as long as the cavern pressure is smaller than the geostatic pressure, the behaviour is considered to be thermo-elasto-viscoplastic and when the cavern pressure becomes larger than geostatic pressure, the behaviour is considered to be thermo-elastic.

Geological Setting

The 1000-m thick exploitable layer consists of approximately 50% rock salt and 50% nonsoluble material (anhydrite and clay). This salt deposit belongs to the northwestern part of the onshore Camargue Basin bounded by the major extensional, SE-dipping Nimes fault. This basin results from the oligo-aquitanian rifting at the "Golfe du Lion". It contains more than 4000 meters of syn-rift sediments which overlay Mesozoic carbonates (1st unconformity) and are covered by the transgressive Burdigalien marine sediments (2nd unconformity).

The structural interpretation reveals an abnormal superposition of three halite series. Only the lower series is autochtonous and affected by the N-W dipping normal faults (Oligocene extension). The underlying allochtonous series above a lower thrust surface (D1) is severely folded (Figure 1). The upper allochtonous series above the upper thrust surface (D2), accommodates the deformation by SE dipping listric extensional normal faults (Figure 1), which also accommodates the extension since the upper Aquitanian series (VALETTE, 1991 and VALETTE *et al.*, 1994).

Seismic Monitoring Network

From 1992 to 1994, we carried out a series of tests to evaluate the feasibility of seismic monitoring for the early detection of mechanical readjustment occurring at

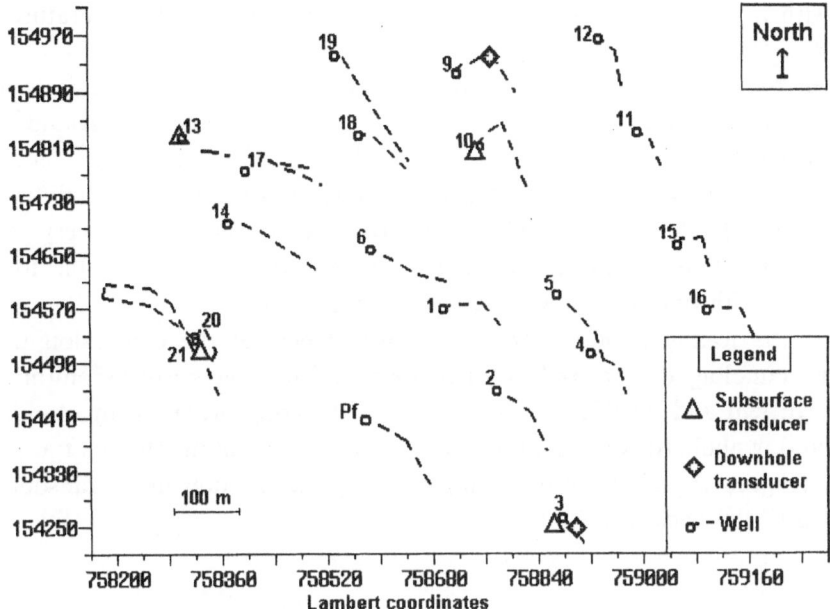

Figure 4
Plan view of the ELF-ATOCHEM seismic monitoring network.

a depth of around 2000 m. The objectives were to evaluate the level of seismicity induced by leaching and to define the network needed to properly identify wells (i.e., caverns) affected by the seismicity. For this period (MAISONS et al., 1994), the microseismic monitoring network consisted of only one downhole triaxial geophone anchored in well number 3 at a depth of 1800 m and four subsurface triaxial geophones cemented in small shallow boreholes 25 m below the surface (Figure 4).

These tests confirmed the accuracy of seismic monitoring for controlling the behaviour of the site in response to stress readjustments induced by leaching. However, the very small magnitude (seismic moments of 10^8 Nm to 10^{10} Nm) and the depth (>2000 m) precluded the recording of most of the events by surface geophones. The detection of seismicity was mainly ensured by the downhole transducer. This situation led to making full use of deep downhole triaxial transducers. Thus, during he testing period the evaluation and improvement of the reliability and the survivability of permanent transducers in high temperature (100°C) conditions became the priority.

The downhole probe used is anchored inside the casing and is retrievable. The anchoring force is provided by two spring blades and the geophones are directly welded on one of the spring blades. The geophones are directly coupled to the casing and uncoupled from the main body of the tool which includes an electric DC motor (to open and close the tool) and a preamplifier unit (with fixed or variable gain of up to 100 dB). The high anchoring force and the design of the tool are responsible for a spurious-free frequency response. This tool is claimed to work for years with a pressure rating up to 700 bar and an operating temperature range between −55°C and 125°C.

From the experience gained in the field, we now believe that this tool does operate efficiently in field conditions on a long-term basis. The background noise is very low and the signals recorded are effectively free of any resonance. The overall quality of downhole signals recorded in Vauvert allows a reliable analysis of the seismic waves. Under these conditions, the processing of the signals is very practical (no additional noise analysis and filtering is needed) and we are able to locate events even with only one downhole triaxial geophone.

After the testing period (1992–1995), we arrived at the conclusion that the existing technology can be applied for permanent downhole seismic monitoring and can be efficient and reliable for a long-term monitoring program. In March 1995, a second downhole triaxial geophone was anchored in well number 9 at a depth of 1800 m (Figure 4) and at the beginning of 1997, another downhole transducer will be added to the network.

Microseismic Location

The microseismic events recorded on site are analysed, classified and located later on with the Seistool software, developed by Géostock. This software allows us to locate events using either a classical multi-station seismic network (at least 4 independent stations are required to calculate the hypocentre) or using the triaxial hodogram method with only one triaxial station at a time. Relative locations are under development and will be applied to both methods.

In the case study presented here, the location of the seismic events is determined using only triaxial downhole tools. The triaxial hodogram method has been extensively used although the relative methods should appreciably improve the spatial distribution of seismicity. The relative location approach should also provide a means to locate most of the events in real time.

Triaxial Hodogram Method

For a given triaxial transducer, the triaxial hodogram method consists of determining the *P*-wave arrival direction and the source-transducer distance. The three-component recording makes it possible to determine the actual motion of the ground (often called particle trajectory). The direction in space of the velocity vector for a given linearly polarised *P* wave can be characterised by two angles: azimuth and inclination. The best way to study this trajectory is to model it by computing the covariance matrix over a time window and then calculate the Eulerian angles (CLIET and DUBESSET, 1988). The source to geophone distance is computed, from the time delay between *P* and *S* waves.

In our applications, the time window used to analyse particle trajectory, and thus to calculate azimuth and inclination, is systematically fixed to the first period of the *P*-wave onset. The *S*-wave arrival time is picked up when the radial-radial angle between the *P*-wave velocity vector (reference) and a moving velocity vector (Figure 5a) is greater than 80°. In addition, the polarisation analysis allows for clear identification of *S*-waves splitting (Figure 5b). The overall procedure applied is aimed at avoiding as much as possible human bias often encountered in seismic wave analysis.

The hypocentre location is then estimated for every triaxial transducer from azimuth, inclination and time delay parameters. Ray tracing could be applied for any given velocity model. However, due to lack of information, the velocity model is often a very simple homogeneous half space estimated from a calibration shot. In this case the calibration shot was produced at 2000 m of depth in a well situated near one of the clusters of seismicity to be studied. The seismic source and the downhole transducer sited at similar depth in the same geological formation; the velocity model is considered to be rather adequate at this stage. The location of seismicity is coherent with general geological features and allows the discrimination of reactions between wells. The overall uncertainty on source location (between 10 m and 50 m) depends on the distance between source and transducer. The uncertainty assumption is on the order of 5% of the distance, which was mainly made from the calibration shot, using uncertainties on the estimation of angles and distance.

Since the velocity model cannot easily be improved and the objective is to follow the trends of seismicity, our approach should pay particular attention to the relative distribution of seismicity. A more frequent use of the relative methods to locate and to analyse induced seismicity will become the priority.

Doublets and Relative Method

The induced seismicity we encountered in this study is characterised by the existence of different families of similar events called doublets (Figure 6a). Taking into account the great similarity between doublets inside the same family, it is assumed that they have all been generated by the same geological discontinuity. The seismic hypocentre location is based on absolute independent triaxial hodogram analysis. In the case of doublets (high level of coherence), the events can be located with a relative method, whereas inside a cluster each event is located with respect to a reference one (FORTIER *et al.*, 1996).

This method consists of computing the relative distance and the relative direction between the two sources. The relative distance is determined from the difference between *P-S* arrival times, using cross correlation peak determination. In addition, the correlation functions are oversampled to improve the accuracy of time delays. The relative angles of incidence can be basically estimated using in-phase

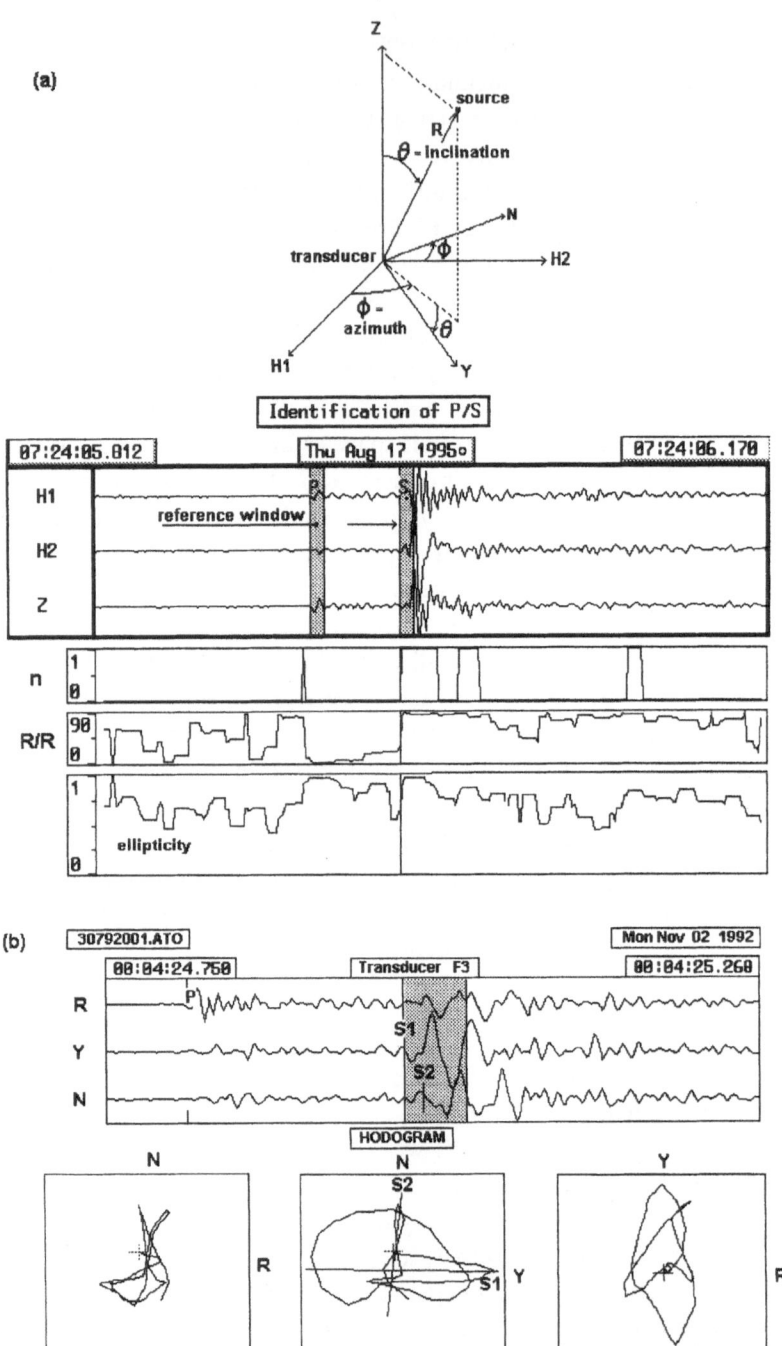

Figure 5

Example of relative polarisation analysis; (a) identification of S-wave polarisation with respect to P-wave polarisation (reference); S-wave polarisation type is indicated by a factor $n = 1$ (in that case $n = 1$ when ellipticity > 0.8 and radial/radial angle $> 80°$); (b) Example of microseismic event with shear-wave splitting. The hodograms are established from the signals in the grey window. The hodogram in the plane normal to the raypath (N, Y) shows the two split shear-wave phases (S1 and S2).

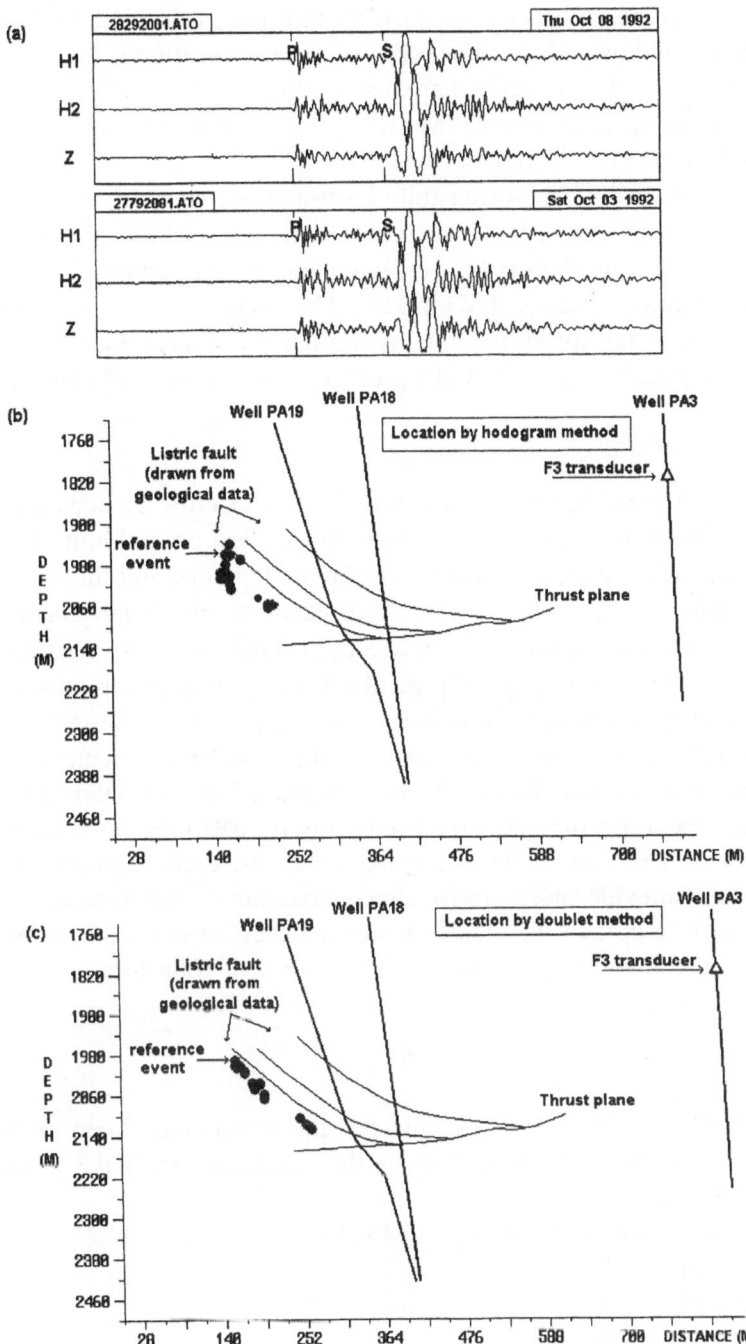

Figure 6
ELF-ATOCHEM microseismicity; (a) Example of two similar events detected by the *F*3 transducer;
Comparison between source locations from (b) the hodogram method and (c) the doublet method.

signals by the difference between the incident directions calculated for each event in the same time window and the same phase of the *P*-wave onsets. The relative angles of incidence could also be derived from the difference in the *P*-wave polarisation directions, using spectral matrix (MORIYA *et al.*, 1994) or "Olindes-Rodrigues" equation (FROISSART and MECHLER, 1993).

The analysis of doublets being fully computer driven, prevents any subjective bias from occurring and provides a precise relative location of events inside a cluster. The relative location is not yet systematically used and its integration in an easy to use software is under development. The results obtained on a few families of doublets have improved the quality of relative source locations, and have allowed us to specify more clearly the geological discontinuity (Figures 6b and 6c).

Seismicity Map

To establish consistency, the epicentre maps and cross sections presented for microseismicity analysis are all taken from the classical triaxial hodogram location data base. In such a monitoring network using initially one and then two downhole triaxial transducers, the seismic level of detection is not homogeneous over the whole site. Therefore, in order to analyse the overall seismicity recorded we have decided to combine three types of general seismicity map presentations. Figures 7 and 8 show plots where each event is represented by a dot. Figure 9a is a density map on which the contours are based on the number of events per unit area (10×10 m in this case). Figure 9b is a "relative" energy map, on which the contours are based on the mean relative energy (in dB) released in each unit area (10×10 m) with respect to the energy of the lowest event recorded. These three representations provide more precise characterization of the induced seismicity in each zone with respect to the others. Moreover, they illustrate the variation of the detection level due to the geometry of the network of transducers.

Results

From 1992 to April 1996, 2133 microseismic events have been recorded from which 1057 have been located. Among the seismicity, we have identified four geographical zones.
— the region around wells 20–21 and 13–14–17,
— around wells 18–19,
— around wells 9–10,
— around wells 11–12 and 15–16.
The seismicity encountered corresponds to three different states of well activity:
— wells used for leaching process,
— wells used to recycle the brine,
— wells in arrest (closed).

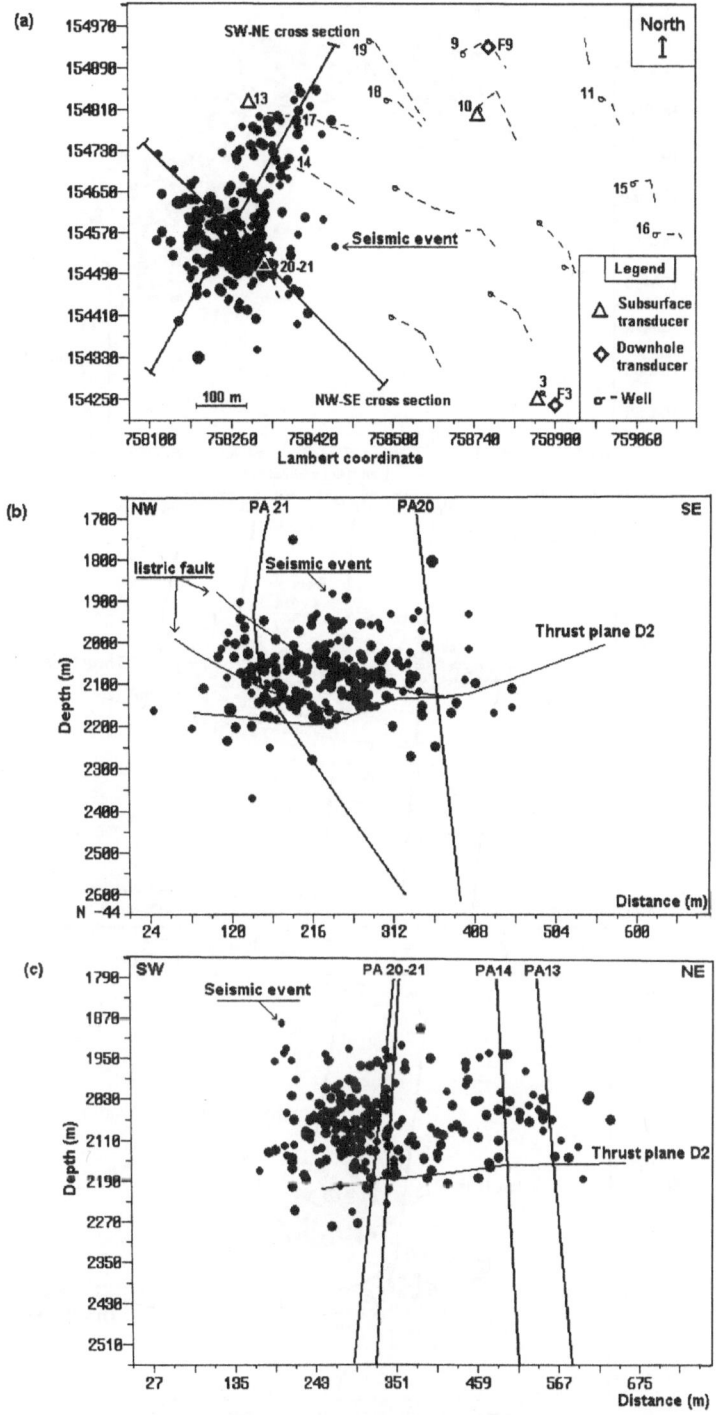

Figure 7
Microseismic activity of set 20–21 (1992–1996); (a) Plan view, (b) NW-SE section, (c) SW-NE section.

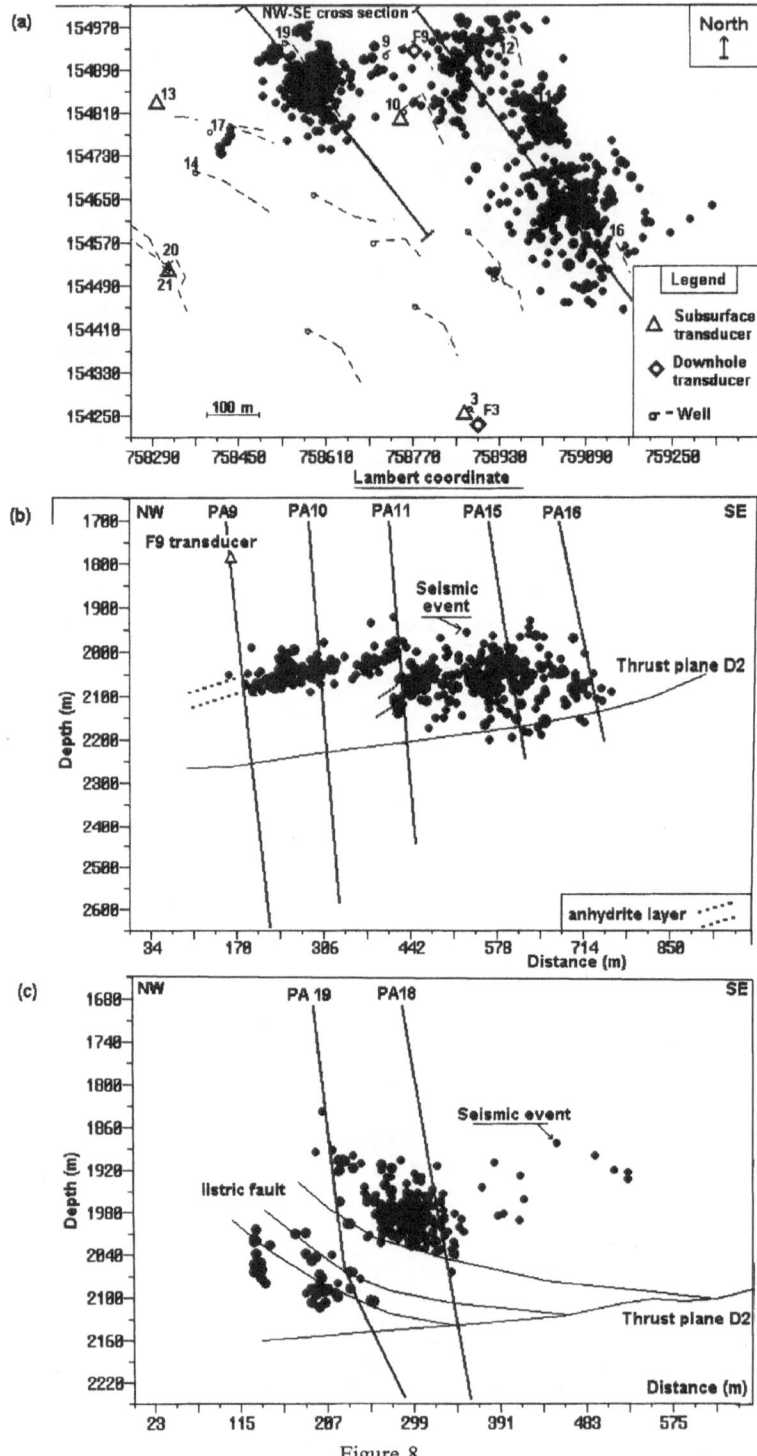

Figure 8

(a) Seismicity map (1992–1996); NW-SE sections of (b) set 15–16 and (c) set 18–19.

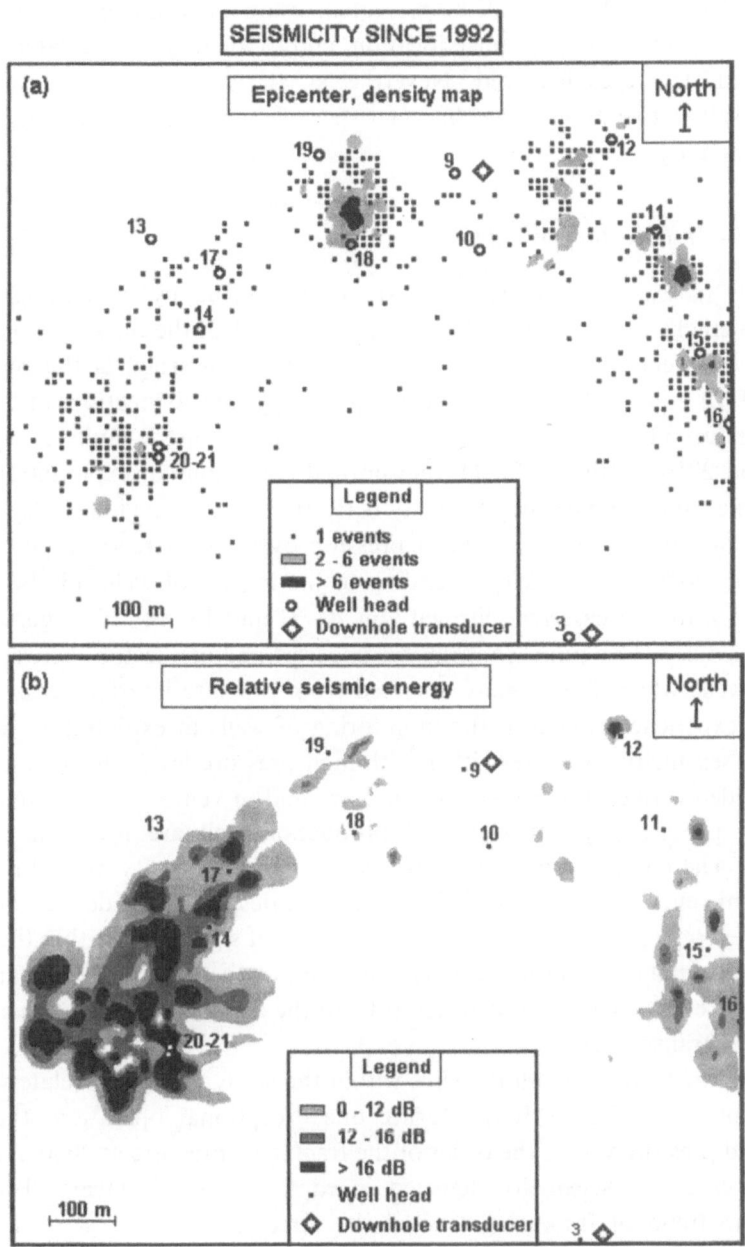

Figure 9
(a) Density map representing the number of events per unit area (10 m × 10 m) recorded since 1992; (b) Relative energy map representing the ratio between the cumulative seismic energy released since 1992 per unit area (10 m × 10 m) and the smallest seismic energy recorded (in dB).

Among all the swarms of microseismicity, we have identified families of similar events. Those families will all be relocated, using the relative doublet method, in order to improve the hypocentre location, and to identify with greater accuracy the geological structures involved. Nevertheless, the results collected with standard analyses allow to extract some characteristics of the seismicity induced by each state of well activity.

Wells 20–21 Used for Leaching

Wells in exploitation are used to produce brine by circulating fresh water through cracks between the wells. Except during 1992, the only set which produced brine were wells 20–21. The geographical zone concerned is the region around wells 20–21 and 13–14–17. For this zone, the hypocentre locations of the microseismicity induced by exploitation are situated around the listric faults at depth between 1850 and 2250 m (Figure 7). During normal conditions of exploitation (well-head pressure around 80 bar) the seismicity is restricted to the region of 20–21, however a high state of pressure (well-head pressure over 200 bar, see Figure 10a) in set 20–21 has induced seismicity extending to the triplets of wells 13–14–17 (Figures 7a and 7c). In that case the seismicity on 20–21 and 13–14–17 (Figure 10a) occurs simultaneously. This point confirms an underground connection between those two sets of wells which was already assumed from pressure interferences.

The experience gained in the monitoring of wells in exploitation demonstrates that the seismicity correlates either with high pressure levels in the injector well or with sudden variation of pressure (shut down and/or venting) due to inversion of the leaching process (injector well and production well are inverted, see the next section). The comparison between the seismic events density map (Figure 9a) and the seismic energy map (Figure 9b) shows that despite a low density of events, the seismic energy released in region 20–21 is the most significant within the study area. The small number of seismic events compared to the other zones means that only the largest events have been detected (due to the distance between this zone and the downhole transducers).

The largest events of all the sites within the study period correlate with a sharp rise in pressure (Figure 10a). During one exceptional operation, the maximum well-head pressure was in the order of the fracturing pressure at 2000 m depth. Even in that case, the seismicity detected is confined to salt layers. The seismicity shows no trend of fracturing toward the upper layers. The general distribution (spatial and temporal) of seismicity is from the southwest to the northeast and the maximum extent of seismicity in that period is in the order of 200 m. The west boundary of the seismicity corresponds very well to the assumed limit of exploitable salt layers.

Most of the microseismic events in this region present shear wave splitting (Figure 5b). The *S*-waves splitting is azimuth dependent, thus it must be related to

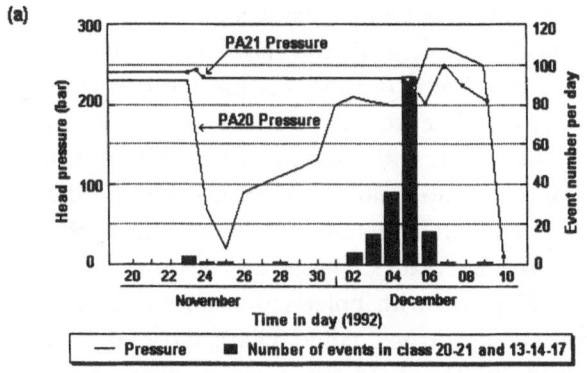

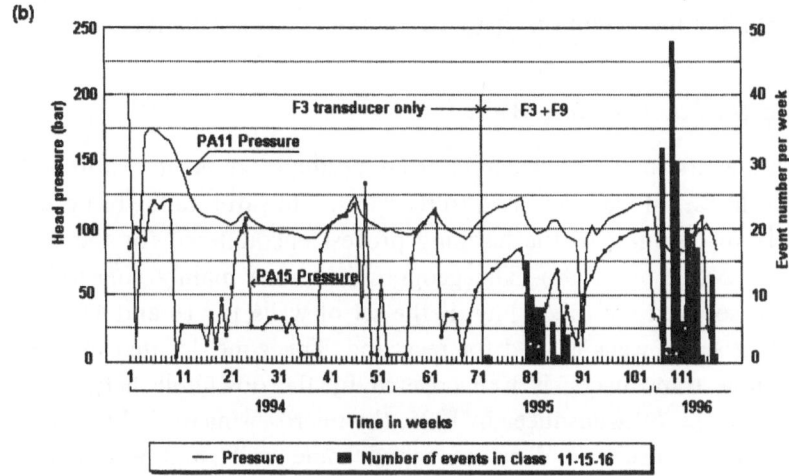

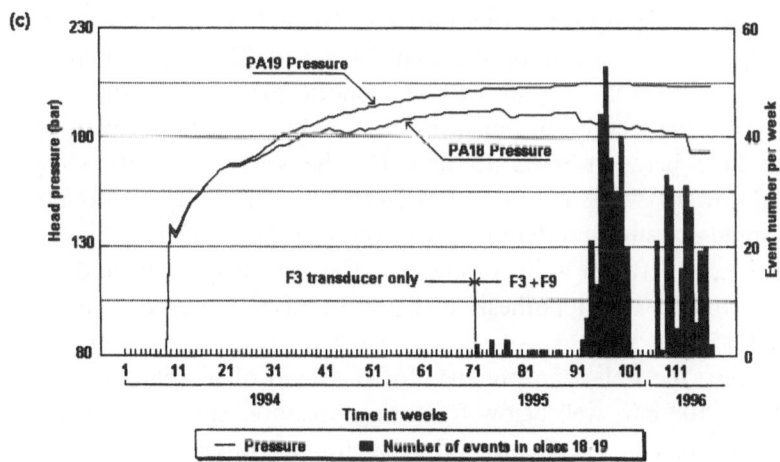

Figure 10
Correlation between well-head pressure and seismicity during; (a) Special operation before leaching; (b)
Recycling process; (c) Closure of a set of wells.

the propagation path. For this region, the azimuth (view from $F3$ sensor) of the seismicity varies from 5° to 25°, while the shear wave velocity anisotropy V_{S1}/V_{S2} varies from 1.10 to 1.06 (where V_{s1} and V_{s2} are the two split shear waves). The shear-wave splitting is mainly encountered for the events located on the west part of this region, and some events with the same azimuth but located on the east part of this seismic region show no clear shear-wave splitting. Shear-wave splitting is known to be associated with fractured rocks in which the $S1$ phase with a polarisation parallel to the cracks, meets less acoustic impedance, and is less attenuated than the $S2$ phase with polarisation normal to the cracks (LIU *et al.*, 1989). The shear-wave anisotropy could characterise the fracturation ratio of rockmass around wells 20, 21, 13, 14 and 17. The shear anisotropy might then be used to monitor the evolution of the fracture system in this region.

Set 15–16 Used as Recycling Wells

The brine produced by leaching from a single set is not saturated enough, therefore other set of wells are used to recycle and to saturate the brine. The process involved is very similar to the leaching process encountered for the set 18–19 in 1992 (MAISONS *et al.*, 1994). The geographical zone mainly concerned with the seismicity associated with recycling, is the set of wells 15–16 and 11–12, of which only the first set of wells is used for recycling. This zone, located just between the two downhole transducers, is well covered by the downhole network. Since the installation of the $F9$ transducer in 1995, the microseismicity induced is characterised by high microseismic density around some wells and a rather important spreading of seismic energy released.

A high-density spot of microseismicity is situated around well 15. This seismicity is restricted to a depth of between 2000 and 2100 m (Figure 9). Another dense spot of seismicity is near well 11, concentrated in two 100-m thick layers, parallel and in the vicinity of anhydrite layers above the thrust plane $D2$. Despite the brine flow between wells 15 and 16, the seismic activity seems to grow northward toward wells 11 and 12 (Figure 8). Well 11 is closed nonetheless it presents some pressure interferences with the recycling well 15 (Figure 10b). Brine flows must exist between wells 15 and 11. The seismicity maps and cross sections (Figure 8) confirm the hypothesis of some connection between the sets of wells 11–12 and 15–16.

In Vauvert, the well-head pressure for this kind of recycling process is usually around 80 to 100 bar, well below fracturing pressure. Since the installation of the $F9$ transducer we can see (Figure 10b) that for this process the seismicity appears mainly during the pressure decrease. The seismicity could be induced by the closure of fractures. We have encountered the same behaviour in 1992 for the leaching process of set 18–19, but at that time, with only the $F3$ transducer, it was not so clear (MAISONS *et al.*, 1992). In both cases the seismicity is clearly characterised by the evidence of different families of similar events and by a kind of repetition of the seismic activity (same families of similar events are recorded from one "shut in" period to another). The process of microseismic event generation must be different here from that encountered during high levels of pressure, as observed in set 20–21. It seems that, even during the recycling process the seismicity, induced by a fall in pressure, highlights the features which could be reactivated after closure due to a rise of pressure. This point is based on observations between the seismicity recorded around set 18–19 during a short period of leaching and during the closure test described in the next paragraph (the network of transducer was upgraded during these 2 periods). This point will be experimentally re-examined subsequently.

Test of Closure on Wells 18–19

Two other microseismic zones are located around the sets of wells 18–19 and 9–10. The proximity of the zone concerned with the downhole transducer F9 means that very tiny events can be detected. The result is a high density of seismic events combined with very low seismic energies. Figures 8 and 9 show that the seismicity is mainly located in the vicinity of well 18.

The set of wells 18–19 is closed and the pressure rises naturally. The seismicity is, as in the other case, restricted to a subhorizontal layer 100 m thick. Set 18–19 presents an increase of well-head pressure since March 1994 (i.e., the date of closure). Figure 10c shows the correlation between the pressure rise and induced seismicity. The seismicity seems to be triggered when the pressure of well 18 reaches a limit beyond which it begins to slowly decrease. One may assume that the opening of fractures might readjust the pressure of the cavity. The same evolution is observed on set 9–10.

Considering the depth of seismicity which roughly corresponds to the top of salt leached cavities, the well-head pressure indicates that the downhole pressure which has triggered the seismicity is between 80% and 90% of the geostatic pressure. This is far below the fracturing pressure estimated from hydrofrac data collected during the initial connection between the wells (110% of the geostatic pressure). On the one hand, the seismicity could indicate that pre-existing fractures have been reopened. On the other hand, the pressure trend could indicate that brine has been injected into the salt formation. The relationship between microseismic events and the volume of brine drained into the salt layers will be studied and should be relevant to the final monitoring. It must be kept in mind that microseismic activity is possible without "significant" fluid injections. Regardless, field observation and seismic monitoring results show that the first critical pressure that must be taken into account in the conceptual models is significantly below the fracturing pressure and could depend on the history of each set of wells. This point is currently under investigation.

Conclusion

French authorities must be assured that there is a procedure for monitoring rock fracturing associated with the closure of brine production wells so that any massive fracturing that could lead to the release of brine into aquifers could be detected and prevented. The seismicity recorded confirms the accuracy of seismic monitoring to observe such leached caverns after closure.

The induced seismicity recorded since 1992 is of a very small magnitude and is mainly horizontally scattered. The vertical extension seems to be very restricted and at this stage there is no indication of fracturation growing vertically toward the upper zones. The rather good agreement between seismic location maps and the overall geological features allows us to put a high degree of confidence in the analysis carried out. In addition, the results obtained lead us to orient our future work toward relative analysis (i.e., relative seismic signature characterisation and real time relative location versus pressure data) which should furthermore improve the interpretation.

Seismic monitoring has yielded useful data on the behaviour of salt layers with respect to normal salt mining operations, or partial closure of the cavern. Initially, seismicity and pressure interferences show that, in the case of Vauvert, the induced seismicity is associated with brine flows. The seismic monitoring during exploitation (leaching or recycling) clearly highlights the weak zones that could be reactivated by cavern pressure after closure. Furthermore this monitoring illustrates that the mechanical behaviour of a cavern cannot be studied separately from the others. The monitoring of a closed pair of wells shows that the microseismicity can be triggered when the pressure inside the caverns rises to a critical pressure below the geostatic pressure. The seismicity is then associated with a release of the cavern pressure. The release of pressure associated with the occurrence of seismicity suggests a release of brine and not a creep effect. Thus, when wells are closed, a relationship between the seismicity and the venting of brine outside the caverns does exist and is of great interest for the closure procedure which is under development. However, the results also show that in order to be truly efficient the final network must display a rather uniform sensitivity over the whole site. The minimum network needed should include at least four downhole transducers.

Indeed, in such experiments, the microseismic monitoring has a much wider aim than the simple detection of fracturing and potential aquifer pollution, which would be very limited. The monitoring of induced seismicity constitutes an early warning system which aims at prediction and prevention rather than merely reporting events after they have occurred. In the different steps involved in the development of such a closure procedure, seismic monitoring is a method which can allow the confirmation and validating of recommended closure criteria and which can also be applied for the post-sealing monitoring. In this case, seismic monitoring is one of the key points in the validating process of the closure procedure.

REFERENCES

CLIET, C., and DUBESSET, M. (1988), *Polarisation Analysis in Three-component Seismics*, Geophysical Transactions *34*, 101–119.

FORTIER, E., MAISONS, C., MECHLER, P., and VALETTE, M. (1996), *Seismic Monitoring: Improvement in Hypocentre Location Using a Doublet Technique*, Eurock'1996, Torino, Italy.

FROISSART, C., and MECHLER, P. (1993), *Online Polynomial Path Planning in Cartesian Space for Robot Manipulators*, Robotica *11*, 245–251.

LIU, E., CRAMPIN, S., and BOOTH, D. (1989), *Shear-wave Splitting in Cross-hole Surveys: Modelling*, Geophysics *54* (1), 57–65.

MAISONS, C., VALETTE, M., and FORTIER, E. (1994), *Use of Great Depth Permanent Borehole 3-axis Geophones for Induced Microseismicity Monitoring*, Eurock'94, Delft, The Netherlands.

MORIYA, H., NAGANO, K., and NIITSUMA, H. (1994), *Precise Source Location of AE Doublets by Spectral Matrix Analysis of Triaxial Hodogram*, Geophysics *59*, 36–45.

VALETTE, M. (1991), *Etude structurale du gisement salifère Oligocène de Vauvert* (Gard), Thèse doct. Lab. Géol. Struct. UGTL, Montpellier II.

VALETTE, M., BENEDICTO, A., LABAUME, P., and IBS-GULF OF LION WORKING GROUP (1994), *Gravity Thrust Halotectonics in the Extensional Camargue Basin (Gulf of Lion Margin, SE France)*, EAPG-6th Conference and Technical Exhibition P804—Vienna, Austria.

YOU, T., MAISONS, C., and VALETTE, M. (1994), *Experimental Procedure for the Closure of the Brine Production Caverns on the "Saline de Vauvert" Site*, SMRI-Fall Meeting, Hannover, Germany.

(Received September 17, 1996, accepted April 2, 1997)

Pure appl. geophys. 150 (1997) 605–625
0033–4553/97/040605–21 $ 1.50 + 0.20/0

⌐ Pure and Applied Geophysics

Three Seismic Nondestructive Methods Used to Monitor Concrete Slab Injection Tests

Kaveh Saleh,[1] Ferri P. Hassani,[2] Philippe Guevremont,[2] Afshin Sadri,[2]
Richard Lapointe,[3] Gérard Ballivy,[4] Jamal Rhazi,[4] and
Yahya Kharrat[4]

Abstract—A long term research program to determine the effectiveness of cement grouts injected in concrete dams was initiated by the Institute of Research in Electricity of Hydro-Quebec (IREQ) in 1986. To investigate the effectiveness of the grouts, three nondestructive methods based on seismic wave propagation were recently applied to various concrete slabs. The Impact-Echo method was used to determine the crack profile before the injection process began in one of the slabs. This method was used successfully and the detected internal crack depths agreed well with crack depths measured on the sides of the concrete slab. Acoustic Emission was used to monitor the penetration of the grout inside the crack. This method also allowed the researchers to determine at which moment, during the injection test, the cement grout mixture needed to be changed. Sonic Tomography was used to produce a tomographic image of the internal seismic wave velocities which traveled through the slab before and after the injection tests. Two different transmitter and receiver arrangements were used to determine the best measurement configuration. This paper illustrates the usefulness of combining various seismic wave based nondestructive methods to obtain a better knowledge of fracture detection and cement grout propagation, which can eventually lead to practical applications on concrete structures and hence on concrete dams.

Key words: Seismic testing, nondestructive testing, Impact-Echo, acoustic emissions, sonic tomography, concrete slabs, cement grout injection.

Introduction

The development of cracks in hydraulic structures is an important preoccupation for the owners of concrete dams. Thermal, physical, and hydrostatic cycles contribute to the propagation of these cracks which, to some extent, weaken dams.

[1]Hydro-Quebec, Institute of Research in Electricity of Hydro-Quebec (IREQ), Concrete Laboratory, 1800 montée Sainte-Julie, Varennes, Quebec, Canada, J3X 1S1.
[2]McGill University, Department of Mining and Metallurgical Engineering, 3450 University Street, FDA 20, Montreal, Quebec, Canada H3A 2A7.
[3]McGill University, Department of Civil Engineering and Applied Mechanics, 817 Sherbrooke Street West, Montreal, Quebec, Canada H3A 2K6.
[4]Sherbrooke University, Department of Civil Engineering, Faculty of Applied Sciences, Sherbrooke, Quebec, Canada J1K 2R1.

One current rehabilitation method is the use of injection techniques which fill the fissures with a bonding material. The typical method consists of injecting a cement grout (or chemical products) which will assist the hydraulic structure to regain its original structural integrity once the product has hardened, and bonded to crack surfaces. An injection is executed under pressure and is usually performed in existing boreholes, cored during investigations of a structure.

The principle of an injection technique is somewhat straightforward, however, the progression of the injection grout still remains an unanswered question. This is due to a lack of understanding of all the factors influencing the progression of grout in a crack (WEAVER, 1991). In order to successfully accomplish a grouting campaign, it is important to know that the grout has been injected in all areas of the crack, regardless of the crack thickness. It is sometimes necessary to use high injection pressures for the preceding purpose, but at all times maintaining the pressure below a level which would induce hydraulic fracturing of concrete. It is also important to ensure that the injection grout has impermeability or consolidation properties. These factors are not always easy to satisfy due to the function of a number of parameters which are linked to the rheology of the injection product, the morphology of the crack, and the *in situ* conditions which vary from one case to another.

It is therefore important to develop a control method which will ensure that the injection has been carried out successfully. To do so, an injection test has been developed jointly by the Institute of Research in Electricity of Hydro-Quebec (IREQ) and the University of Sherbrooke, in the province of Quebec, Canada. This test is primarily used to investigate the performance of various injection products during the grouting process. In essence, a concrete slab is hydraulically fractured along its length. The slab is dotted with linear variable displacement transducers (LVDTs) and pressure meters. The pressure meters are placed on a grid, on top of the slab and are used to measure the variations in pressure at different points as the crack is being injected. By analyzing the pressure variations and the grout take, one can determine if the grout hole has been completely injected. However, it may be difficult to determine if the grout has traveled between the pressure meters.

To address this problem, three nondestructive techniques were used before, during, and after the injection process, namely the Impact-Echo method, Acoustic Emission, and Sonic Tomography. These three nondestructive tests are partially based on seismic theory, and were applied in a controlled laboratory environment. These three methods were applied on a numerous concrete injection slabs of the same dimensions and properties.

This paper is the result of a joint effort initiated by Hydro-Quebec and the involvement of two Quebec universities. The presented research briefly underlines current work performed in the area of nondestructive testing of concrete in Canadian institutions.

Cracking Problems in Concrete Dams

Concrete is essentially a man-made material consisting of cement, water, sand, aggregates of various sizes, air and occasionally chemical additives. Therefore, concrete can be subjected to many types of physical and chemical deterioration processes. It is a well known fact that many hydaulic structures around the world suffer from one or many forms of deterioration. Concrete dams in the province of Quebec are no exception. Often, problems associated with cracking in these dams are caused by many factors acting simultaneously. For example, initial shrinkage provokes microcracking, thermal changes cause dilation and contraction of concrete, and freeze thaw cycles can cause pop-outs in concrete with high porosity. Other deterioration processes also of importance are: reinforcement corrosion, surface erosion, carbonation, sulfate attack, and alkali-aggregate reaction. All these detrimental effects can lead to substantial cracking in concrete dams. Another physical cause of cracking is differential settlement of structures, which can be caused by inadequate foundation design unobserved subterranean geologic abnormalities, and variations in water table levels.

Presently, only coring has been used to detect these problems. Coring is a standardized and widely accepted investigation technique, which is used to determine concrete quality and its mechanical properties, along with observing the presence and orientation of cracks (Houlsby, 1990). This is a very time consuming technique if one takes into account the time needed to retrieve and analyze the samples. Economics is a very important factor when rehabilitation of a structure is considered, and in the case of dams high investigation costs occur when this technique is applied to these large-scale structures. It is obvious that substantially coring a dam is not an ideal procedure due to structural considerations, therefore in some instances it is necessary to extrapolate results between boreholes. This brings about a certain degree of uncertainty in the results, however coring remains today the best way to investigate structures such as concrete dams.

Due to high coring costs and a requirement to interpolate results, Hydro-Quebec has proceeded to use nondestructive techniques which would increase the accuracy of the results and help ensure a better structural analysis of dams afflicted with cracks. It was determined that seismic techniques would be best suited for these situations because of their sensitivity to changes in material densities and elastic constants.

As mentioned in the introduction, grouting is a method that can be adopted to inject cracks in concrete hydraulic structures. The method has been shown to be effective in restoring the structural integrity of dams. However, an injection is only beneficial if the selected injection grout is adequate for a given situation, and if a certain control method exists to determine if the grout has been injected in all areas of a crack. Hence, it is important to experimentally determine which injection product to use and, to some degree, predict the propagation and the effect of the grout inside the crack.

Injection Test Setup

The purpose of this laboratory project is to simulate the injection of fine cracks in concrete dams. The grouting products and equipment were studied first to select those with the best performance. These are subsequently tested in an injection simulation at Hydro-Quebec's grouting centre. Inaugurated in 1987, this facility possesses a grouting test bench placed in a climatic chamber with controlled temperature and relative humidity levels, which allows the cracking and injection of large concrete slabs at a laboratory scale. The research program consists of testing the products and injecting the concrete slabs in the laboratory to assess grouting equipment and methods.

The first injection products to be used were clay-based grouts followed by cement-based and chemical products. Currently, it is recommended to use products featuring improved physical and mechanical properties, however the behavior of these new products still requires investigation. In fact, it is critical to determine their rheological, physical and mechanical properties and to compare these results with those of the conventional products before they are used on a wider scale. Also, since these products are used to repair cracked structures subjected to large temperature changes, and since their physical and mechanical behavior in response to these changes is not well known, such investigations would provide information to choose products that perform well for dam repairs.

In order to assess the performance as well as the relevance of the grouting equipment and procedures, it was decided to inject concrete slabs in the laboratory where it is possible to simulate accurately the real conditions at concrete dams.

The Test Slabs

Large concrete slabs (2.62 m × 1.42 m × 0.40 m) were cast for the simulation. They comprise two kinds of concrete: a high-strength concrete (50 MPa) with reinforcing fibers for the top and the bottom layers, and a low-strength concrete (15 MPa) for the middle layer where a horizontal crack should propagate (BALLIVY *et al.*, 1992). The slabs are placed in the test bench, a rigid metallic frame, where the initial crack opening and displacements caused during the injection test can be instrumentally controlled (Fig. 1). Usually a force in the order of 7200 kN is applied to the frame and the slab to keep any uplift displacements below 0.05 mm. The test bench allows the slab to be inclined to simulate a crack. The experimental slab injections take place in a climatic chamber where the temperature is set at 4°C and the relative humidity level at 100%, representing the conditions encountered upstream of a concrete dam.

In order to monitor the progress of the injection, the slabs are instrumented with 24 pressure sensors, 8 loading cells and 12 displacement sensors (Fig. 2). The pressure sensors serve mainly to establish the pressure profile and the grout penetration within the crack. The loading cells and displacement sensors are used to evaluate the effect of pressure on the crack opening and on the reactions on the test bench. Other sensors read the flow, the pressure and the temperature of the grout in the incoming line. All sensors and cells are connected to a data acquisition system, which helps to control the different injection parameters.

Once the slab is correctly installed in the metallic frame, the crack is initiated by a hydraulic force. This splitting force is the result of a generalized pressure increase obtained when a high flow of water is pumped toward the crack initiator (sand layer, see Fig. 2), located at one end of the slab, while all other points are kept closed. The slab generally cracks horizontally along the plane of the cold joint.

The next step consists of a water test to saturate the slab and assess the crack pattern. This test is performed using the grouting equipment and procedures but replacing the grout by water. The injection itself follows the water test immediately after the whole system has been drained. The choice of an initial injection point is based on the results of the water test: it is usually central and shows a relatively good absorption rate. The slab is injected with a grout whose water to cement ratio (W/C) varies depending on the product used. The spread of grout is then observed. Sometimes it is necessary to modify the W/C according to the absorption rate or to change the injection point if the flow decreases to a low value while the pressure

Figure 1
Concrete injection slab placed in the metal test frame at IREQ. This picture of a recently tested slab is identical to older slabs.

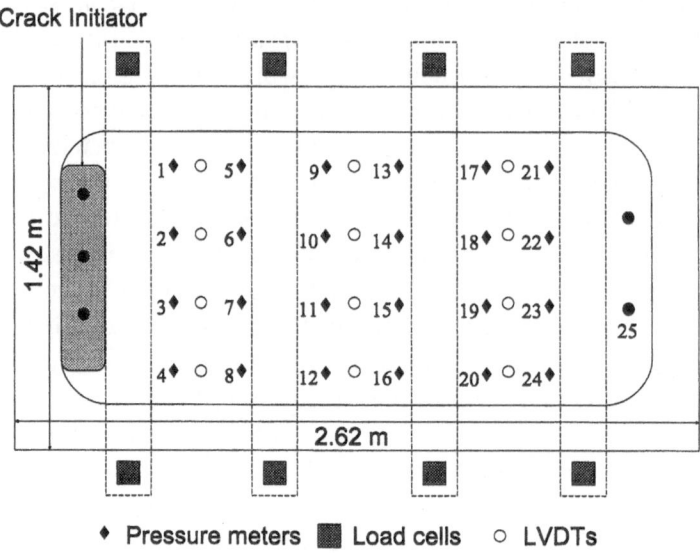

Figure 2
Top view of test slab and position of instrumentation.

stays high or increases. This procedure must be followed until complete grouting of the crack is achieved.

The grouted slab must then cure for 28 days in the climatic chamber still at 4°C and 100% of relative humidity. Coring is then accomplished over the entire slab. The cores are useful for determining the quality of the injection either by visual inspection or by other relevant tests such as tensile and compressive testing. The slab is also split along the crack to assess the spread, bonding and hydration of the grout. Figure 3 shows a typical surface profile of a grouted crack.

Nondestructive Testing

Nondestructive testing has been gaining a wider acceptance in the concrete industry due to the use of stress wave propagation methods. Such methods are used to determine layer profiles (SASW method), defect locations (Impact-Echo), material quality (MSR and Sonic Tomography), and also crack propagation and orientation detection (Acoustic Emission). There are many nondestructive techniques based on stress waves, however they typically vary with the equipment used and data analysis techniques.

For the concrete slabs discussed in this paper, three techniques were deemed capable of enhancing the information retrieved during the grout injection tests. Three techniques were used, namely the Impact-Echo method, Acoustic Emission, and Sonic Tomography. The Impact-Echo tests were performed by McGill University's Geomechanics Laboratory from the Department of Mining and Metallurgical Engineering. The Acoustic Emission and Sonic Tomography tests were conducted by Sherbrooke University's Rock Mechanics Laboratory from the Department of Civil Engineering in the Faculty of Applied Sciences. The following sections describe the various nondestructive tests and discuss results obtained before, during and after injection tests performed on some of the concrete slabs.

Impact-Echo Tests

The Impact-Echo method is a microseismic technique which was developed for the detection of defects in thin concrete structures. Initial work on this method was carried out at the National Bureau of Standards in the United States in the mid 1980s (SANSALONE and CARINO, 1986).

The principle of the Impact-Echo method is based on seismic reflection. A spherically tipped impact source generates body waves at the surface of the material. These waves travel through concrete and are reflected at interfaces inside the medium. When the distance between the impact source and the receiver is small

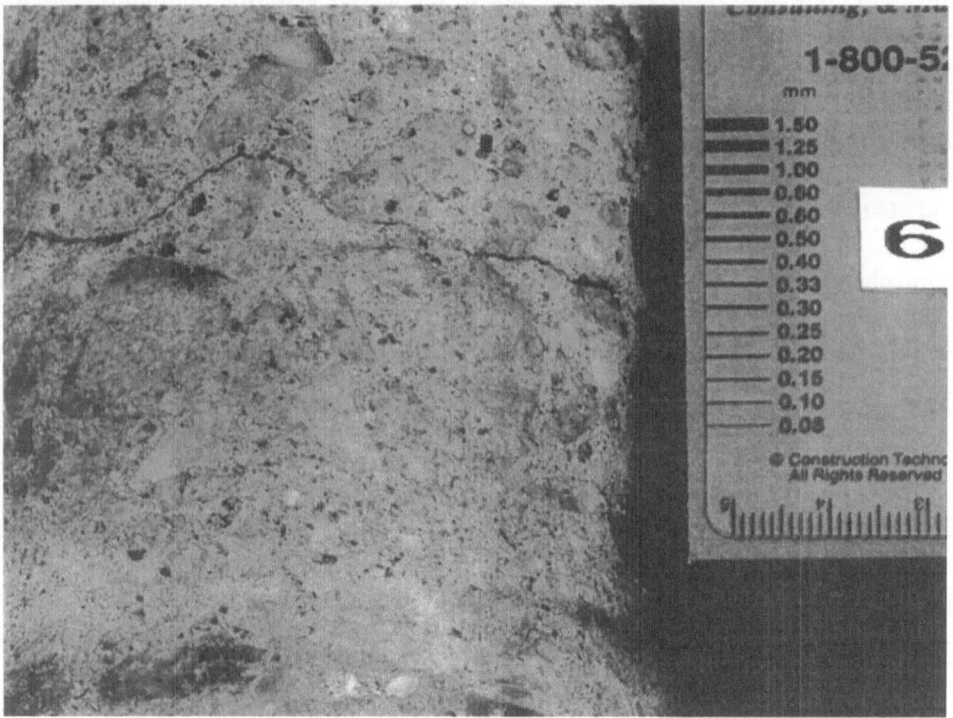

Figure 3
Typical crack profile observed on a core from one of the grout injected slabs.

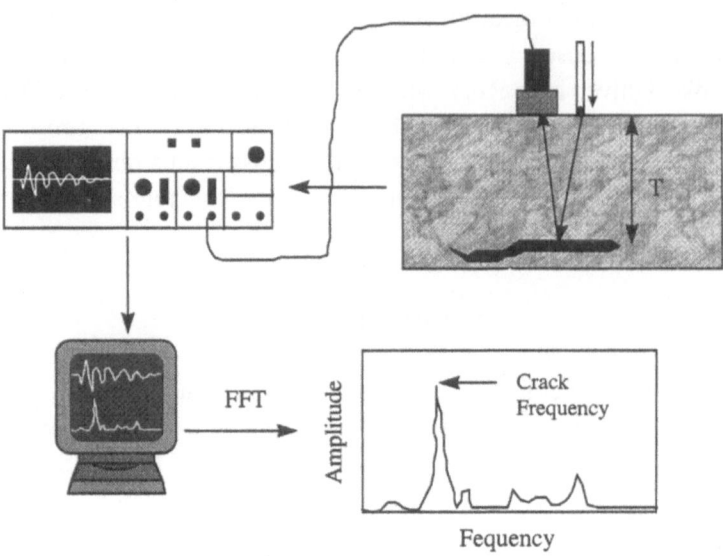

Figure 4
Schematic of Impact-Echo test procedure.

and both are placed on a solid plate such as a concrete slab, the dominant response is due to P-wave reflections between the top and bottom of the plate. Defects within the plate are detected as a result of a higher frequency resonance between the top surface of the defect and the top of the plate as seen in Figure 4 (CARINO and SANSALONE, 1988).

The reflected pulse travels back and forth between the two reflecting surfaces and creates a periodic wave form at the top surface. Shear waves (S waves) are also generated when the impact source hits the plate's surface. S waves have the same hemispherical wave front characteristics as P waves (TELFORD *et al.*, 1990). The amplitude of the reflected wave forms are mainly controlled by the elastic properties of the material. The maximum reflection amplitude for the P wave is directly beneath the impact tip. The P wave is characterized by the propagation of compression and tension waves between two surfaces. The compression wave produces a downward vertical particle motion at the surface, while the tension wave produces an upward particle motion at the surface. A P wave changes mode (i.e., compression to tension or *vice versa*) when it reflects back from a concrete to air interface. The result on the surface of the slab is a periodic sinusoidal displacement at the surface, which can be detected by a vertical displacement piezoelectric transducer. With this instrument it is possible to obtain the maximum amplitude of the reflected P wave.

The mechanical impact on the surface of a solid plate generates stress waves that have a wide frequency content. By changing the impact tip one can increase or decrease the frequency content of a stress wave. This effect is the result of varying the contact time of the sphere on the test surface. The wavelength generated by an impact determines the type, depth, and dimension of defect which can be detected from the test surface. For example, a short contact time will generate a high frequency signal and a short wavelength which is useful for detecting small and shallow effects. Correspondingly, a long contact time will allow detection of defects at greater depths. For depth calculations, a dominant frequency is selected in the frequency spectrum. The amplitude of the dominant frequency is partially dependent upon a reflection coefficient which is a factor of the acoustic impedance of two different materials forming an interface. The acoustic impedance is equal to the density of a material multiplied by the P-wave velocity in the same material. The reflection coefficient is the difference between the acoustic impedance of both materials, divided by their sum. The dominant frequency is found by applying a Fast Fourier Transform (FFT) to the wave form recorded by the oscilloscope and this dominant frequency is used with the following equation:

$$T = Cp/(2 \times f),\qquad\qquad(1)$$

where $T =$ the depth from the surface to the crack (m), Cp is the P-wave velocity (m/s) in the material, and f is the dominant frequency (Hz) related to the P wave in the frequency spectrum. The factor of $1/2$ is added because the actual depth of the crack is one half the travel distance of the wave.

The system used for the Impact-Echo test was assembled at McGill University by the Geomechanics Laboratory. This unit includes a spherical tip spring loaded impact device, one broadband vertical displacement transducer, a digital oscilloscope and a personal computer for data analysis. The analysis software program was developed on the GAUSS mathematical and statistical system, using the time series analysis package to identify and measure P-wave frequencies and velocities. Figure 4 is an illustration of the Impact-Echo test equipment.

Impact-Echo: Experimental Setup and Discussion of Results

The aim of this test was to measure the distance of the hydraulically induced fracture from the top surface of a concrete slab. To achieve this goal, a grid of 20×40 cm was drawn on the top surface of the slab. The grid was drawn to cover as much area as possible by the Impact-Echo scans (see Fig. 5). The average P-wave velocity in the concrete slab was found to be 4300 m/s. The results are given in Table 1, where the crack depths were computed from equation (1) and Cp is the average velocity. In general, there is a good agreement with the actual crack depth, as determined by visual inspection (see Table 2). At some impact locations, such as point $C5$, there is a difference of approximately 7 cm between adjoining points. A depth of 17 cm could indicate a lack of bond between the top layer of concrete and the middle layer of concrete, however, it was not possible to physically investigate this explanation since it was not feasible to core a concrete sample at that location on the slab.

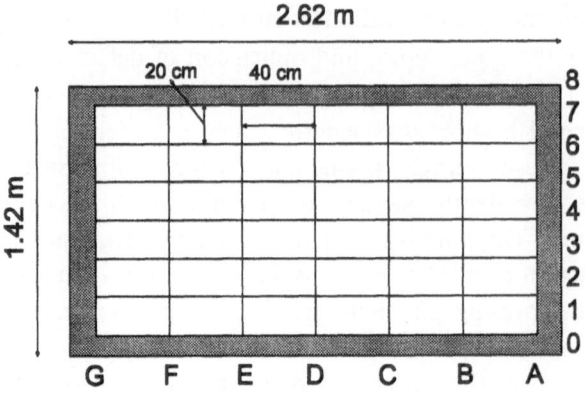

Figure 5
Layout of Impact-Echo test grid on the slab.

As can be seen in Table 1, the Impact-Echo method offers a good identification of the crack profile in the slab. Note that there is consistency in the data with respect to the observed crack depths at the surfaces and within the concrete. For

Table 1

Crack depths from a concrete injection slab determined by the Impact-Echo method

Impact position	Frequency of P wave to crack (Hz)	Thickness to crack from top surface (cm)	Impact position	Frequency of P wave to crack (Hz)	Thickness to crack from top surface (cm)
A1	9131	23.6	E1	10009	21.5
A3	9961	21.6	E2	9814	21.9
A6	10107	21.3	E3	10009	21.5
B1	9375	22.9	E5	8594	25.0
B2	9277	23.2	E6	10498	20.5
B3	9424	22.8	E7	12061	17.8
B7	7764	27.7	F1	8203	26.2
C1	8300	25.9	F2	7031	30.6
C2	9082	23.7	F3	7373	29.2
C3	9277	23.2	F4	7275	29.6
C4	8740	24.6	F5	8350	25.8
C5	12000	17.9	G1	8203	26.2
C6	8789	24.5	G2	8398	25.6
C7	8350	25.8	G3	7666	28.1
D1	8252	26.5	G4	7666	28.1
D2	8887	24.2	G5	8250	26.1
D3	10010	21.5	G6	7861	27.4
D4	8838	24.3	G7	7715	27.9
D5	7470	28.8			
D6	7471	28.8			
D7	8301	25.9			

Table 2

Crack depths from the top surface observed on the 2.60 m sides of the slab by visual inspection

Position	Depth (cm)	Position	Depth (cm)
$A0$	25.5	$A8$	25.3
$B0$	25.6	$B8$	24.9
$C0$	25.6	$C8$	24.9
$D0$	25.7	$D8$	24.5
$E0$	25.9	$E8$	25.8
$F0$	25.9	$F8$	25.8
$G0$	26.0	$G8$	27.0

example, if test line D is considered (Fig. 5), one can see that the depth difference between the point $D0$ and $D1$ is 0.8 cm which shows that the investigators can have confidence in the results obtained inside the slab by the Impact-Echo method. The depth differences between the cracks visually observed at each of the 2.62 m sides and the Impact-Echo results along rows 1 and 7 are displayed comparatively in Figure 6.

The slab was also scanned by Impact-Echo after the injection. It is important to mention that these tests were performed on a slab with a thickness of 46 cm. The results are given in Table 3. As shown, it was difficult to accurately determine the thickness of the slab with this method after the injection process. The Impact-Echo method is strongly dependent on P-wave velocities inside the slab. The use of an average P-wave velocity for depth calculation at the various impact points may account for the discrepancies in the preceding results.

With the accumulated data, it is possible to draw a crack profile along the test lines as set up on the grid. The crack profile can then be interpolated between the test lines, allowing for a three-dimensional view of the top surface of the crack to be obtained as shown in Figure 7. However, it is important to note that this would not generate a true three-dimensional image, since no tests were carried out on the bottom surface of the slab (i.e., the crack width cannot be determined in this case).

Acoustic Emission Tests

Acoustic Emission (AE) tests were performed to evaluate the injection process. Tests were conducted during the grouting periods. When concrete elements are subjected to dynamic or static loads and plastically deform, they generate sounds. It is possible to locate the origin of these sounds and determine the progression of cracks or of an injection grout. Technological advances in transducers, amplifiers, and oscilloscopes have allowed researchers to detect sounds which emanate from a concrete element loaded to 10% of its ultimate capacity. Acoustic emissions from concrete occur over a wide range of frequencies (from approximately 2 to 400 kHz) and it has been determined that there is no precise frequency range to investigate while testing concrete (MINDESS, 1991).

On some of the concrete slabs, tests were performed with a SPARTAN data acquisition system manufactured by the "Physical Acoustic Corporation". It is used for locating and analyzing, in real time, acoustic emission events. It is a high speed acquisition system which is capable of a sampling rate up to 10 MHz. This system includes a PAC 3000 computer, which works in the CMP environment. It is also a multi-channel acquisition system which can be extended to 128 channels. As just mentioned, the analysis of AE signals is conducted in real time and involves a statistical analysis of parameters such as: amplitude, duration, rise time, energy analysis, and event counting. Three options are available for AE source location: linear, plane, and three-dimensional. In a series of tests on two concrete injection slabs, it was necesary to determine the frequency range needed to detect the AE signals generated during grouting. It was found that 50 kHz transducers were best suited to conduct AE investigations during injection tests (RHAZI and BALLIVY,

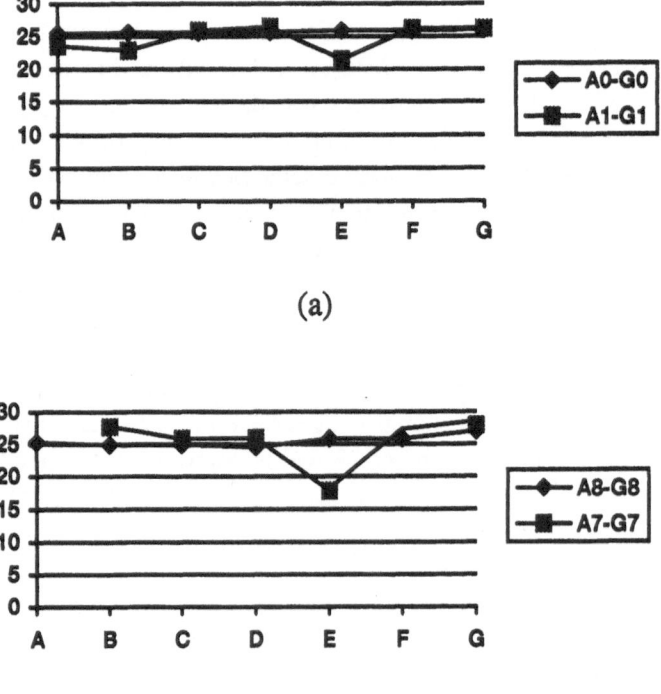

(a)

(b)

Figure 6
Depth comparison between the visually observed depths on the long sides of the slab and the first internal test points: (a) Row $A0-G0$ compared to row $A1-G1$, (b) Row $A8-G8$ compared to row $A7-G7$.

Table 3

Thickness to bottom of the concrete slab as determined by the Impact-Echo method

Impact position	Frequency of P wave to crack (Hz)	Thickness to bottom (cm)	Impact position	Frequency of P wave to crack (Hz)	Thickness to bottom (cm)
A1	7397	29.1	E1	6226	34.5
A2	4541	47.4	E2	4395	48.9
A3	5029	42.8	E3	6207	34.6
A4	5127	41.9	E4	4492	47.9
A5	5151	41.7	E6	4492	47.9
A6	6787	31.7	E7	4395	48.9
B1	5286	40.7	F1	4411	48.7
B2	4999	43.0	F2	3564	60.3
B3	5664	38.0	F3	4614	46.6
B7	4194	51.3	F4	5459	39.4
C3	6185	34.8	F6	4443	48.4
C4	5151	41.7	F7	5078	42.3
C5	4443	48.4	G1	4590	46.8
C6	4492	47.9	G2	4919	43.7
C7	6346	49.5	G3	4639	46.4
D1	5518	39.0	G4	4248	50.6
D2	4614	46.6	G5	3874	55.5
D3	5778	37.2	G6	4639	46.4
D4	4932	43.6			
D5	4102	52.4			
D6	4541	47.4			
D7	4565	47.1			

1993a). In this study, the analysis showed that the 50 kHz transducers are more sensitive to the hydraulic noise, induced by the flow phenomenon. Therefore, the 50 kHz transducers are more appropriate for injection monitoring as compared with 135

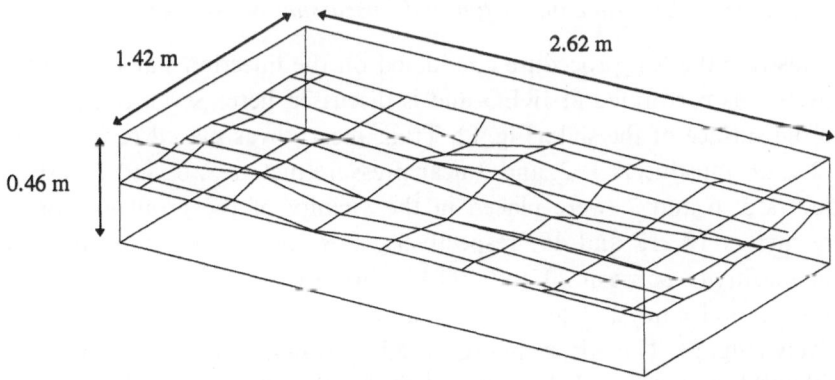

Figure 7
Three-dimensional view of the crack surface as determined by the Impact-Echo method.

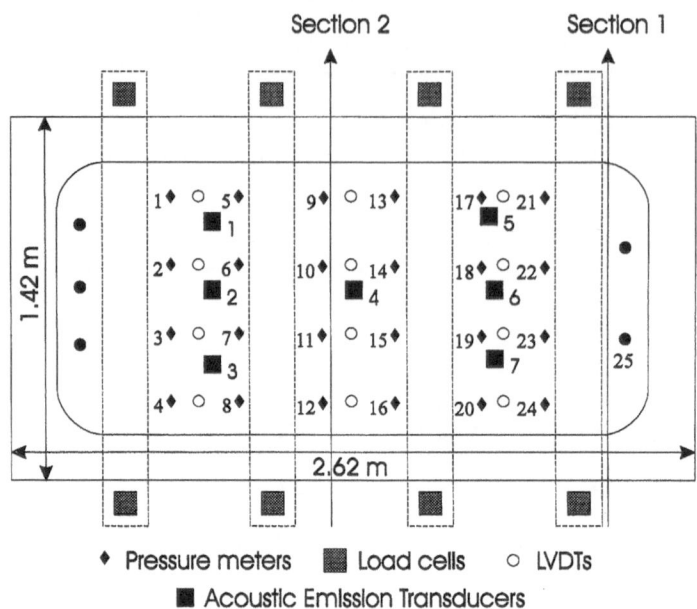

Figure 8
Layout of AE transducers and positions of sections 1 and 2 used for Sonic Tomography testing.

kHz transducers. If lower frequency transducers were available, the sensitivity of the test would be better. In order to correlate AE activity to flow parameters, the evolution of the AE signals was compared to water pressure during a water test and at different crack positions. It was found that the detected AE signals are not due to constant pressure flow, but to the acceleration and deceleration of the injected product, and to the mechanisms resulting from the variations of pressure and flow in the fluid.

Acoustic Emission: Experimental Setup and Discussion of Results

To illustrate the test procedure conducted on the injection slabs, a recent series of AE tests was performed at IREQ and is discussed here. Seven transducers were fixed to the surface of the slab (Fig. 8). Transducer 4 was placed near the injection point 12, and transducers 1, 2, and 3 near pressure meters 5, 6, and 7, respectively. Transducers 5, 6 and 7, were placed in the vicinity of the grout exit point near pressure meters 17, 18 and 19, respectively. For these tests, transducers with a constant sensitivity between 20 and 100 kHz were used.

Transducers 1 and 2 detected a small amount of activity, 2.3% and 3.9%, respectively (Fig. 9). The AE events recorded by transducers 3 (6.8%) and 4 (7.4%) are comparable but remain below 10% of the total activity. Transducers 5 (19.6%), 6 (25%), and 7 (35%) were placed in an area where 80% of the activity occurred.

Figure 9 shows the contributions from each transducer. Figure 10 is a graphical representation of the cumulative number of events for transducers 1, 4, and 6 with respect to time.

The number of AE events from each transducer gives an indication of the success of the injection in each area of the slab and also offers a qualitative analysis of the injection product. In this case, it seems that the flow of the grout was easier in the direction of the grout exit point 25. This tends to correspond with the grout injection procedure.

The general evolution of the signals indicates an increase in AE activity during approximately the first 90 minutes of the test. After this period, the activity generally decreases with only occasional increases. This activity tends to be representative of the injection procedure used during the tests (SALEH et al., 1995).

As previously mentioned, transducer 4 was placed in proximity to the injection point. From Figure 10(b), it can be seen that the AE activity constantly increased which indicated a normal injection process. The events stabilized around the 5000 sec (83 minutes) mark after the beginning of the injection test. This demonstrates that there was a decrease in penetration flow at that time. This was also noted on the injection grout discharge charts and led to a change in the grout mix a few minutes later. Although at that time the flow remained quite low, there was significant AE activity. The activity shown between 8000 seconds (133 minutes) and 10000 seconds (166 minutes) is not completely due to the injected grout but to a break in the injection main tube. However, it can be seen that the AE method could help to determine when to change the injection mix and where the injection grout has penetrated.

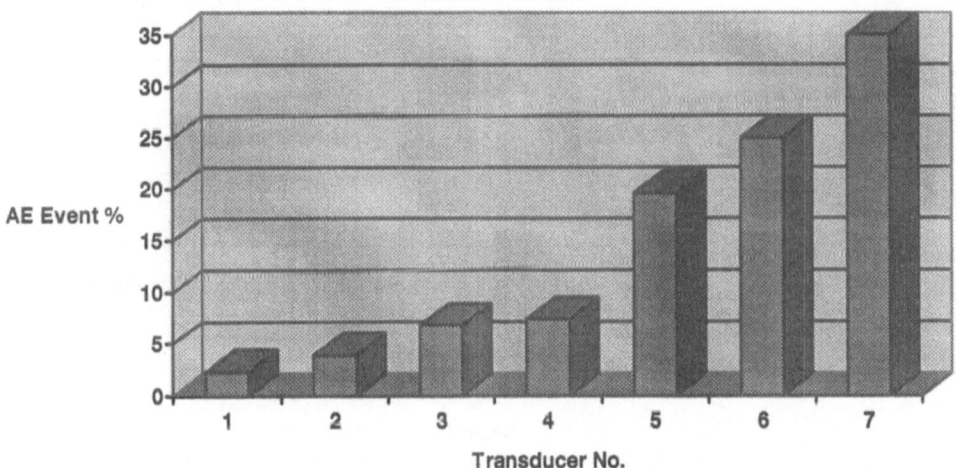

Number of AE Events In %

Figure 9
Contribution in percentage (%) of each AE sensor.

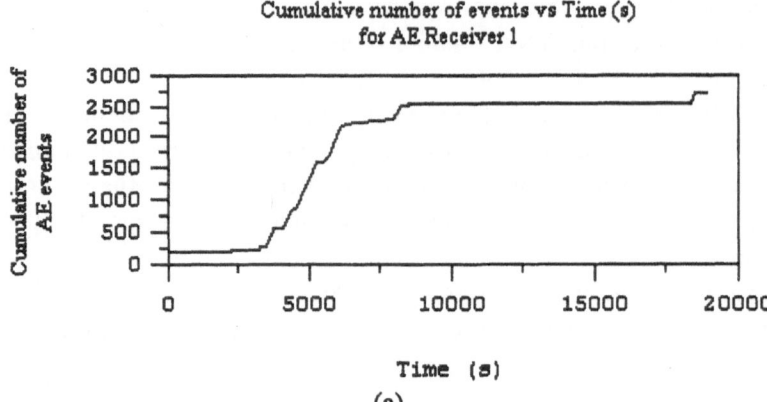

(a)

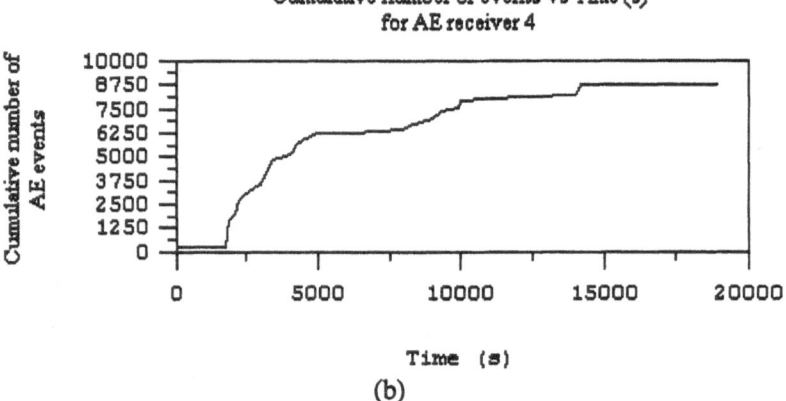

(b)

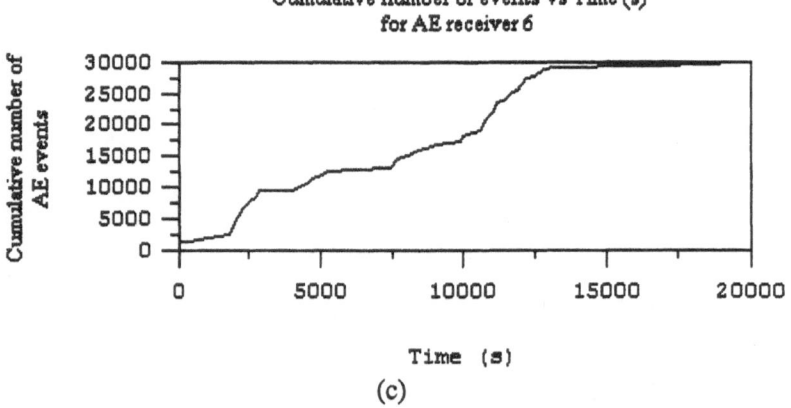

(c)

Figure 10
Cumulative number of AE events: a) Receiver 1, b) Receiver 4, and c) Receiver 6.

It is possible to locate events with the AE method by applying what is called linear location. This technique helps determine what is happening between two receivers, and allows the users to define the linear distribution of the grout penetration into the slab. This is accomplished by analyzing information retrieved between two adjacent receivers. The information was analyzed and it was found that most of the penetration occurred near receivers 5, 6, and 7. The most intense activity zone was determined to be in the plane of receivers 6–7 and 4–7. Therefore, it seems that during this injection test most of the grout penetration occurred between the injection point (no. 12), and the grout exit point (no. 25).

Since the AE events are essentially due to the hydraulic noise, the detected signals are of a continuous type. However, in this case, it is not a major problem for event location by travel time. Indeed, AE transducers are not only sensitive to the variation of flow rate and therefore, the response of the transducer is not constant but periodical. A detailed explanation of this aspect is given by RHAZI and BALLIVY (1993b).

Sonic Tomography Tests

Sonic Tomography is an imaging technique based on the analysis of seismic P-wave velocities. The principle of the technique is similar to the ultrasonic pulse velocity technique in that both methods require access to more than one side of the test sample, and both study the velocity of a seismic wave traveling through a medium. Sonic Tomography can use numerous transmitters and receivers simultaneously. This technique is used in geophysics and mining engineering applications and has recently been used on civil engineering structures such as dams (KHARRAT *et al.*, 1993).

Due to the high sensitivity of this method, researchers can differentiate internal anomalies which possess various seismic wave velocity characteristics. Sonic Tomography tests can be conducted on concrete structures a couple of meters thick to tens of meters thick depending on the resolution requirements and the objectives of the tests.

The instrumentation consists of a multitude of receptors and transmitters, data acquisition cards and a portable computer. This method is essentially a numerical technique for data analysis. Each transmitter is triggered consecutively and recorded by the receivers. The P-wave travel times are obtained by using all possible P-wave path combinations between the transmitters and the receivers. The data are then processed by an inversion software which is generally based on an iterative process. The geometric iterative reconstruction techniques have been found to be faster than other methods for this type of analysis. All the data undergoes a tomographic reconstruction, and the P-wave velocity tomography of the section is displayed on a computer screen. The velocity distribution is in direct relation with the distribution of the discontinuities of the medium being tested.

Sonic Tomography: Experimental Setup and Discussion of Results

For a particular test on one of the injection slabs, two test configurations were used. Both configurations consisted of placing the receivers and the transmitters in plane with the 1.42 m side of the slab. Both transmitter and receiver configurations were placed along sections 1 and 2 as shown in Figure 8. The first configuration consisted of sounding the slab with sensors on each side. The second configuration consisted of sounding the slab with sensors placed on only three faces because the bottom face was not accessible.

The first configuration placed along section 1, used 31 transmitter and 15 receiver positions. This generated 350 ray combinations between receivers and transmitters. For the second configuration placed along section 2, 18 transmitters and 10 receivers were used. This configuration allowed for 155 ray combinations. Note that the second configuration provides a less dense ray path field in the bottom of the section. Both transmitter-receiver configurations are shown in Figure 11 along with the ray paths.

The equipment used for these tests consisted of a 10-channel data acquisition system with a 2 MHz sampling frequency, and B & K Detlatron accelerometers with a resonance frequency of 28 kHz. The first sensor was always placed next to the transmission point in order to trigger the data acquisition for the other nine accelerometers. Seismic waves were generated by an air gun with a 2 mm diameter marble impact object.

The tomographic images for sections 1 and 2 are given in Figures 12 and 13, respectively. Figure 12(a) shows the tomographic image of section 1 before grouting, and Figure 12(b) after grouting. As it can be seen in Figure 12(a), the *P*-wave velocity varies from 2,500 to 4,000 m/s. In the center of the slab, a horizontal weak area can be singled out. This confirms the presence of the weak compressive resistance concrete in the middle layer and possibly the influence of crack. Figure 12(b) demonstrates the ability of this technique to experimentally determine the homogeneous nature of the slab after the injection test, thus confirming that the injection has proceeded successfully.

Section 2 tomographic images are shown in Figure 13. As above, Figure 13(a) illustrates the *P*-wave velocities before grouting. Note that in the middle of the section, where velocities are higher than 4,000 m/s, a high strength area is present. It is well known, however, that this area is weaker due to a weaker concrete in the center of the slab and to the presence of a horizontal crack in that region. This discrepancy is attributed to the lower ray path density used on section 2. The same discrepancy is illustrated in Figure 13(b), but to a lesser extent.

Therefore, the configuration of the transmitters and receivers plays an integral part in the evaluation of concrete elements by Sonic Tomography. The density of the ray paths is critical for satisfactory data analysis. It is thought that by increasing the density of the ray paths up to a certain limit, one could eventually generate a more accurate representation of the crack profile.

Conclusions

This paper discussed the information retrieved by nondestructive techniques used to evaluate a grouting test in a concrete slab, which is used to simulate the injection process of cracks in concrete dams. The Impact-Echo method was employed to determine the crack profile, and demonstrated that the induced crack propagated throughout the slab, however the thickness of the slab after injection

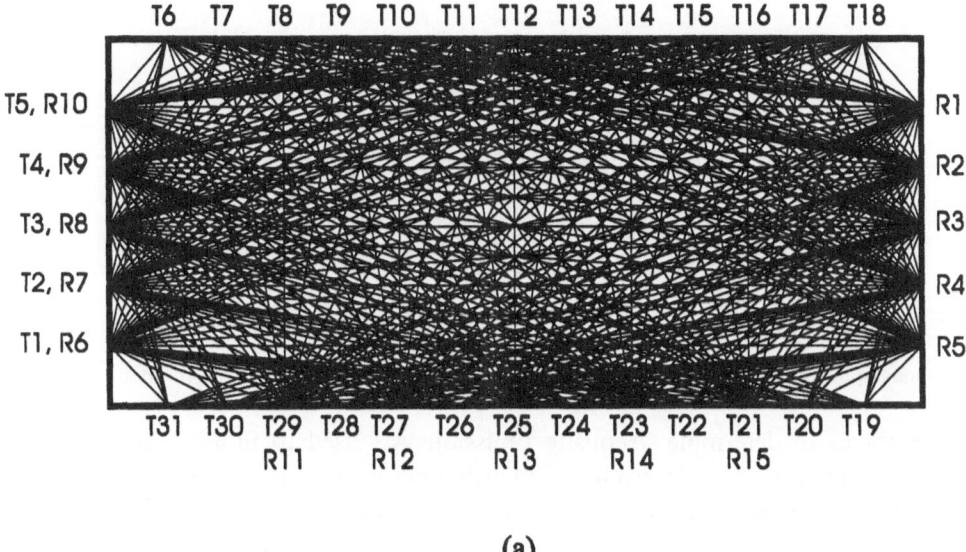

(a)

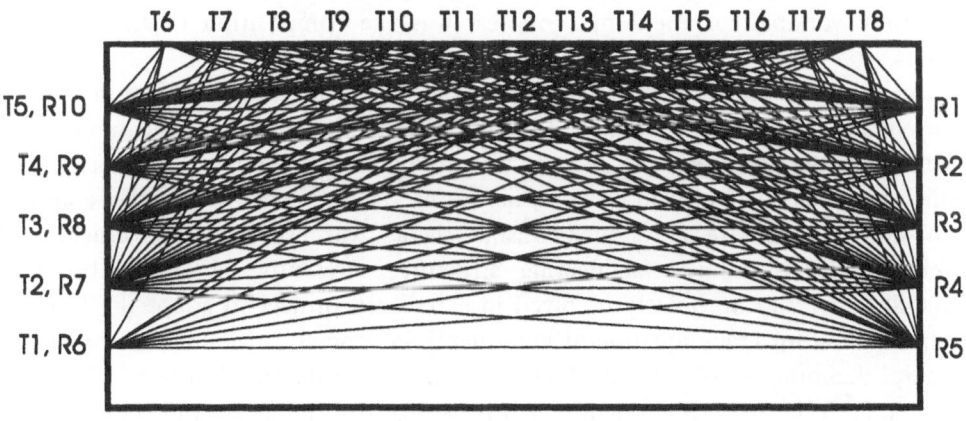

(b)

Figure 11
Ray paths for sonic tomography tests: a) section 1, and b) section 2.

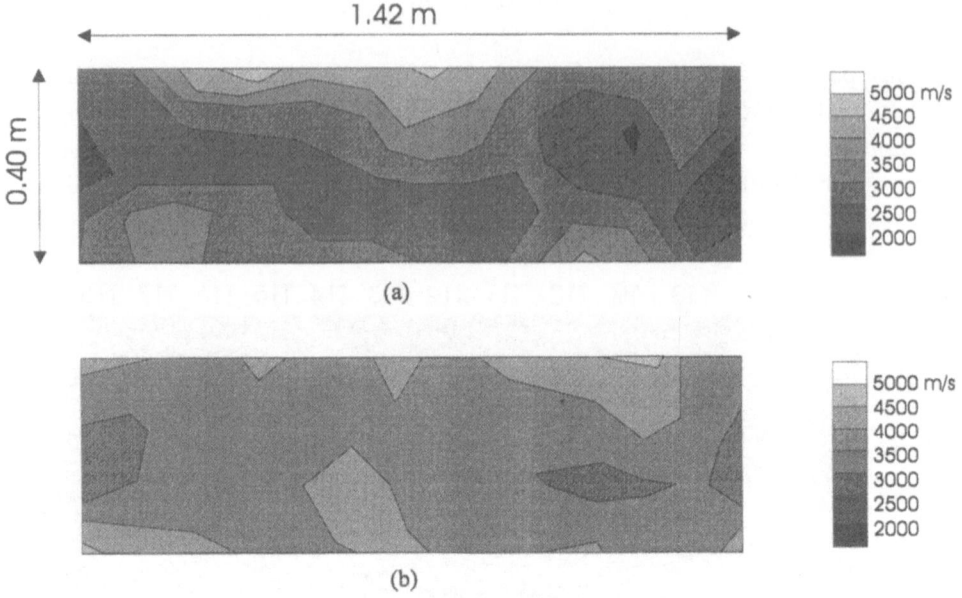

Figure 12
Tomographic image of *P*-wave velocities for section 1: a) before grouting, and b) after grouting.

was difficult to determine. Acoustic Emission was used to monitor the injection process, and helped to determine at which time a change in the injection grout composition was needed. Sonic Tomography is an emerging technique which was used here to provide an internal image of two sections of an injection slab before and after the injection. It is important to bear in mind that not all three methods were applied on the same slab, but to slabs constructed to identical specifications. The only variable for each injection test was the composition of the injection grouts.

In conclusion, the nondestructive techniques based on seismic wave propagation can enhance the results obtained by traditional methods, and at the same time, facilitate better decisions. Moreover, this study outlined a trend in research in which methods developed in certain sciences are being borrowed to solve complex problems in other areas. As such, seismic theory was used to detect cracks and other anomalies in civil engineering structures. Presently, Hydro-Quebec and McGill University are jointly involved in a research project to adapt the Impact-Echo method for crack location at great depth in concrete structures. The Geomechanics Laboratory of McGill University has recently developed a Miniature Seismic Reflection (MSR) system which is used to determine dynamic constants of materials such as Young's Modulus and Poisson's Ratio. Research at Sherbrooke University is presently concentrated on attenuation tomography of sections by the addition of the quality factor distribution (Q) of a medium. In this case, rise time is used instead of travel time of the waves.

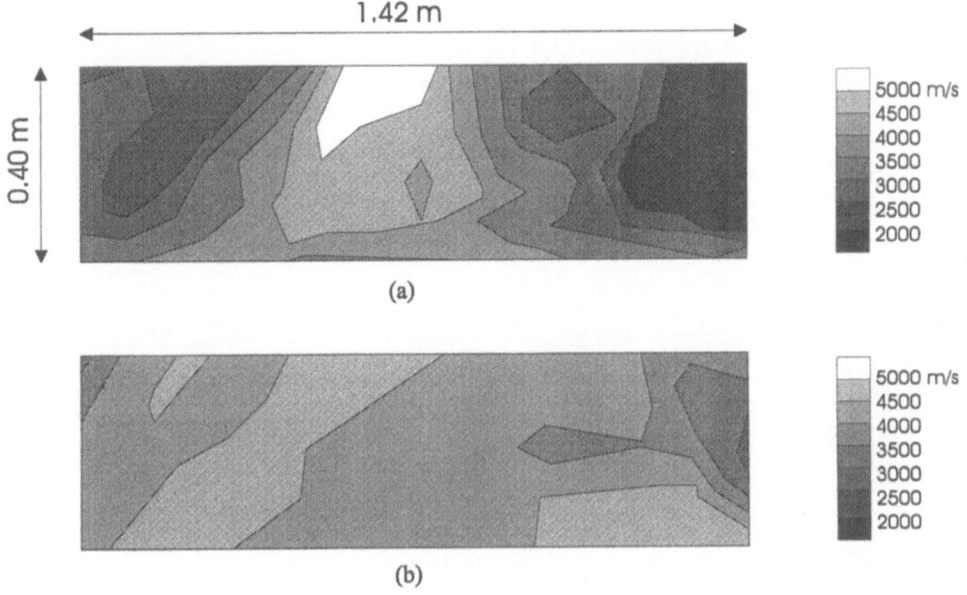

Figure 13
Tomographic image of *P*-wave velocities for section 2: a) before grouting, and b) after grouting.

REFERENCES

CARINO, N. J., and SANSALONE, M. (1988), *Impact-Echo: A new method for inspecting construction materials*. In *NDT & E for Manufacturing and Construction Conference*, p. 15.

BALLIVY, G., SALEH, K., MNIF, T., MANIEZ, J., LANDRY, L. M., and NADEAU, M. (1992), *Rehabilitation of Concrete Dams: Laboratory Simulation of Cracking and Injectibility*, Specialty Conference on Grouting, Soil Improvement, and Geosynthetics, American Society of Civil Engineering, New Orleans.

HOULSBY, A. C., *Construction and Design of Cement Grouting: A Guide to Grouting in Rock Foundations* (John Wiley and Sons, Inc., New York, 1990) pp. 211–215.

KHARRAT, Y., CÔTÉ, P., and BALLIVY, G. (1993), *Essai de mesures tomographiques sur le barrage de Sorin (France)*, University of Sherbrooke, Canada, Report no. GR 93-01-02.

MINDESS, S., *Acoustic Emission Methods*. In *CRC Handbook on Nondestructive Testing of Concrete* (ed. Malhotra, V. M., and Carino, N. J.) (CRC Press, 1991) pp. 317–333.

MOMAYEZ, M., SADRI, A., and HASSANI, F. P., *Impact-Echo: A technique for determining the mechanical properties of rocks*. In *Rock Mechanics* (Balkema, Rotterdam, 1995) pp. 843–848.

RHAZI, J., and BALLIVY, G. (1993a). *Suivi de l'injection au moyen de l'émission acoustique*, Can. Geotechn. J. *30*, 965–973.

RHAZI, J., and BALLIVY, G. (1993b), *Damage Study by Acoustic Emission: The Role of the Transducer*, Canadian Acoustics *21*(4), 9–14.

SALEH, K., TREMBLAY, S., and MNIF, K. (1995), *Injection de la dalle de béton no 30 par le coulis de ciment Portland type 10*, Report # IREQ-95-209, 37 pages.

SANSALONE, M., and CARINO, N. J. (1986), *Impact-Echo: A Method for Flaw Detection in Concrete Using Transient Stress Waves*, National Bureau of Standards, U.S.A., NBSIR 86–3452.

TELFORD, W. M., GELDART, L. P., and SHERIFF, R. E. *Applied Geophysics* (Cambridge University Press, 2nd Edition, 1990).

WEAVER, K., *Dam Foundation Grouting* (American Society of Civil Engineers, 1991).

(Received September 3, 1996, accepted January 24, 1997)

Pure appl. geophys. 150 (1997) 627–646
0033–4553/97/040627–20 $ 1.50 + 0.20/0

⌐ Pure and Applied Geophysics

Effect of Injected Water on Hydraulic Fracturing Deduced from Acoustic Emission Monitoring

T. Ishida,[1] Q. Chen[1] and Y. Mizuta[1]

Abstract—In order to investigate the effects of injected water in hydraulic fracturing, experiments were conducted on cubic granite specimens, comparing fracturings induced by conventional water injection with those induced by pressurization of a urethane sleeve, thereby realizing "hydraulic fracturing" without the use of fracturing fluid. In both experiments, a shear type mechanism was found to be dominant in fault plane solutions of AE events. However, in the case of water injection, cracks extended rapidly with large drops in hole water pressure and bursts of AE, whereas in pressurization by the urethane sleeve, cracks extended stepwise with no such large drops in hole pressure and no bursts of AE. The difference in crack extension in the two experiments can be analyzed by comparing relations between crack length and stress intensity factor of mode I at a crack tip. The observation and analysis indicate that existence of fracturing fluid like water helps initiated cracks to extend rapidly and widely in hydraulic fracturing in actual HDR fields.

Key words: Hydraulic fracturing, acoustic emission, fault plane solution, fracturing mechanism.

1. Introduction

In order to extract geothermal energy from a HDR (hot dry rock) mass, a water injection well must be connected to a production well through cracks created by hydraulic fracturing. Therefore, clarifying the mechanism of hydraulic fracturing is important for constructing efficient HDR geothermal systems. HUBBERT and WILLIS (1957) theoretically showed that these cracks are created by tensile fracture and extend along the direction of maximum compressive stress, if a well is bored in one of the directions of principal stress and the surrounding rock is isotropic, homogeneous and impermeable. Based on this theory, methods of measuring stress *in situ* have been developed and applied to problems in mining engineering, civil engineering and geophysics (for examples, see HAIMSON, 1978 and MIZUTA *et al.*, 1987). However, most of the fault plane solutions of AE (Acoustic Emission) events recorded in the fields of hydraulic fracturing, ranging from small scale at 10 m depth to large scale at 4,000 m depth, have been classified as quadrant type, thereby suggesting shear fracture (BARIA and GREEN, 1986; SASAKI, 1995; TALEBI and

---

[1] Department of Civil Engineering, Yamaguchi University, Tokiwadai, Ube, 755 Japan.

CORNET, 1987). The conflict between tensile fracture suggested by theory and shear fracture deduced from recorded wave forms has not yet been solved.

On the other hand, the authors conducted a hydraulic fracturing experiment on cubic granite specimens, to clarify the viscous effect of fracturing fluid (ISHIDA *et al.*, 1997). In this experiment, the viscosity was varied by 1 cP water and 80 cP oil. Fault plane solutions of recorded AE events indicated the dominance of shear fracture in the water injection, while that of tensile fracture in the oil injection. In addition to this, viscous oil injection tended to create thick and plane cracks with few branches, while water injection tended to generate thin and wavelike cracks with many branches of a network. Also in an actual HDR project, viscous fluid was used to create fractures (BARIA and GREEN, 1986; BARIA *et al.*, 1989). Therefore, the experimental results suggested that changing the viscosity of fracturing fluid can control not only the fracture mechanism but also the fracture patterns even in actual HDR fields.

Not only to solve conflict between tensile fracture suggested by theory and shear fracture deduced from recorded wave forms but also to control fracture patterns by changing the viscosity of fracturing fluid, it is important to clarify the effect of fluid existence itself on hydraulic fracturing. For this purpose, in this study, the authors conducted experiments to compare fractures induced by conventional water injection with those induced by pressurization of a urethane sleeve, thereby realizing "hydraulic fracturing" without the use of fracturing fluid.

2. *Experimental Method*

2.1 *Specimen*

A 19-cm cubic block of Kurokami-jima granite with a 2-cm diameter hole, as shown in Figure 1, was employed as a specimen. In both experiments, the rift plane of granite was oriented so as to correspond to the YZ plane in the Cartesian coordinate system. P-wave velocities along the X, Y and Z directions of a specimen for conventional water injection were 4.93, 5.14 and 5.20 km/s, respectively, whereas, those of a specimen for pressurization by a urethane sleeve were 4.98, 5.16 and 5.27 km/s. Two horizontal confining pressures, 3 MPa in the X direction and 6 MPa in the Y direction, were applied to the specimen through flat jacks, using a loading system as shown in Figure 2. Under the confining pressures, from the solution by KIRSCH (1898), as shown in Figure 3(a), the tangential compressive stress on the hole wall has a maximum value of 15 MPa at the two points, A_1 and A_2, whose X coordinates represent the maximum and the minimum, and the minimum value of 3 MPa arises at the two points, B_1 and B_2, whose Y coordinates represent the maximum and the minimum. Therefore, without hole pressure, tangential compressive stress exceeds 3 MPa at all points on the hole wall. How-

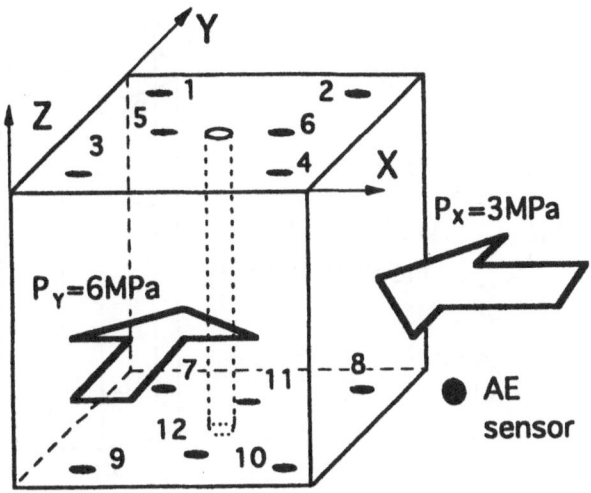

Figure 1
Specimen, loading condition, setting positions of AE sensors, and coordinate system.

ever, once the hole pressure exceeds 3 MPa, the maximum tensile stress should arise at the two points on the XY plane; in other words, along the two lines, B_1B_1' and B_2B_2' in Figure 3(b), on the hole wall whose Y coordinates represent the maximum and the minimum. In addition to this, the rift plane of the granite was oriented parallel to the YZ plane. Therefore, it should be easiest to initiate cracks at the two lines, B_1B_1' and B_2B_2', in the Y direction on the hole wall by tensile fracture, and the cracks should extend along the YZ plane.

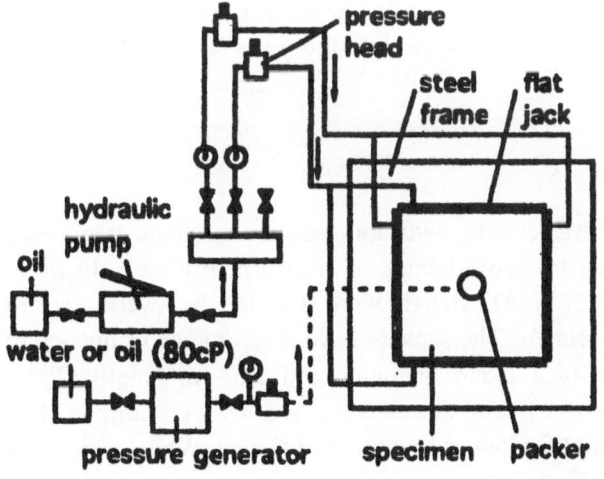

Figure 2
Diagram of loading system.

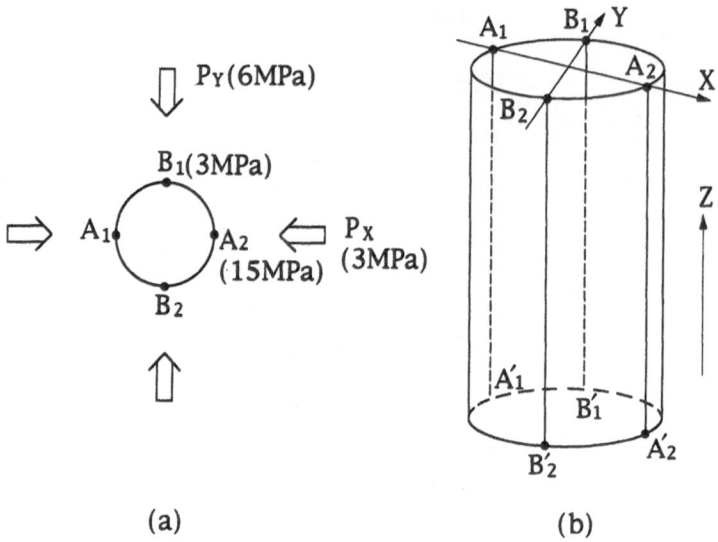

(a)   (b)

Figure 3
Stress condition on the borehole wall on the XY plane, (a), and on a bird's-eye view, (b).

2.2 Water Injection and Pressurization by a Urethane Sleeve

A mechanical type double packer shown in Figure 4(a) was used to inject water at a flow rate of 10 cm³ per minute. This packer applies water pressure on a 6.0 cm section of the hole wall. In contrast, to realize "hydraulic fracturing" without the use of fracturing fluid, a urethane sleeve shown in Figure 4(b) was inserted into the hole and inflated by oil pressure. A-10 cm section of the inflated urethane sleeve applied uniform pressure to a wall of the hole.

Water injected into the hole was pressurized by a screw pump and oil inside the urethane sleeve was pressurized by a hand pump, while a pressure meter was monitored. The pressures at two-second intervals were recorded in a personal computer.

2.3 AE Monitoring System

To monitor AE events, two independent systems were employed, one for recording and one for reproducing, as shown in Figure 5. In the recording system shown in Figure 5(a), AE signals were detected by twelve sensors placed on the surface of the specimen. The sensors have a resonance frequency of 150 kHz and a cylindrical shape of 1.74 cm diameter and 1.63 cm length (Type R-15, Physical Acoustic Corporation). The detected AE signals were amplified by 70 dB in total through a pre-amplifier and a signal conditioner. When a signal from one among the three sensors Nos. 1, 6, and 10 shown in Figure 1 exceeded a threshold level, a wave memory was triggered. For each channel, the wave memory has a record length of 2 kwords and 8 bit resolution in amplitude. In the wave memory, after the signal was digitized using a sampling interval of 0.2 μs, and analog signal was produced again using a sampling interval of 5 μs. The analog signals produced in the wave memory were recorded on tape in a FM recorder. This procedure was employed in order to avoid dead time needed for directly storing digital data and to capture as many AE events as possible in burst occurrences. Through this procedure, the time axis of the signal wave forms was enlarged 25 ($= 5\,\mu$s$/0.2\,\mu$s) times, and a 150 kHz wave of the sensor resonance frequency was converted to 6 kHz so that it could be recorded on the tape, as the frequency band of the recorder was from DC to 20 kHz at its highest tape speed of 76 cm/s.

After the experiments, the signals were reproduced from the tape, using the system shown in Figure 5(b). Tape speed of the reproducing system was 1.2 cm/s, whereas that of the recording system was 76 cm/s. The time axis of the wave form was thereby further enlarged, by 64 ($= 76/1.2$) times, in addition to the 25 times enlargement during the course of the A/D and D/A procedure performed in the wave memory. Therefore, the reproduced signals were 1,600 ($= 25 \times 64$) times slower than the real wave forms. The reproduced signals were digitized in a wave memory, using a sampling interval of 0.2 ms, and stored in floppy disks through a GP-IB cable. A signal conditioner and a wave memory for the reproducing system have low frequency characteristics, and they are different from those used in the recording system.

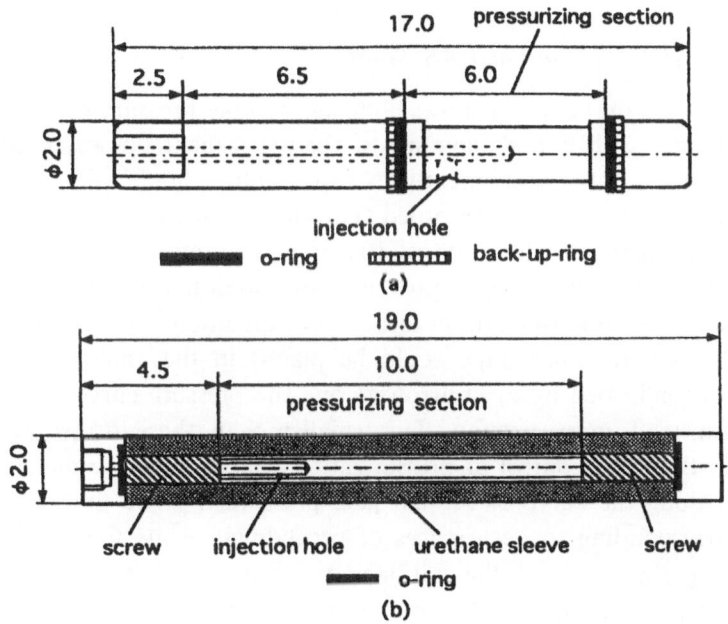

Figure 4
Packer and urethane sleeve inserted into the boreholes, (unit: cm) (a) Packer to inject water. (b) A urethane sleeve to realize "hydraulic fracturing" without fracturing fluid.

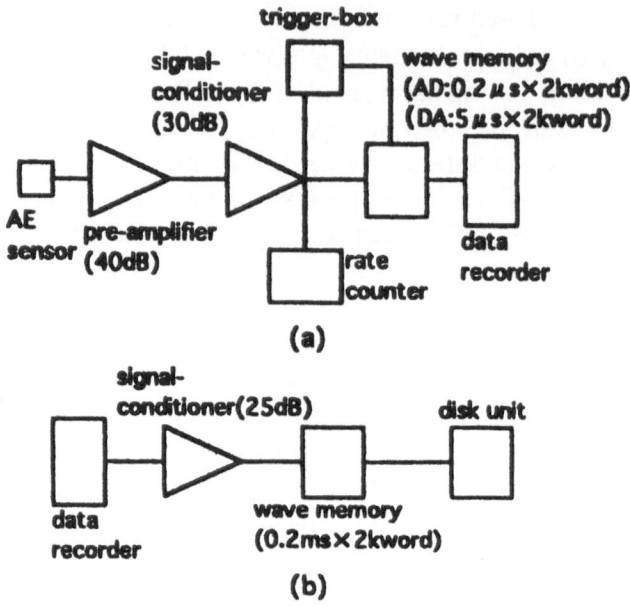

Figure 5
Diagram of AE monitoring system. (a) Recording system. (b) Reproducing system.

3. Results

3.1 Change of Hole Pressure and AE Rates

Each of Figure 6(a) and (b) shows a change of hole pressure with a thick solid line and AE rate with a bar diagram. In the experiment on water injection, as shown in Figure 6(a), the experiment was stopped after three rapid pressure drops, indicated by solid arrows, were recorded. Through the use of a screw pump, water was injected by hand at a low rate of 10 cm^3 per minute. Since the screw pump contained only a small water tank, pressurization often had to be stopped so as to supply water to the tank. In addition to this, pressurization had to be stopped every 330 seconds so that a new tape could be placed in the analog data recorder. Pressure drops indicated by open arrows along the pressure curve were caused by these stoppages of pressurization. Jags smaller than these drops were due to variations in the hand control of pressurization while a pressure meter was being monitored. Along the curve of the flat jack pressure, P_X, we find small pressure increases corresponding to occurrences of breakdown, while there is no pressure increase along the curve of the flat jack pressure, P_Y. Upon inducing cracks extending in the Y-direction and seeping into apertures of the induced cracks, the injected water would by expected to open the cracks with displacements. This would expand the specimen in the X-direction and the specimen surfaces would push back on the flat jacks.

A pressure curve in Figure 6(b) is shown with a lateral axis twice as large and a longitudinal axis three times as large as those in Figure 6(a). Jags along this curve were most likely due to variations in the hand control of pressurization and stepwise crack extensions in the specimen. The curve indicates oil pressure applied

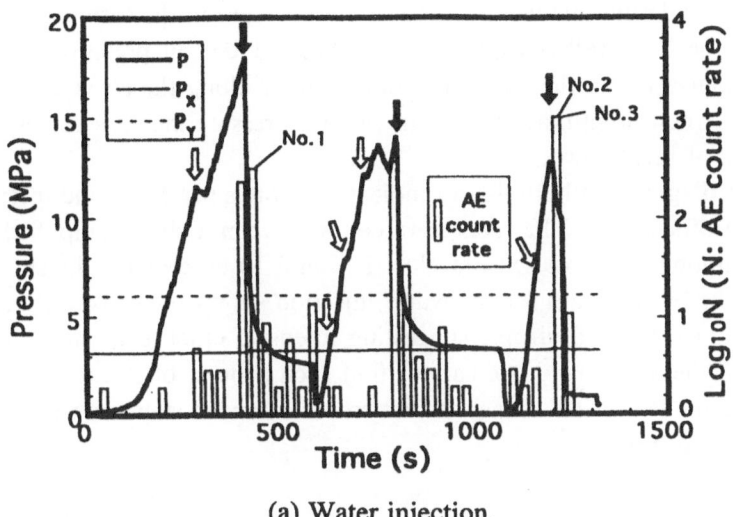

(a) Water injection.

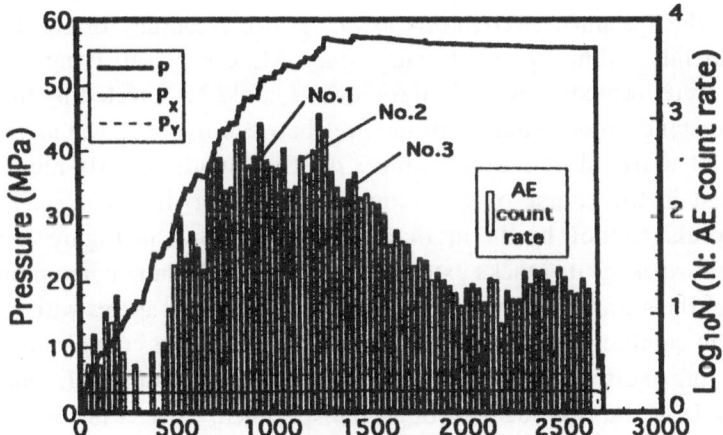

(b) Pressurization by the urethane sleeve.

Figure 6
Hole pressure and AE rate vs. elapsed time. Numbers 1, 2 and 3 in each figure indicate recorded time of AE events whose fault plane solutions are shown in Figures 10 and 11. (a) Water injection. (b) Pressurization by the urethane sleeve.

inside the urethane sleeve, which has a stiffness. CROUCH and STARFIELD (1983) obtained a formula for the stress field in an annulus described by the relation $a \leqq r \leqq b$ and having Poisson's ratio v_1 and the modulus of rigidity G_1 inside a circular hole of radius $r = b$ in a large plate with v_2 and G_2. Since Young's modulus of the urethane sleeve is 96 MPa and this is much smaller than those of the granite specimens, we can assume that $G_1/G_2 = 0$. Substituting 0.47 for v_1, 5.5 and 10 mm for the inner and outer diameters of the urethane sleeve in the formulation, pressure applied to the hole wall is expected to be 88% of the oil pressure shown in the figure. In contrast with the water injection, neither along the curve of the flat jack pressure, P_X, nor along that of P_Y, can we find pressure increase corresponding to occurrences of breakdown.

The bar diagrams of both figures indicate AE rates per 30 second interval, using a semilogarithmic scale. The rates were counted when a signal amplified by 70 dB from one among the three sensors Nos. 1, 6 and 10 exceeded a threshold level of 1.2 volts. The bar diagram for the water injection shows bursts of AE corresponding to the three pressure drops, suggesting intensive crack extension within short periods. On the other hand, we cannot find such a burst of AE for pressurization by the urethane sleeve. This suggests that cracks extended stepwise with increases of the applied pressure.

3.2 Located AE Sources and Observed Visible Cracks

AE sources were located while taking into account anisotropy of P-wave velocity in the specimen (ROTHMAN *et al.*, 1974). Accuracy of the location was expected to fall within 10 mm, because only AE events satisfying the following conditions were located; P-wave initial motions could be detected by more than six sensors and standard deviation of differences between the observed arrival time and back-calculated arrival time were within 3 μs. One hundred thirty-nine AE sources were located in the course of water injection and two hundred sixty-two in the course of pressurization by the urethane sleeve, as shown in Figure 7. We can see that in both experiments cracks extended along the YZ plane corresponding to the rift plane of the granite specimen. The crack direction also agrees with the direction of maximum compressive stress expected, based on the stress condition on the hole.

Comparing the two figures, in the case of water injection, AE sources spread from the hole to the two edges through the specimen, as seen in the projection on XY plane. On the other hand, in the case of pressurization by the urethane sleeve, the AE sources did not spread to the edges but were distributed only in the vicinity of the hole.

Figure 8 shows visible cracks observed on surfaces of the specimens and on surfaces of cores bored around the hole for the purpose of inspecting cracks extending through the specimens. We can see that the visible cracks correspond to the located AE sources.

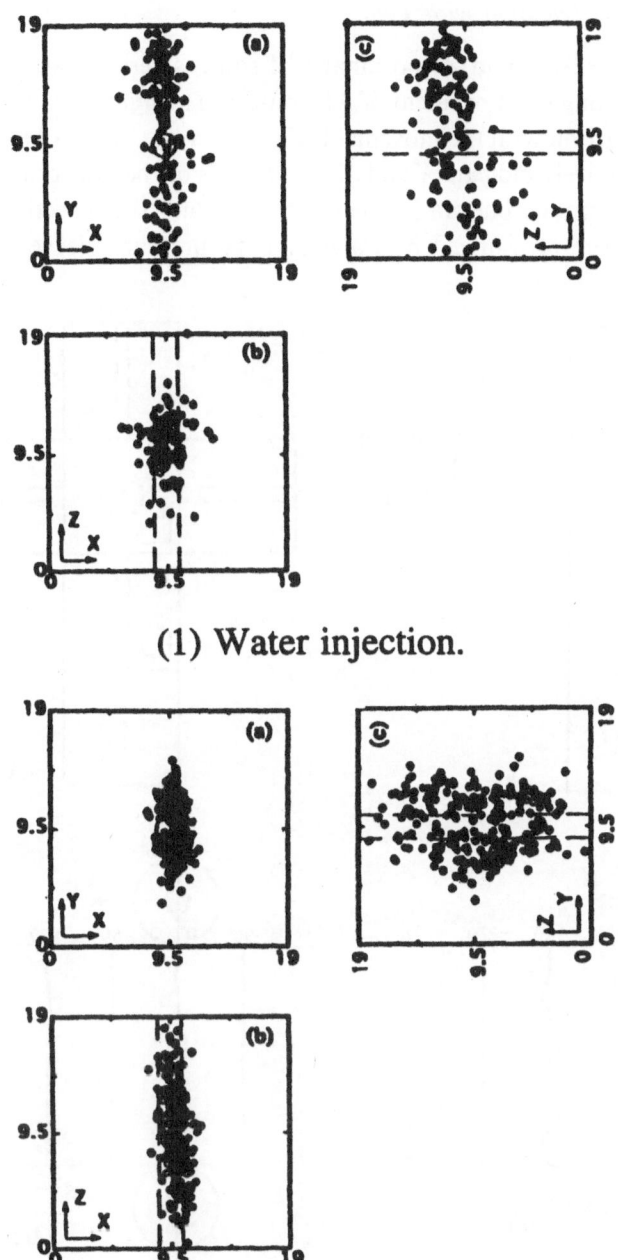

(1) Water injection.

(2) Pressurization by the urethane sleeve.

Figure 7
Located AE sources projected on *XY*, *YZ* and *ZX* planes. Broken lines indicate positions of the borehole. (1) Water injection. (2) Pressurization by the urethane sleeve.

3.3 *Change of AE Location with Time*

Figure 9 shows X, Y and Z coordinates of sources versus elapsed time, so as to facilitate determining the movement of AE sources. The figure also provides pressure curves for comparison with the movement of AE sources. The broken lines in figures (a) and (b) respectively indicate X and Y coordinates of the hole center. The sections between the two broken lines in figures (c) correspond to the sections in which the borehole walls were pressurized by water and the urethane sleeve, respectively.

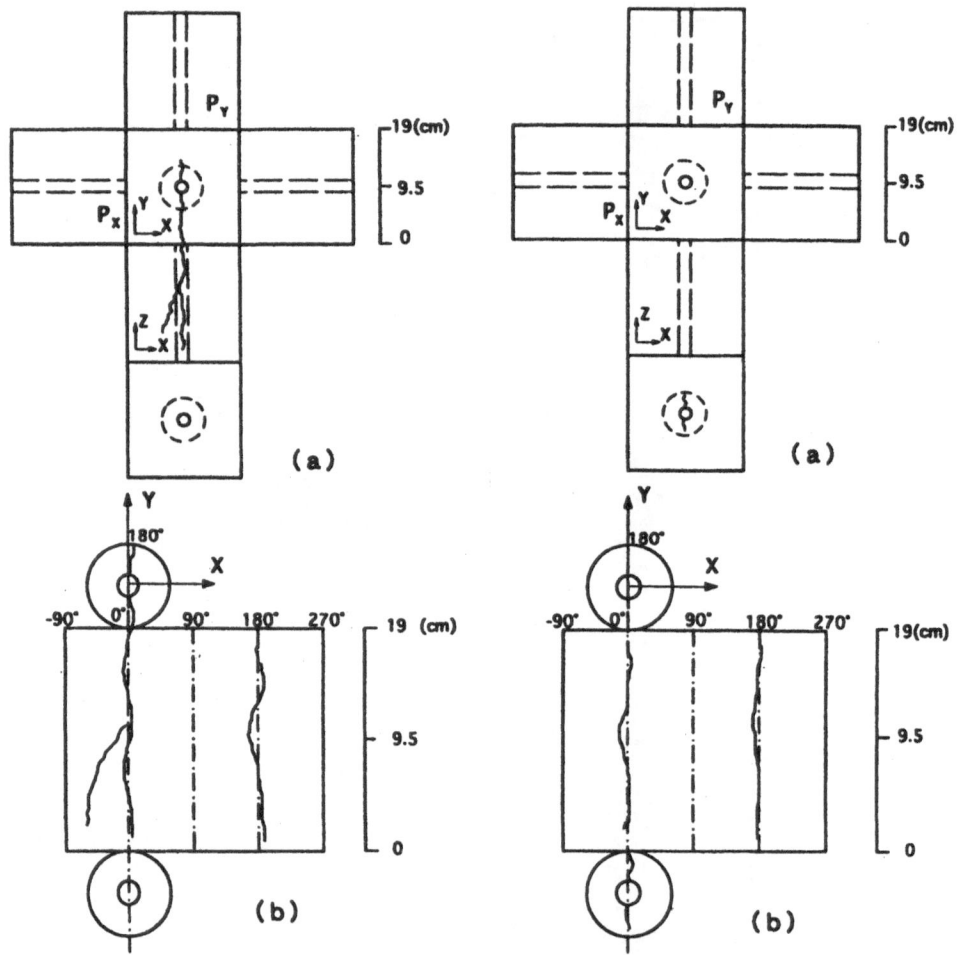

(1) **Water injection.**   (2) **Pressurization by the urethane sleeve.**

Figure 8

Sketch of visible cracks shown on unfolded planes, observed after the experiments. Upper figures, (a), show cracks observed on the surfaces of the specimens. Broken-line circles indicate the positions of cores bored for inspection of cracks in the specimens. Lower figures, (b), show cracks observed on the surfaces of the bored cores. (1) Water injection. (2) Pressurization by the urethane sleeve.

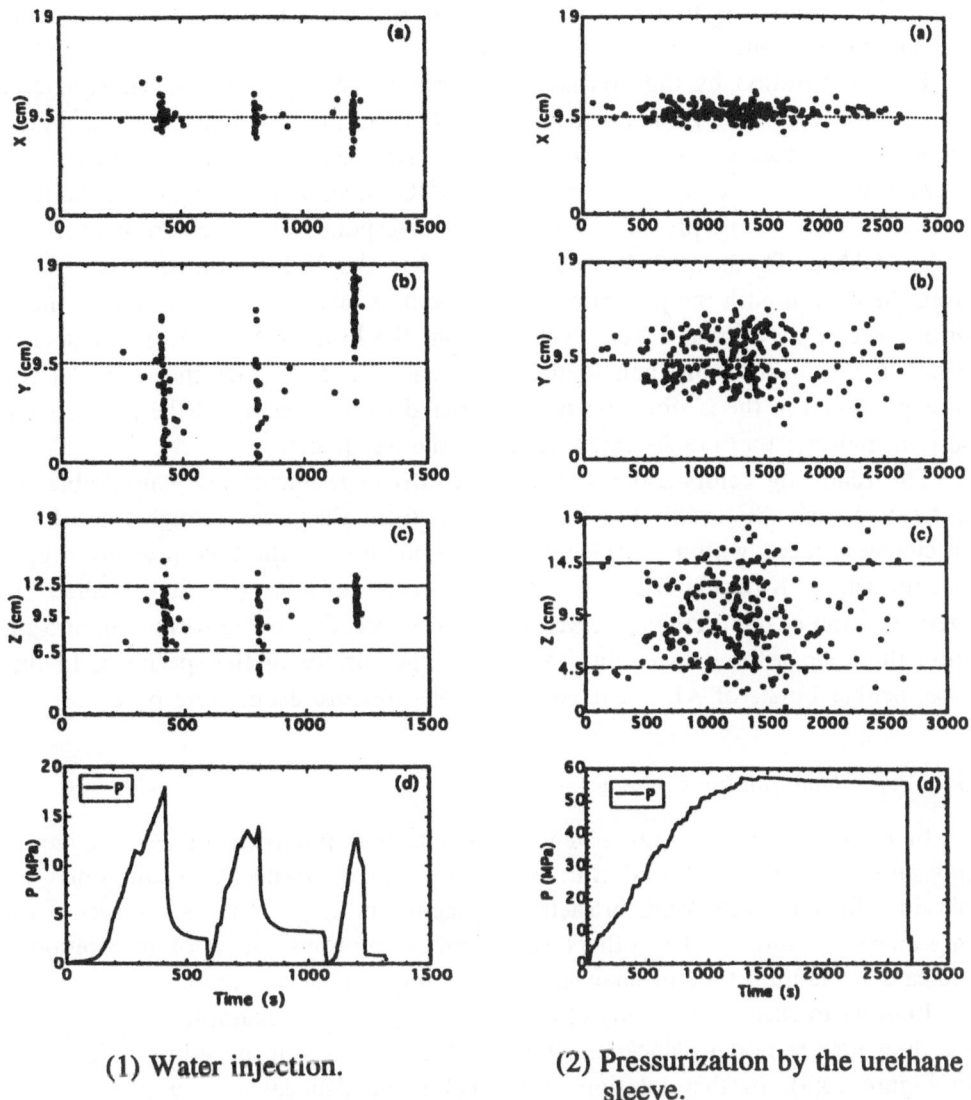

(1) Water injection.

(2) Pressurization by the urethane sleeve.

Figure 9

Time-space distributions of AE hypocenters. In each experiment, three figures (a), (b) and (c) respectively show movements of AE hypocenters in x, y and z directions with respect to elapsed time. The figures (d) show pressure vs. elapsed time, which are the same as in Figure 6, for facilitating comparison with the movements of AE hypocenters. (1) Water injection. (2) Pressurization by the urethane sleeve.

In the water injection displayed in Figure 9 (1), we can easily find that located AE sources clustered with the three pressure drops. The located AE sources moved away from the hole through one of the specimen surfaces, in the minus Y direction with the first and second pressure drops, and in the plus Y direction with the third drop. In contrast to this, movement in the X direction was restricted to the vicinity of the hole, and that in the Z direction was restricted to the proximity of the pressurizing section.

In pressurization by the urethane sleeve shown in Figure 9 (2), whose lateral axes are all twice as large as those in Figure 9 (1), located sources were distributed along the time axes with no burst occurrence of AE. Many AE sources were located between 600 and 1400 seconds. The time of 1,400 seconds corresponds to the time to stop to increase the pressure. This period corresponds to a pressure between 35 and 57.5 MPa. The located AE sources moved in the Y direction to around 5 cm from the hole in both the plus and minus directions, whereas those moved in the X direction to only around 1.5 cm. The fact that the located AE sources are spread wider in the Y direction than in the X direction, and the fact that most of the sources spread in the Z direction were restricted to the vicinity of the pressurizing section, indicate that cracks extended along the YZ plane.

The following comparisons between the two experiments are remarkable: In water injection, AE events burst and spread from the hole through one of the specimen surfaces, within short periods corresponding to the hole pressure drops. In contrast, in "hydraulic fracturing" without the use of fracturing fluid, realized by pressurization of the urethane sleeve, AE sources spread stepwise within about 5 cm from the hole, without extending through to the surface of the specimen. In this case, neither bursts of AE occurrence nor hole pressure drops were observed.

3.4 Fault Plane Solutions of AE Events

In each experiment, only seven AE events had polarities of P-wave initial motion that were easily found at more than six sensors. Fault plane solutions were obtained for the seven events in each experiment, and three of the seven are shown in Figures 10 and 11. The others yield similar solutions. Most of the solutions indicate a quadrant-type mechanism, thereby implying shear fracture.

In order to check the validity of the solution, as the first example, let us examine the directions of nodal planes and those of P and T axes in the solution displayed in Figure 10(a). In this solution, a vertical plane delineated as a line segment labelled "A" and a horizontal plane labelled "B" could be chosen as the two nodal planes. The alternative of the two corresponds to a real fracture plane causing the AE event. Although the macroscopic crack extended along a vertical YZ plane, it is actually composed of many small jagged planes. Upon consideration that a real fracture plane causing the AE event should be one of these planes, the vertical plane, A, forming a 38 degree angle in the plus Y direction, seems to be suitable for the real fracture plane.

The positions of the letters P and T indicate the directions of the P and T axes, respectively. In this experiment, a pressure of 6 MPa was applied in the Y direction, that of 3 MPa in the X direction and no pressure in the Z direction. If a size of the specimen was very large in comparison with the borehole diameter and a rift plane

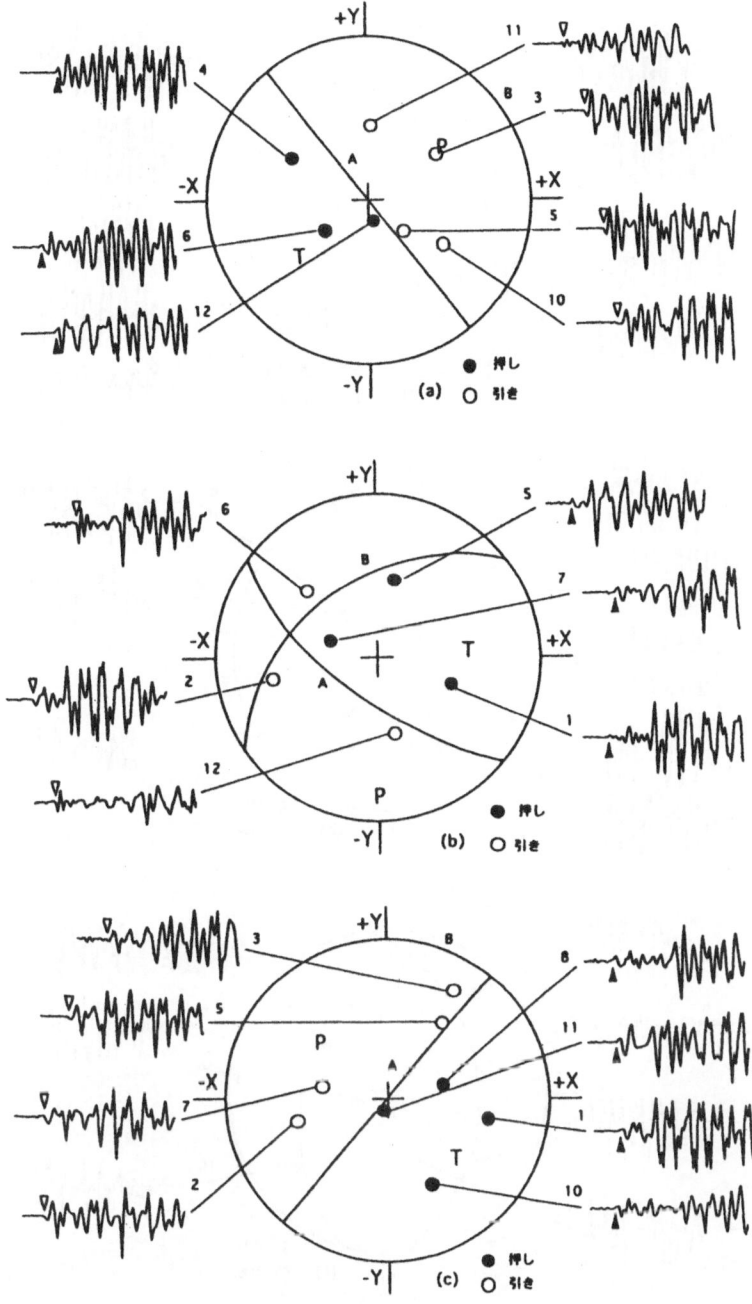

Figure 10

Fault-plane solutions of AE events associated with the water injection, projected onto a lower hemisphere of a Schmidt net. A solid circle indicates compression in P-wave initial motion, whereas an open circle indicates dilatation. (a) No.1 event, $(X, Y, Z; T) = (10.33, 4.27, 10.99; 424.92)$. (b) No. 2 event, $(X, Y, Z; T) = (9.16, 15.58, 11.44; 1207.92)$. (c) No. 3 event, $(X, Y, Z; T) = (10.00, 15.46, 12.43; 1209.04)$. The X, Y and Z coordinates indicate the location of an AE source (unit: cm), and T indicates recorded elapsed time (unit: s).

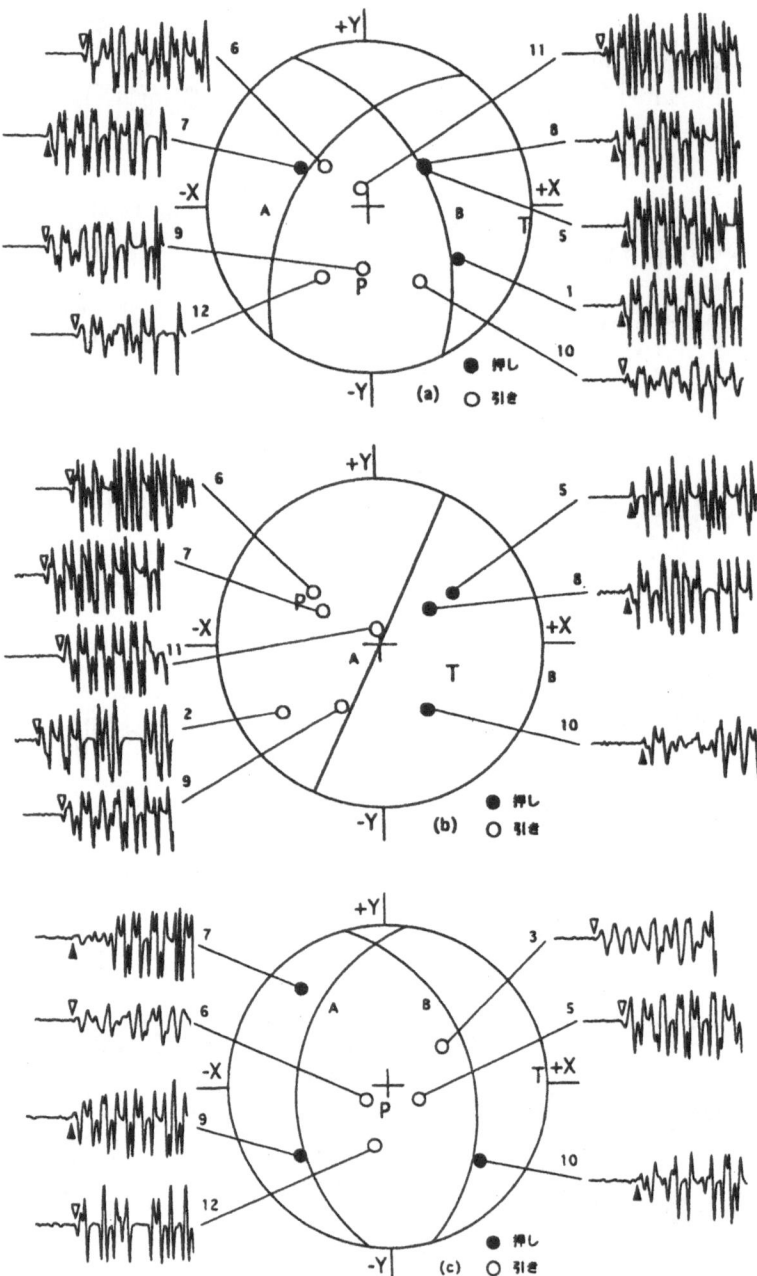

Figure 11

Fault-plane solutions of AE events associated with pressurization by the urethane sleeve, projected on a lower hemisphere of a Schmidt net. A solid circle indicates compression in P-wave initial motion, whereas an open circle indicates dilatation. (a) No. 1 event, $(X, Y, Z; T) = (9.75, 12.22, 11.68; 890.37)$. (b) No. 2 event, $(X, Y, Z; T) = (9.53, 12.23, 14.39; 1480.03)$. (c) No. 3 event, $(X, Y, Z; T) = (9.73, 7.40, 4.51; 1169.50)$. The X, Y and Z coordinates indicate the location of and AE source (unit: cm), and T indicates recorded elapsed time (unit: s).

did not affect crack extension, a crack initiating along a vertical YZ plane in the vicinity of the borehole wall should bend at a distance from the borehole and extend along a horizontal XY plane normal to the minimum principal stress, Pz, being zero. However, since the specimen size was limited and the lift plane was seated in the vertical YZ plane, a crack extended along the YZ plane to the lateral surfaces of the specimen without bending. Therefore, directions of the P and T axes of the solution do not coincide with the maximum and minimum compressive directions of the applied stress. From this point of view, the T axis seems to more reasonably correspond to the X direction rather than to the Z direction. Additionally, considering that the real fracture plane corresponds to one of the jagged planes composing the macroscopic fracture plane, differences in the directions between the P axis and the Y axis and between the T axis and the X axis should exist. Through these considerations, the directions of the P and T axes shown in Figure 10(a) seem to be reasonable.

Although, from the theory, it is expected to find very weak initial motions near the nodal planes, with stronger first motions away from these planes, wave forms attached in the solution may run counter to the expectation. While a method of sensitivity calibration for AE sensors attached to a much smaller specimen (48 × 48 × 90 mm) was found (SHAH and LABUZ, 1995), that for a larger specimen has not yet been found. In other words, the amplitude of each wave form cannot be compared to that of another wave form with sufficient accuracy. For this reason and to show polarities of the first motions as clearly as possible, amplitudes of wave forms attached in the solution were normalized with the maximum amplitude of each wave form. Therefore, the amplitude of each wave form is independent and first motion amplitudes counter to the expectation could exist.

In the solution shown in Figure 10(b), plane "A" seems to correspond to a real shear plane, because its dip is closer to vertical than that of plane "B" although the angles between their strikes and the Y direction are similar. The P axis is close to the Y direction, in addition to this, the T axis is not distant from the X direction although it is at an angle of 45 degrees from the horizontal direction. Based on these observations, the solution seems to be reasonable as the mechanism of microfractures composing the macroscopic fracture.

Through discussion similar to that for Figure 10(a), we can rationalize the solution shown in Figure 10(c) as well.

Fault plane solutions obtained for the AE events in pressurization by the urethane sleeve show a quadrant-type mechanism just as do those in the water injection. In the solution shown in Figure 11(a), since both nodal planes are nearly parallel to a YZ plane along which a macroscopic fracture was generated, either can correspond to a real fracture plane. Although pressurization by the urethane sleeve made cracks only near the borehole wall, the cracks extended along a vertical YZ plane without bending toward a horizontal XY plane normal to the minimum principal stress, Pz, being zero. Therefore, as well as in the water injection, directions of the P and T axes of the solution do not coincide with the maximum and minimum compressive directions of the applied stress, and the T axis seems to more reasonably correspond to the X direction rather than to the Z direction. From this point of view, directions of the P and T axes of the solution seem to be reasonable.

In the solution shown in Figure 11(b), the nodal plane A is a vertical plane forming a 24-degree angle in the plus Y direction, and the nodal plane B is a horizontal plane. Upon comparing it to the direction of the macroscopic fracture, and upon consideration that the AE event was accompanied by one of the jagged microfractures composing the macroscopic fracture plane, the nodal plane A should correspond to the real fracture plane. However, the directions of P and T axes may be slightly different from those expected from the consideration mentioned. Although the difference may suggest the fracture includes some components beside shear, it appears certain that shear is dominant in the fracture.

The solution in Figure 11(c) is similar to that in Figure 11(a), and either of the nodal planes can correspond to a real fracture plane. Although the direction of the T axis is consistent, the direction of the P axis is inconsistent. The difference may suggest the fracture includes some components beside shear. Consequently, it appears certain that shear is dominant in the fracture, as well as in the solution of Figure 11(b).

Based on the fault plane solutions, we can conclude that, in both the water injection and pressurization by the urethane sleeve, all AE events, or at least all large AE events whose polarities of P-wave motions could be easily observed at more than six sensors, were associated with fracture in which shear mode is dominant.

4. Discussion

Comparison of the results of the two experiments leads to the following two conclusions. First, cracks extend rapidly in fracturing induced by injection of water, while cracks extend stepwise when no fluid is injected as observed in pressurization by the urethane sleeve. Second, in both water injection and pressurization by the urethane sleeve, the shear type mechanism appears to be dominant in fracturing.

The first conclusion can be analyzed by comparing the stress intensity factor at a crack tip (ZOBACK and POLLARD, 1978), as shown in Figure 12. The figure on the left depicts the case in which hydrostatic pressure, σ, is applied throughout an ellipsoidal crack surface. In this case, $K_1 = \sigma(\pi l)^{1/2}$, where K_1 is the stress intensity factor of mode I at a crack tip and l is half of the crack length. According to this relation, the stress intensity factor at a crack tip increases with crack length; in other words, a crack never stops once it starts to extend. The figure on the right depicts the case in which a pair of point loads, F, is applied at the center of an ellipsoidal crack. In this case, $K_1 = F(\pi l)^{-1/2}$, which indicates that the stress intensity factor at a crack tip decreases with crack length; that is, a crack never extends without an additional increase in applied loads, even after the crack starts to extend. Needless to say, the case depicted in the left figure corresponds to fracturing with fracturing fluid such as water, whereas the case depicted in the right figure corresponds to fracturing with pressurization by the urethane sleeve. Additionally, the presence of water may promote crack extension, because water can induce a chemical reaction in a crack tip in rock, a phenomenon known as stress corrosion. Therefore, existence of fracturing fluid like water should help initiated cracks to extend rapidly and widely, also in hydraulic fracturing in actual HDR fields. Another experiment by the authors (1997) suggested that more viscous fracturing fluid like oil makes thicker and planar cracks with fewer branches than water. Further investigation on the effects of fracturing fluid would make it possible to control the fracture patterns in actual HDR fields.

Regarding the second conclusion, shear fracture induced by water injection may be explained by using a model for volcanic earthquake swarms. Although many volcanic seismic events characterized by magma intrusions or eruptions show compression in P-wave initial motions on seismographs of all stations, other events display standard quadrant-type distribution of P-wave initial motions, thereby

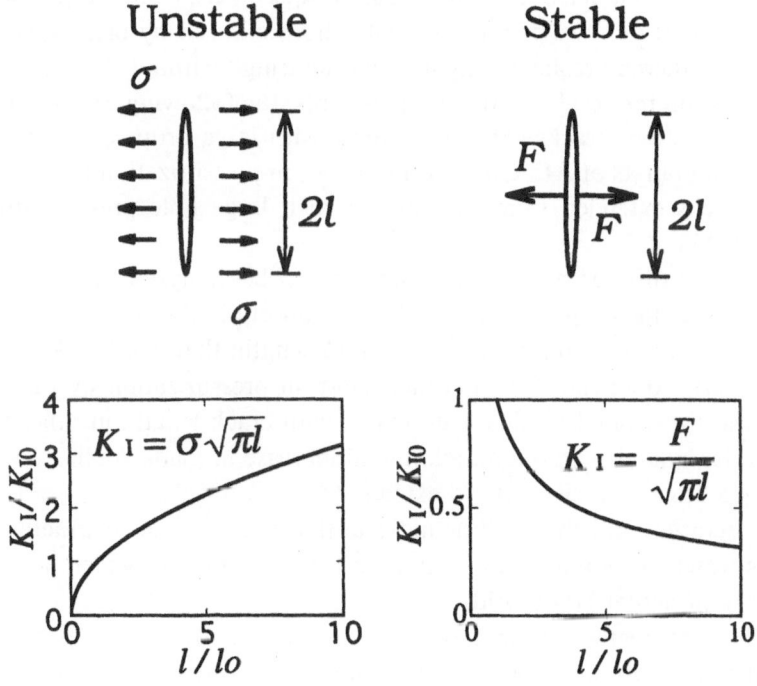

Figure 12
Stable and unstable crack growth from a viewpoint of the stress intensity factor of mode I at a crack tip.

implying shear fracture. HILL (1977) suggested, based on a model wherein preexisting weak planes lie along the direction of maximum stress, that shear fracture should occur with fracture connecting between the tips of the two neighboring weak planes, in addition to tensile fracture accompanying magma intrusion into the weak planes. Such a type of shear fracture is believed to have been dominant in the experiment involving water injection, because the specimen had many defects along the rift plane in the direction of maximum compressive stress. On the other hand, in pressurization by the urethane sleeve, injection of a fluid such as water or magma never affected the fracturing. Based on this observation, the fracturing mechanism in pressurization by the urethane sleeve seems to correspond to usual seismic events, because these events are caused by compressive stresses without the effects of fluid injection and exhibit a normal quadrant-type distribution of P-wave initial motions, thereby implying shear fracture. Understanding the fracturing mechanism of injected fluid like this would help to realize fracture control with the use of various fracturing fluid in field operations of hydraulic fracturing.

5. Conclusions

In order to investigate the effects of injected water in hydraulic fracturing, experiments were conducted on cubic granite specimens, comparing fracturings induced by conventional water injection with those induced by pressurization of a urethane sleeve, thereby realizing "hydraulic fracturing" without the use of fracturing fluid. Based on the results of these experiments, the following can be concluded:

(1) In water injection, cracks extended rapidly with large drops of the hole water pressure and bursts of AE. On the other hand, in pressurization by the urethane sleeve, cracks extended stepwise with no such large hole pressure drops nor bursts of AE.

(2) The observed difference in crack extension can be analyzed by comparing the stress intensity factor at a crack tip. In water injection, the stress intensity factor of mode I at a crack tip increases with crack length; that is, a crack never stops once it starts to extend. On the other hand, in pressurization by the urethane sleeve, the stress intensity factor decreases with crack length; in other words, a crack never extends without an additional increase in loads applied on the hole wall, even after a crack starts to extend.

(3) The observation and the analysis indicate that the existence of fracturing fluid such as water helps initiated cracks to extend rapidly and widely in hydraulic fracturing in actual HDR fields.

(4) In both experiments, shear type mechanism was dominant in fault plane solutions of AE events. The shear fracture in the water injection is believed to have occurred through the connection of tips of preexisting defects lying in the direction of maximum compressive stress along a rift plane of the granite

specimen, based on a model for explaining shear mechanism seismicity often observed in volcanic earthquake swarms. Conversely, the shear fracture induced by pressurization by the urethane sleeve is believed to correspond to usual seismic events, because these events are caused by compressive stresses without the effects of fluid injection and manifest a normal quadrant-type distribution of P-wave initial motions, thereby implying shear fracture.

(5) Another experiment by the authors (1997) suggested that more viscous fracturing fluid like oil makes thicker and planar cracks with fewer branches than water. Therefore, understanding the fracturing mechanism of injected fluid like this would help to realize fracture control with using various fracturing fluid in field operations of hydraulic fracturing.

Acknowledgments

We wish to acknowledge valuable discussions with Dr. S. Sasaki, Central Research Institute of Electric Power Industry, Japan and Mr. I. Matsunaga, National Research Institute for Resources and Environment, Japan. We also would like to thank Professor J.-C. Roegiers, The University of Oklahoma, USA and Professor F. H. Cornet, Institut de Physique du Globe de Paris, France, for their valuable comments and encouragement. This research could not be completed without the support of Dr. O. Sano and Mr. Y. Ohnish, Yamaguchi University, Japan. We are grateful to them as well. This research was supported by The Ministry of Education, Science and Culture under Grant-in-Aid for Exploratory Research (Grant No. 08875202).

REFERENCES

BARIA, R., and GREEN, A. S. P. (1986), *Seismicity Induced during a Viscous Stimulation at the Camborne School of Mines Hot Dry Rock Geothermal Energy Project in Cornwall, England*, Progress in Acoustic Emission, The Japanese Society of NDI, 1986.

BARIA, R., GREEN. A. S. P., and JONES, R. H. (1989), *Anomalous Seismic Events Observed at the CSM HDR Project, U. K.*, Int. J. Rock Mech. Min. Sci. and Geomech. Abstr. 26, 257–269.

CROUCH, S. L., and STARFIELD, A. M., *Boundary Element Methods in Solid Mechanics* (Uniwin Hyman, 1983).

HAIMSON, B. C. (1978), *The Hydrofracturing Stress Measuring Method and Recent Field Results*, Int. J. Rock Mech. Min. Sci. and Geomech. Abstr. *15*, 167–178.

HILL, D. P. (1977), *A Model for Earthquake Swarms*, J. Geophy. Res. *82*, 1347–1352.

HUBBERT, M. K., and WILLIS, D. G. (1957), *Mechanics of Hydraulic Fracturing*, Petrol. Transact. Am. Soc. Min. Eng. *210*, 153–168.

ISHIDA, T., CHEN, Q., and MIZUTA, Y., *Effects of Fluid Viscosity in Hydraulic Fracturing Deduced from Acoustic Emission Monitoring*. In *Proc of Fourth International Symposium on Rockburst and Seismicity in Mines*, Kraków, August 1997 (ed. Gibowicz, S. J.) (Balkema 1997, in print).

KIRSCH, C. (1898), *Die Theorie der Elastizität und die Bedürfnisse der Festigkeitslehre*, Zeitschrift des Vereines Deutscher Ingenieure *42*, 797–807.

MIZUTA, Y., SANO, O., OGINO, S., and KATOH, H. (1987), *Three-dimensional Stress Determination by Hydraulic Fracturing for Underground Excavation Design*, Int. J. Rock Mech. Min. Sci and Geomech. Abstr. *24*, 15–29.

ROTHMAN, R. L., GREENFIELD, R. J., and HARDY, Jr. H. R. (1974), *Errors in Hypocenter Location due to Velocity Anisotropy*, Bull. Seismol. Soc. Am. *64*, 1933–1966.

SASAKI, S., *Study on Characteristics and Mechanism of Acoustic Emission Induced by Hydraulic Fracturing* (PhD. Thesis, Tohoku University, 1995) in Japanese.

SHAH, K. R., and LABUZ, J. F. (1995), *Damage Mechanisms in Stressed Rock from Acoustic Emission*, J. Geophys. Res. *100* (B8), 15,527–15,539.

TALEBI, S., and CORNET, F. H. (1987), *Analysis of the Microseismicity Induced by a Fluid Injection in a Granitic Rock Mass*, Geophys. Res. Lett. *14*, 227–230.

ZOBACK, M. D., and POLLARD, D. D. (1978), *Hydraulic Fracture Propagation and the Interpretation of Pressure Time Records for In Situ Stress Determination*, Proc. of 19th US Rock Mech. Symp., Mackay School of Mines, Reno, Nevada, 14–22.

(Received September 12, 1996, accepted January 24, 1997)

Pure appl. geophys. 150 (1997) 647–659
0033–4553/97/040647–13 $ 1.50 + 0.20/0

❙ Pure and Applied Geophysics

An Advanced Nonlinear Signal Analysis Method For Damage Detection in Geomaterials

B. BAZARGAN-SABET,[1] H. LIU,[1] and S. CHANCHOLE[1]

Abstract—An algorithm is described which enables us to evaluate the Volterra kernels and the corresponding transfer functions. This method is then used to detect the cracking threshold of the geomaterials under loading. The responses of a sample of sandstone under axial leading and subjected to ultrasonic excitation are analyzed. The occurrence of microcracking is characterized by the changes in the linear and nonlinear parts of the measured signal energy.

Key words: Damage detection, signal processing, Volterra series, nonlinear analysis.

1. Introduction

Determining the cracking threshold in rock samples under compressive loading is very important for the study of rock behavior. Standard methods used for this purpose are essentially based on the measurement of the global strains (WOLTERS, 1971). However, this kind of method is not very sensitive. The density of cracks must reach relatively high levels before it produces measurable effects on deformations.

More sophisticated methods of detection use ultrasounds. Damage is thus detected by a modification in the characteristics of the output signal (GHOREYCHI, 1978). However, the results obtained by most of these methods are no more accurate than those obtained by the standard method of strain measurements. The reason is that signal analysis is generally restricted to the assumption of a linear system (governed by a linear convolution between the input signal and the impulse response). Since the process of microcracking is inelastic (GLADWIN and SACEY, 1974), linear processing can only provide partial information. In order to identify the damage threshold, it is therefore more appropriate to analyze the occurrence of nonlinearity in the output signal. Several possible methods exist for achieving this objective, among which the multidimensional treatment by the Volterra functional series seems to be particularly promising.

---

[1] Groupement pour l'étude des Structures souterraines de Stockage (G.3S), CNRS URA 317 (LMS)—Ecole Polytechnique, 91128 PALAISEAU Cedex, France.

Although the Volterra series have been the subject of active research (BARRETT, 1963; CHOI and WARREN, 1978; HUNG *et al.*, 1977) their use as part of a method of analysis for practical problems is still rare. Therefore we will begin by presenting an original algorithm which we have developed to separate the linear and nonlinear parts of the signal. We will then use this tool to detect the cracking threshold in geomaterials under compressive loading. Finally, we compare our results with those obtained by the standard global strain measurement method.

2. *Methodology*

Let $x(t)$ and $y(t)$ be respectively the input and output measurements for a dynamic system invariable in time. For example, the response $y(t)$ of a rock subjected to an ultrasonic excitation $x(t)$. The relationship between the two signals defined by the Volterra series is the following;

$$y(t) = y_1(t) + y_2(t) + \cdots + y_n(t), \tag{1}$$

where

$$y_1(t) = \int_{-\infty}^{\infty} h_1(\tau)x(t - \tau)\, d\tau, \tag{2}$$

$$y_2(t) = \int_{-\infty}^{+\infty} \int_{-\infty}^{+\infty} h_2(\tau_1, \tau_2)x(t - \tau_1)x(t - \tau_2)\, d\tau_1\, d\tau_2, \tag{3}$$

$$y_n(t) = \int_{-\infty}^{+\infty} \int_{-\infty}^{+\infty} h_n(\tau_1, \tau_2, \ldots, \tau_n) \prod_{i=1}^{n} x(t - \tau_i)\, d\tau_i. \tag{4}$$

Here the first term $y_1(t)$ is a linear term. It is obtained by the convolution product between the first order kernel, $h_1(t)$, and the excitation function $x(t)$. The nonlinear behavior is evident from the presence of higher order terms $y_2(t)$, $y_3(t)$, ... etc.

The number of possible terms in (1) is in theory infinite. In practice n is chosen such that the difference between the recorded signal and the signal calculated based on the n first components is reasonably small. This margin of error is fixed according to the equipment used for each application and the precision of the measurement.

The problem we wish to solve is the calculation of the kernels, given excitation $x(t)$ and output signal $y(t)$. To obtain large order kernels $h_n(t_1, t_2, \ldots, t_n)$, multidimensional deconvolutions are necessary, which are generally very difficult to achieve. However, in some cases it is possible to obtain solutions in which the input signal is assumed to be of a particular type. For example VINH (1987) uses excitation by shock (modeled by a Dirac distribution) which allows him to directly deconvolute the Volterra kernels of different orders. Others, like BOYD *et al.* (1983), use multi-harmonic excitations.

In the case of ultrasounds, the input signal has a complex form. Indeed current transducer technology does not allow the generation of signals assimilable to Dirac distributions or of sufficiently pure harmonic waves. Here we propose a method for calculating Volterra kernels regardless of the form of the input signal. To do this we first express equation (1) in the frequency domain. Using the multidimensional Fourier transform and by extending Planchenel's theorem to dimensions greater than one, equation (1) can be written as:

$$Y(f) = \sum_{i=1}^{n} Y_i(f) = \sum_{i=1}^{n} H_1(f_1, f_2, \ldots, f_i) \prod_{k=1}^{i} X(f_k) \quad (f \in f_1, f_2, \ldots, f_n), \quad (5)$$

where $H_i(f_1, \ldots, f_n)$ is the Fourier transform of $h_i(t_1, \ldots, t_n)$. We must therefore calculate different order transfer functions $H_1(f)$, $H_2(f_1, f_2)$, $\ldots$, by solving the equation (5). To illustrate the resolution procedure, let us examine the case where $n = 2$:

$$Y(f) = H_1(f_1)X(f_1) + H_2(f_1, f_2)X(f_2)X(f_1) \qquad (6)$$

or

$$H_2(f_1, f_2) = \frac{Y(f) - H_1(f_1)X(f_1)}{X(f_1)X(f_2)}. \qquad (7)$$

To obtain $H_2(f_1, f_2)$ it is necessary to know the system's response in the plane (f_1, f_2). To achieve this we return to the time domain and artificially reconstitute this response from the measurements of the signals $x(t)$ and $y(t)$. Figure 1 shows this reconstruction procedure. By shifting k times the excitation $x(t)$ in time, we obtain k new input signals $x_i(t)$ such as:

$$x_i(t) = x(t) + x(t - i\Delta t) \quad i = 1, 2, \ldots, k \qquad (8)$$

$k \in \mathbf{N}$ defined by the length of the output signal.

For each $x_i(t)$, the corresponding response is placed parallel to the time axis t in the plane (t_1, t_2) with a shift Δt. We thus obtain $y(t_1, t_2)$ $(t_1 \geq 0, t_2 \geq 0)$. The bidimensional Fourier transform of $y(t_1, t_2)$ provides the system response in the frequency domain, thereby allowing us to solve (7). By reproducing the construction procedure in space (t_1, t_2, t_3) it is possible to extend the method to $n = 3$ and thereafter to $n > 3$.

However, for real applications it is often sufficient to take the case in which the frequency variables in (5) are identical $(f_1 \equiv f_2 \equiv \cdots \equiv f_n \equiv f)$. In other words, we reduce these multivariable functions to functions with a single variable by selecting specific planes of intersection. For example for $H_2(f_1, f_2)$ taking $(f_1 \equiv f_2 \equiv f)$, the second order transfer function is a function situated along the diagonal in the plane (f_1, f_2). In the same way, taking $(f_1 \equiv f_2 \equiv f_n \equiv f)$, the third-order transfer function will be situated along the trisectrice. The information derived from these functions is sufficiently significant to be used in the interpretation of real experiments. This information can be complemented by including the functions on the axes with only one variable (LIU, 1991) (for example $f_1 \equiv f$ and $f_2 \equiv f_3 \equiv 0$ for H_3).

In order to evaluate the transfer functions in the specific case where $f_1 \equiv f_2 \equiv \cdots \equiv f_n \equiv f$, we must use (n) different signals. In practice we use the same input signal with (n) different levels of amplitude. For the mth level of excitation the Fourier transform of the corresponding response is:

$$Y_{(m)}(f) = \sum_{i=1}^{n} H_i(f) X_{(m)}^i(f), \quad m = 1, 2, \ldots, n. \tag{9}$$

We put (9) in matrix form by taking $m = n$ (which means that the number of levels of excitation is equal to the number of terms included in the Volterra series).

$$Y = XH \tag{10}$$

where

$$Y = \{ Y_{(1)}(f) \quad Y_{(2)}(f) \quad \cdots \quad Y_{(n)}(f) \}^T \tag{11}$$

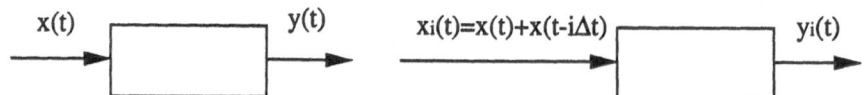

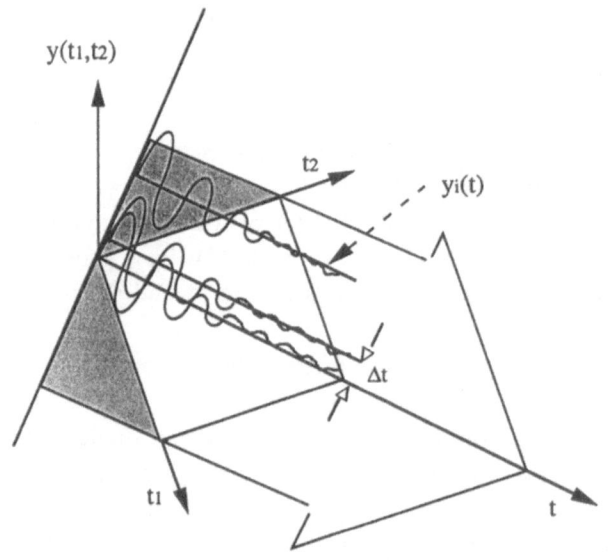

Figure 1
The reconstitution procedures for $y(t_1, t_2)$.

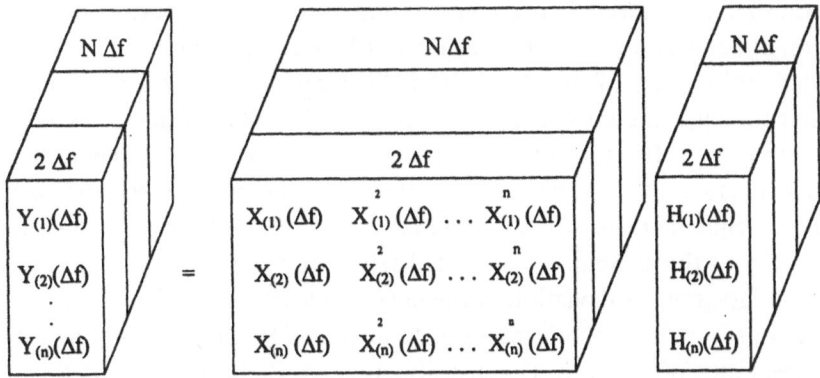

Figure 2
The process for solving equation (10).

$$H = \{H_{(1)}(f) \quad H_{(2)}(f) \quad \cdots \quad H_n(f)\}^T \tag{12}$$

$$X = \begin{bmatrix} X^1_{(1)}(f) \; X^2_{(1)}(f) & \cdots & X^n_{(1)}(f) \\ X^1_{(2)}(f) \; X^2_{(2)}(f) & \cdots & X^n_{(2)}(f) \\ \cdots & \cdots & \cdots & \cdots \\ X^1_{(n)}(f) \; X^2_{(n)}(f) & \cdots & X^n_{(n)}(f) \end{bmatrix}. \tag{13}$$

Figure 2 shows the resolution process schematically.

Equation (10) is solved step by step up to the frequency $f_{max} = N\Delta f$ (N is the number of samples) which must remain inferior or equal to the Nyquist frequency (defined by the sampling theorem).

3. Practical Application

Here we present the practical application of the described algorithm. We use it to detect the occurrence of microcracking in a sandstone sample during a uniaxial compressive test. In order to demonstrate the capacities of our method we have chosen to compare our results with those obtained by the standard global strain measurement method.

The following experiment is one of a group of 10 tests which we carried out on three different kinds of rock, as part of a research contract on radioactive waste storage. The results we obtained from each of these tests are all qualitatively similar. We therefore limit our detailed presentation to one representative test.

The sample studied is a cylinder 75 mm in diameter and 150 mm in height. We placed a broad band (0.2 MHz to 0.8 MHz) transducer on either side of the sample.

The transmitter is connected to an ultrasound excitation generator, enabling the production of acoustic waves of between 0 and 5 kPa of pressure. The axial loading is provided by a press of 1000 kN. The displacement sensors, positioned both axially and radially, allow calculation of the global strains during loading.

Axial stress as a function of axial strain shows very obvious elastic behavior, which is expressed by a linear relation between the axial load and the axial deformation, characteristic of this material (see Fig. 4). Figure 5 illustrates the volumic strain of the sample, reconstituted from the axial and radial deformations versus the axial strain. A deviation in the curve is clearly noticeable from a volumic strain of about 0.68%, which corresponds to an axial load of 420 kN. The threshold of the appearance of cracking is generally situated at the point of this deviation (dilatation—contraction threshold) (BEREST *et al.*, 1989).

We now examine the results obtained from signal analysis. Having separated the output signal into its linear and nonlinear components we will follow the evolution of each part as a function of load. As previously mentioned, the sample must be subjected to several levels of excitation. To ensure that nonlinearity obtained is due to the material and not to peripheral elements, we must use the lowest possible levels of input signal (HEITZ, 1992). For these experiments we selected three levels

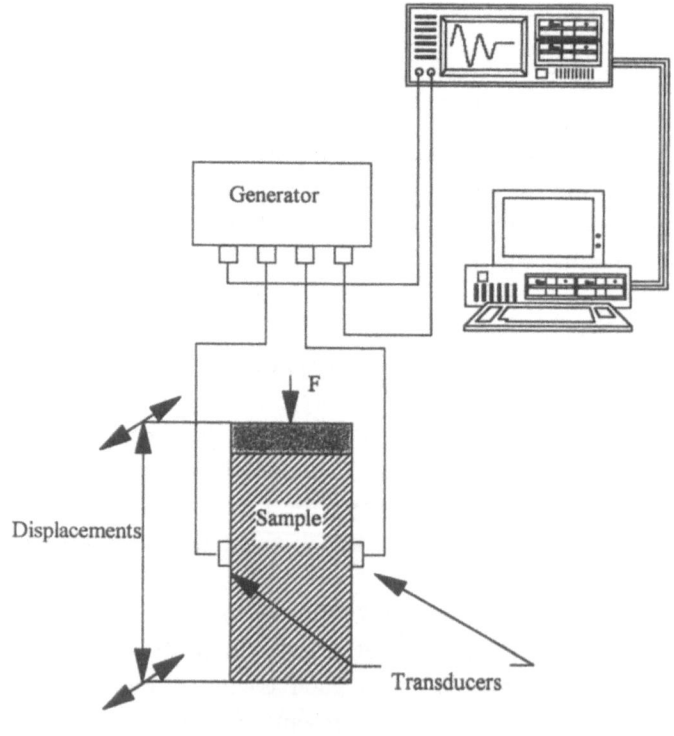

Figure 3
Experimental setting.

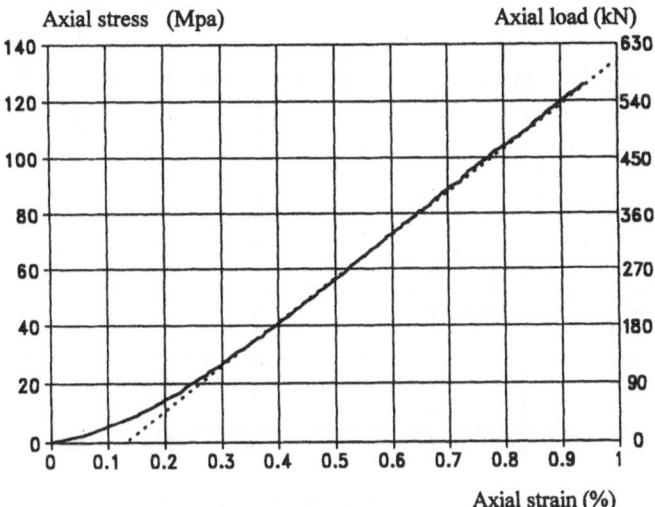

Figure 4
Axial stress versus axial strain.

equivalent to acoustic pressure of 0.25 kPa, 0.5 kPa and 1.0 kPa. In addition, we made sure that the transducers did not bear any loading. In this way the relative signal variations we recorded could not be attributed to the equipment.

Figure 6 shows the excitation signal in the time domain, recorded using a perfectly elastic reference material with properties similar to sandstone. Figure 7

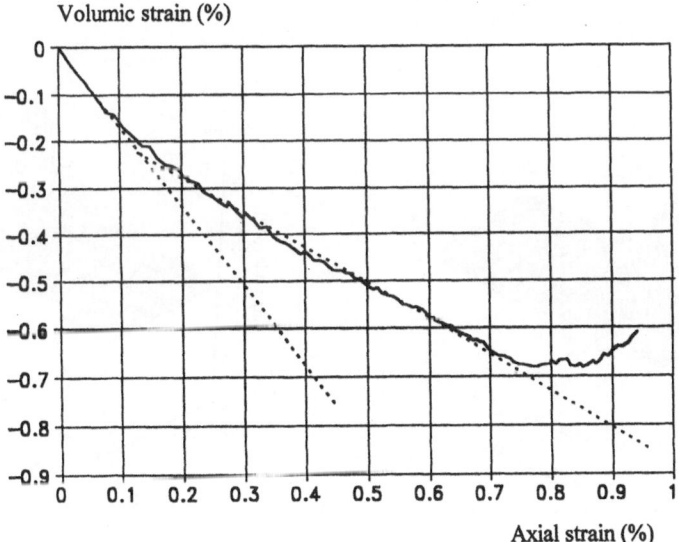

Figure 5
Volumic strain versus axial strain.

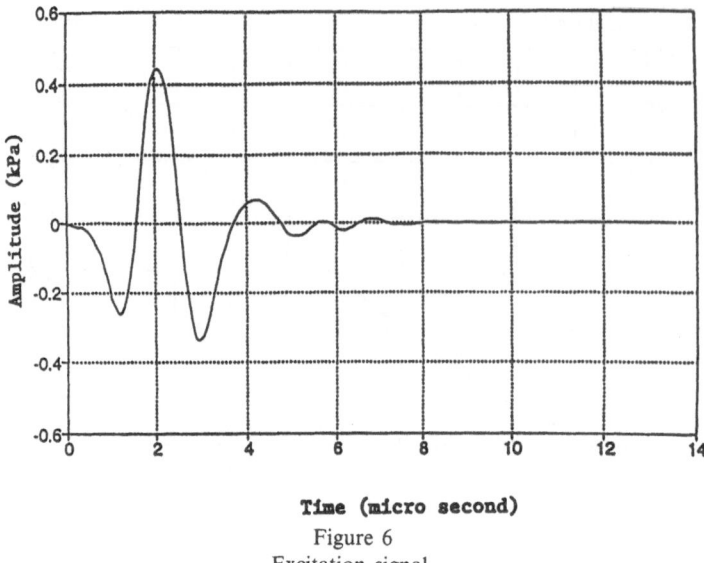

Time (micro second)
Figure 6
Excitation signal.

shows the evolution of sample responses $y(t, \text{load})$ as recorded during loading. We fix the margin of error at 1% of the magnitude of the output signal. With three levels of excitation it is possible to calculate the first three Volterra kernels, so that:

$$y(t) = y_1(t) + y_2(t) + y_3(t). \tag{14}$$

Our experience has shown that for most materials it is sufficient to consider three terms in the Volterra series. However, given the weak level of $y_3(t)$ (see Fig. 8) compared to the two other components $y_1(t)$ and $y_2(t)$, we may simply retain only the first two terms.

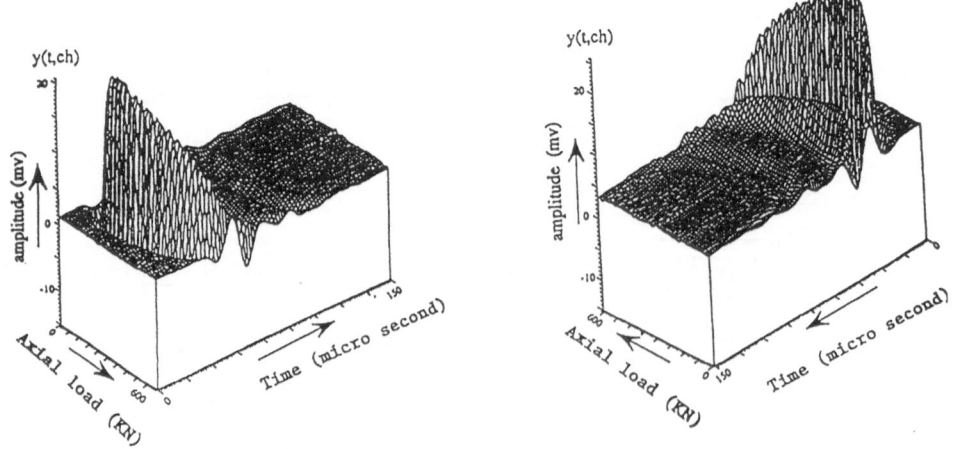

Figure 7
Sandstone responses according to axial deformation.

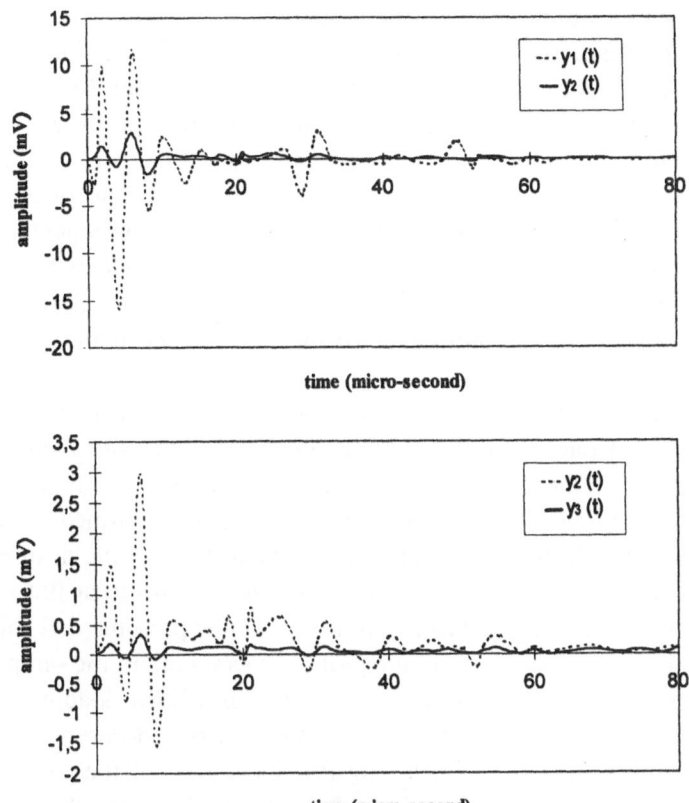

Figure 8
Comparison between the three components of the output signal ($y_1(t)$, $y_2(t)$ and $y_3(t)$) corresponding to 450 kN load.

We take the ratio of the signal energy to the total energy transmitted as the adimensional parameter for which we will study the variations as a function of applied load. By "energy" we mean the quantity represented by:

$$E(y_k(t)) = \int_{-\infty}^{\infty} |y_k(t)|^2 \, dt. \tag{15}$$

It would be possible to follow the variations of other signal parameters, for example the magnitude of the peak around the excitation's main frequency, but the energy parameter, which includes the variations across the entire frequency band, seems to be more appropriate.

Figure 9 shows the energy variation in the linear part of the signal $E(y_1)$ according to the axial load applied to the sample. This curve comprises three noteworthy parts: First, between 0 and 90 kN which shows a fall in the value of $E(y_1)$. Then between 90 and 400 kN showing stability in the value of $E(y_1)$. Finally from 400 kN where we see a sharp rise in $E(y_1)$.

Given the nature of the rock selected (sandstone behaves in an elastic-fragile manner), and the weak acoustic energy levels used, we may hypothesize that the system's behavior depends solely on the stiffness of the material. In other words, we assume that the mass and the viscous damping coefficient of the sample are conservative before the failure point. In that case the first part of the curve (9) corresponds to a tightening in the sample's structure. In this phase the linear stiffness of the rock increases and the linear response $y_1(t)$ consequently diminishes thus keeping with the dynamic equation $\mathbf{K}y(t) = ax(t)$ in which $\mathbf{K}$ is the linear stiffness (BUI, 1993). The stabilization phase corresponds to the elastic response of the material. Finally, the diminution in stiffness, when the material is damaged, causes a rise in $y_1(t)$.

Figure 10 demonstrates the energy variation of the nonlinear part of the signal $E(y_2)$ versus the axial load. It is interesting to note the difference in magnitude between this curve and that of the linear part. Nonlinearity represents only a small part of the total response. Similar to the linear part, this curve is formed of three parts: Initially, a rapid fall at the beginning of the curve which indicates the closure of existing microcracks. Then, a quasi-stability in the loading zone of between 100 and 300 kN. Finally a progressive rise in $E(y_2)$ from 310 kN. We may presume that this variation in nonlinear energy indicates the beginning of microcracking in the sample, the only possible modification of the physical state under compressive loading.

It is interesting to note that this last change in slope occurs well before that observed in the previous curve concerning the linear part of the energy. Similarly we would point out that in the zone preceding the slope change (between 100 and 300 kN), the magnitude of nonlinear energy approaches zero.

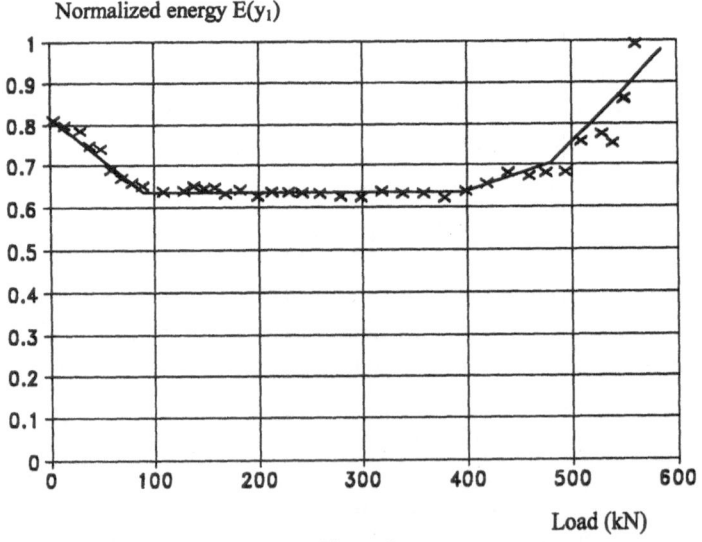

Figure 9
Signal energy variation of the first Volterra kernel versus the axial load.

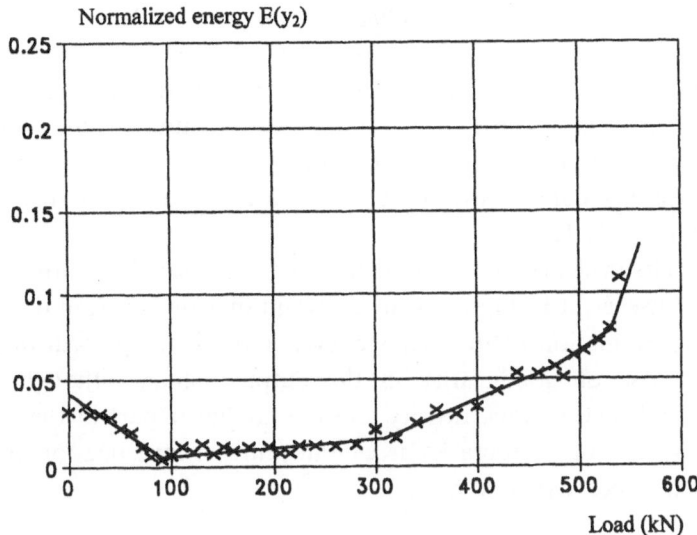

Figure 10
Signal energy variation in relation to the second Volterra kernel versus the axial load.

We now compare our results with those obtained by the standard method (curves in Figs. 4 and 5):

a. The initial closure of the microcracks and the tightening of the material are equally evident in the curves showing the volumic strain and the axial stress versus axial strain (slope change). Reference to the load/strain curve reveals that the end of this initial phase occurs around a stress of 20 MPa, i.e., an axial load of 90 kN. This corresponds to the beginning of the stabilization phase for both the linear and nonlinear energy variation curves.

b. The sample's dilation-contraction threshold shown on the volumic strain curve corresponds to an axial load of approximately 420 kN, equivalent to the value at which we observe a rapid increase in the linear part of the energy $E(y_1)$.

c. The material's behavior is purely elastic between 100 and 300 kN. This is evident in the curves, but is borne out by the fact that in this interval nonlinear energy is almost absent and linear energy remains constant (no variation in stiffness).

d. The rise in energy from a loading threshold of 310 kN has no equivalent on the other curves. This change indicates structural modifications in the sample, probably related to the beginning of microcracking. Only when the density of the microcracks becomes sufficiently high to influence the material's stiffness (from 400 kN), will the effect be visible in the other curves.

4. Other Results

As mentioned, the above experiment is part of a research program. Output signals obtained from 9 other samples were also treated by the Volterra series. Three different kinds of rock were examined. Both uniaxial and triaxial (under confining pressure) compressive tests were carried out. The main results are summarized in Table 1.

As the results evidence, in all cases nonlinear analysis allows forward detection of damage. However, it is important to be aware that this method does not lead to a quantitative estimation of the damage level. Indeed, the growth of microcracks produces the rise of nonlinearity in the output signal, although there is no proportionality between crack density and the nonlinear part of the signal. Determining the quantity of microcracks necessitates further investigation and assumes a modeling of the rock dynamic behavior.

5. Conclusion

Signal analysis using the Volterra series appears to be an efficient tool for the detection of the microcracking threshold in geomaterials under compressive loading. As we have seen the occurrence of microcracks causes the appearance and subsequent increase of the nonlinear component of the Volterra series. Whereas the linear part of the signal produces comparable results to those obtained by the

Table 1

Identification of the damage threshold for three different rocks, using standard and nonlinear methods

Material	Test type	Damage threshold (MPa)		
		Strains analysis	Linear analysis	Nonlinear analysis
sandstone	uniaxial	95	93	70
sandstone	uniaxial	82	80	65
sandstone	**uniaxial**	**93**	**90**	**70**
sandstone	triaxial $(P_c = 5\ \text{MPa})$	170	165	135
sandstone	triaxial $(P_c = 10\ \text{MPa})$	205	200	150
deep marl	uniaxial	10	10	5
deep marl	uniaxial	12	12	8
deep marl	triaxial $(P_c = 5\ \text{MPa})$	20	18	15
granite	uniaxial	80	80	60
granite	triaxial $(P_c = 20\ \text{MPa})$	140	135	110

standard method of strain measurements, the greater sensitivity of the nonlinear part allows forward detection of the damage threshold.

REFERENCES

BARRETT, J. F. (1963), *The Use of Functionals in the Analysis of Nonlinear Physical Systems*, J. Elect. and Control, 567–615.

BEREST, P., BERGUES, J., HOMAND, F., TROALEN, J. P., HENRY, J. P., and IKOGON, S., *Comportement thermique et mécanique du grés de Fontainebleau* (Acute du col. Bilan et perspective du GRECO Géomatériaux 1989) pp. 19–52.

BOYD, S., TANG, Y. S., and CHUA, L. O. (1983), *Measuring Volterra Kernels*, IEEE Transaction on Circuits and Systems *30*, 571–577.

BUI, H. D. (1993), *Introduction aux problèmes inverses en mécanique des matériaux* (n° 83) (EYROLLES 1993) pp. 75–90.

CHOI, C. H., and WARREN, M. E. (1978), *Identification of Nonlinear Discrete Systems* (Proc, South Western Conf. Atlanta 1978), 329–333.

GHOREYCHI, M. (1978), *Mesure de l'endomagement des roches par l'attenuation des ultrasons* (Proc. 2ème colloque Franco-Polonais de géotechnique Nancy 1978) pp. 113–121.

GLADWIN, M. T., and SACEY, F. D. (1974), *Inelastic Degradation of Acoustic Pulses in Rocks*, Ph. of Earth and Planetary Int. *8*, North-Holland Pub. Co.

HEITZ, J. F. (1992), *Prepagation d'ondes en milieu non-linéaire* (Thèse de doctorat Université J. Fourier Grenoble).

HUNG, G., STARK, L., and EYKHOFF, P. (1977), *On the Interpretation of Kernels*, Ann. Biomed. Eng. *5*, 130–143.

LIU, H. (1991), *Multi-dimensional Signal Processing for Nonlinear Structural Dynamics*, Mechan. Syst. and Signal Proc. *5*, 61–80.

VINH, T. (1987), *Techniques for the Identification of Nonlinearity* (Course note Heriot-Watt University, Edinburgh 1987).

WOLTERS, R. (1971), *Initial Fracturing in Rock under Triaxial Loading* (Proc. Symp. of the Int. Soc. for Rock Mech., Nancy 1971), sec. 1.5.

(Received June 18, 1996, accepted March 10, 1997)

Pure appl. geophys. 150 (1997) 661–676
0033–4553/97/040661–16 $ 1.50 + 0.20/0

Pure and Applied Geophysics

Estimation of Deeper Structure at the Soultz Hot Dry Rock Field by Means of Reflection Method Using 3C AE as Wave Source

NOBUKAZU SOMA,[1] HIROAKI NIITSUMA[1] and ROY BARIA[2]

Abstract—We investigate the deep subsurface structure below the artificial reservoir at the Soultz Hot Dry Rock (HDR) site in France by a reflection method which uses acoustic emission (AE) as a wave source. In this method, we can detect reflected waves by examining the linearity of a three-dimensional hodogram. Additionally for imaging a deep subsurface structure, we employ a three-dimensional inversion with a restriction of wave polarization angles and with a compensation for a heterogeneous source distribution.

We analyzed 101 AE wave forms observed at the Soultz site during the hydraulic testing in 1993. Some deep reflectors were revealed by this method. The bottom of the artificial reservoir that is presumed from all of the AE locations in 1993 was delineated at the depth of about 3900 m as a reflector. Other deeper reflectors were detected below the reservoir, which would not have been detected using conventional methods. Furthermore these reflectors agreed with the results of the tri-axial drill-bit VSP (ASANUMA *et al.*, 1996).

Key words: Reflected AE, coda, linearity of three-dimensional hodogram, polarization.

1. Introduction

The Soultz HDR site has been supported by France, Germany, United Kingdom and European Commission (EC) since 1987 (BARIA *et al.*, 1995). An artificial reservoir was made by hydraulic fracturing below the depth of 2000 m, and a second well was drilled and succeeded in developing a circulation system in 1995 (BARIA *et al.*, 1996).

Generally in geothermal fields, the outline of the enhanced fractured reservoir can be estimated by monitoring acoustic emission (AE), also termed microseismicity (NIITSUMA *et al.*, 1985). We can roughly evaluate the shape, extent and permeability of the reservoir. At the Soultz site, the extent of the artificial reservoir was mapped by source locations of numerous AE events (BARIA et al., 1995).

[1] Department of Geoscience and Technology, Graduate School of Tohoku University, Sendai, 980-77 Japan, Tel. & FAX: +81-22-217-7401.

[2] SOCOMINE, Project Géothermique, Route de Kutzenhausen BP39, 67250 Soultz-sous-Forêts, France.

It is essential to investigate deeper subsurface structures not only for the conventional geothermal field but also for HDR field in order to understand the behavior of the reservoir and to increase heat production. The knowledge of the structure below or around the geothermal reservoir which has already been developed is useful and important to understand its character, to estimate a life span for the reservoir and to evaluate the potential of extensive exploitation. Deep geothermal resources contain large amounts of energy which could be more environment-friendly to produce.

Seismic reflection surveys with artificial seismic sources, such as explosive, air gun, and Vibrator, are useful in estimating deep structures in oil exploration. However, it is difficult to obtain detailed structure by such a conventional reflection survey in high-temperature geothermal fields. This is caused by a strong heterogeneous geology and a significant propagation loss at a fractured shallow reservoir. Sometimes, high velocity zones in geothermal fields, such as granitic rocks, seriously disturb the propagation of both direct and reflected waves. Furthermore, the high temperature and high pressure in the field prevent the use of powerful seismic sources such as explosives. The number and spread of the seismic sources and the observation network are often significantly limited by cost consideration and the rugged topography, and it is difficult to get three-dimensional extensive profiles.

Meanwhile, we developed a type of reflection method using acoustic emission as a wave source (the AE reflection method), and succeeded in imaging the deeper reflectors which correspond to other geological estimates in the Kakkonda geothermal field in Japan (SOMA and NIITSUMA, 1997).

The use of AE events as seismic source may provide some advantages for geothermal development. Indeed, the seismic energy of AE events is larger than that of artificial wave sources which we can generally use in geothermal fields (IIO, 1982). The arrangement of wave sources is similar to that in the Inverse Vertical Seismic Profiling (Inverse VSP) method, that is AE sources are located below surface at some depth. Therefore, we can anticipate less attenuation of seismic waves than that of surface seismic sources. Furthermore, AE sources are widely distributed corresponding to the reservoir fracture without restriction of topography. It is easy to make a continuous long-term observation with only a passive measurement system. Although the location of a seismic source itself includes an essential ambiguity about both the direction and the distance from the detector to the source, we can easily use a large number of AE events and reduce the error in the estimates of reflectors.

This paper describes an application of the AE reflection method in which AE events are used as a seismic wave source to the Soultz HDR site in France. First, we present an outline of the AE reflection method using AE events as a seismic source, which consists of the detection of reflected waves by three-dimensional hodogram analyses and the three-dimensional inversion constrained by wave polarization angles for the deep estimation. Thereafter, we show the results for the data observed in the Soultz HDR field in France, and compare with the tri-axial drill-bit VSP (ASANUMA et al., 1996).

2. Outline of AE Reflection Method

2.1 Detection of Reflected Waves by Three-dimensional Hodogram

We developed a method to detect reflected waves by analyzing the linearity of three-dimensional hodogram (a Lissajous figure in three dimensions). The linearity of three-dimensional hodogram changes corresponding to the wave condition (Figure 1). We can infer that the shape of the hodogram becomes elongated when a reflected wave is detected, on the other hand, it changes into a spherical shape when incoherent components such as coda are detected (NAGANO et al., 1986). For the quantitative evaluation of the shape of the hodogram, we use a global polarization coefficient in the principal component analysis (SAMSON, 1977). This coefficient is defined using eigenvalues of variance-covariance matrix considering the three-dimensional hodogram to be a locus of discrete time-series vectors $\mathbf{u}(nT)$:

$$\mathbf{u}(nT) = [S_x(nT), S_y(nT), S_z(nT)], \tag{1}$$

where: S_i $(i = x, y, z)$ are elements of the vector, n is integer, and T is sampling interval.

The variance-covariance matrix of $\mathbf{u}(nT)$ is denoted as

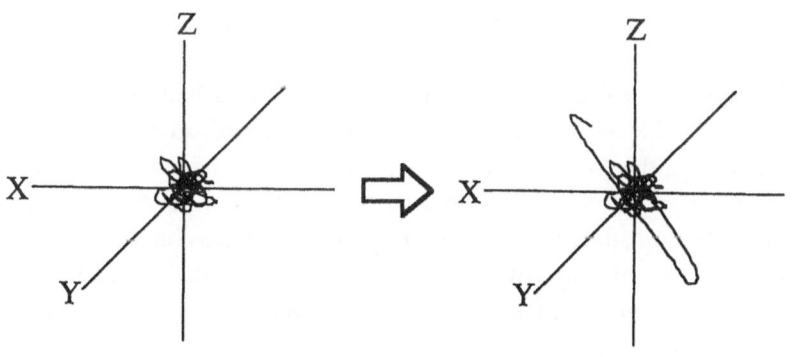

incoherent wave or random noise   coherent wave arrival

(a) random polarization   (b) linear polarization

Figure 1
Concept of linearity of three-dimensional hodogram.

$$V = \begin{pmatrix} C_{xx} & C_{xy} & C_{xz} \\ C_{yx} & C_{yy} & C_{yz} \\ C_{zx} & C_{zy} & C_{zz} \end{pmatrix}, \tag{2}$$

where C_{ii} $(i = x, y, z)$ are variance terms, C_{ij} $(i, j = x, y, z, i \neq j)$ are covariance terms.

A global polarization coefficient (SAMSON, 1977), which represents the linearity of the three-dimensional hodogram, is given by

$$Cp = \frac{(\lambda_1 - \lambda_2)^2 + (\lambda_1 - \lambda_3)^2 + (\lambda_2 - \lambda_3)^2}{2(\lambda_1 + \lambda_2 + \lambda_3)^2}, \tag{3}$$

where: λ_j $(j = 1, 2, 3, \lambda_1 > \lambda_2 > \lambda_3)$ are eigenvalues.

This coefficient changes from 0.0 to 1.0, corresponding to the linearity of the hodogram. Moreover the wave polarization direction is calculated from the first eigenvector of the matrix **V**. In order to detect a reflected wave, the time-variant shape of the hodogram must be examined. Here we use a moving window method and calculate the global polarization coefficient as a function of time. The change in the linearity of the three-dimensional hodogram can be evaluated almost independently of a wave energy using this method.

2.2 3D-inversion for Estimation of Deep Three-dimensional Structures

Because AE has a quasi-random distribution of sources contrary to the systematic source distribution in a conventional survey, the conventional data processing methods of reflection survey cannot be applied directly to estimate deep subsurface structures.

We employ a three-dimensional inversion (3D-inversion) of wave forms which is obtained by the linearity analysis. The concept of the 3D-inversion is shown in Figure 2. This inversion is a type of diffraction stack migration in a conventional reflection survey (BERKHOUT, 1985).

In the present application, amplitude of P wave is usually smaller than that of S wave, and a reflected P wave may be covered by direct and coda of S wave. Therefore, we use the S-wave velocity for the inversion as S-wave reflections make the analysis simple. Converted phases such as P-Sv and Sv-P should be taken in account in reflection survey. However we did not treat the converted waves since we can expect that the influence of converted waves is cancelled in the inversion in the next stage because of the incoherence of wave velocity and a restriction of wave polarization angles.

If we detect a reflected S wave with some delay ΔT relative to the arrival of the direct S wave, then the propagation path length of the reflected wave can be calculated with an assumed velocity. The possible reflection points for a reflected wave of a delay ΔT are distributed on an iso-delay ellipsoid, which is determined by the location of the detector and the source, and the propagation path length of the

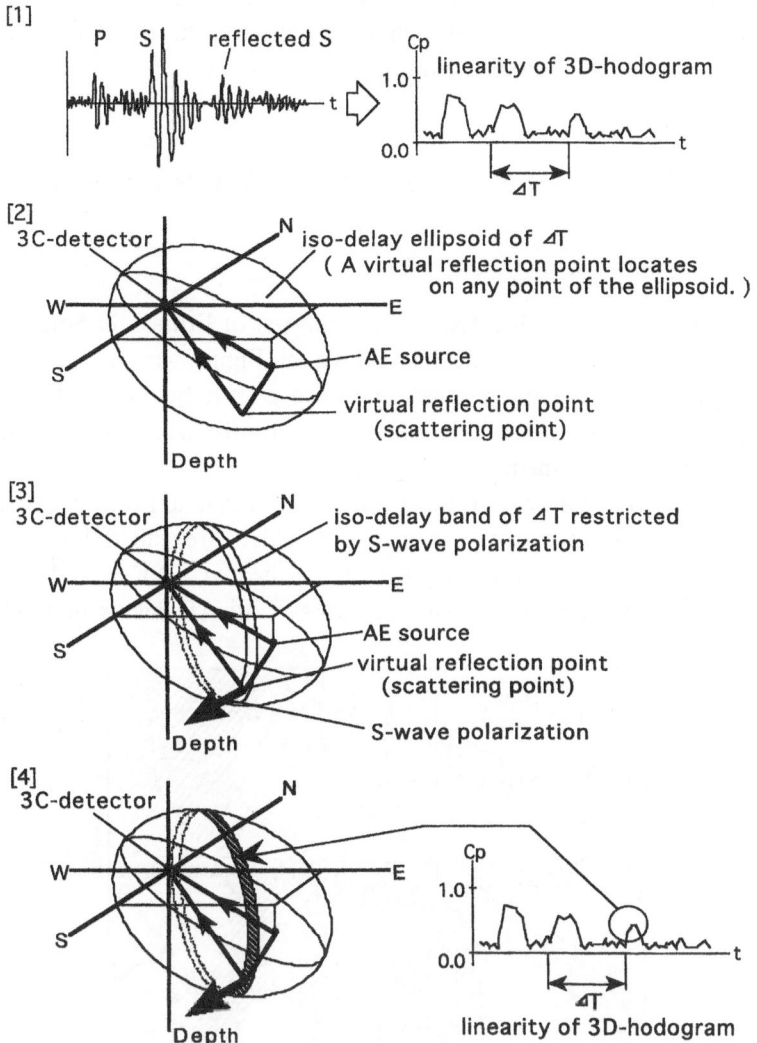

Figure 2
Concept of 3D inversion for the discriminated reflected waves.

reflected wave. Basically, the strength of the linearity signal is plotted in the ellipsoid to image reflectors. Moreover in the inversion, the distribution of possible reflection points is restricted to a narrow band on the iso-delay ellipsoid, by assuming that the direction of the reflected wave polarization is orthogonal to the geometrical ray path of the reflected wave. We plot the strength of the linearity signal of ΔT in the restricted possible reflection zone. We repeat this process over all delays in the signal for all AE events, and stack all the results.

In this inversion, the effect of the heterogeneous distribution of AE sources is suppressed by a normalization of signal strength for the number of adjacent AE events within a radius of 200 m from each AE source.

3. *Estimation of Deeper Subsurface Structure in Soultz*

3.1 *AE Measurement in the Soultz Field*

The Soultz HDR site is located in the northeast of France, in Soultz-sous-Forêts (Fig. 3). A plan view of this field is shown in Figure 4. The geology of Soultz consists of shallow sedimentary layers and deep granite below the depth of 1400 m. The artificial reservoir was made inside the granite which is assumed to be a relatively homogeneous medium.

There is an injection well GPK-1 drilled in 1987, a production well GPK-2 drilled in 1995 and three observations wells which are called E4550, E4601 and E4616 (Fig. 4). A hydraulic testing was performed at GPK-1 in 1993 and about 20,000 AE events were induced. The locations of all AE events in 1993 are shown

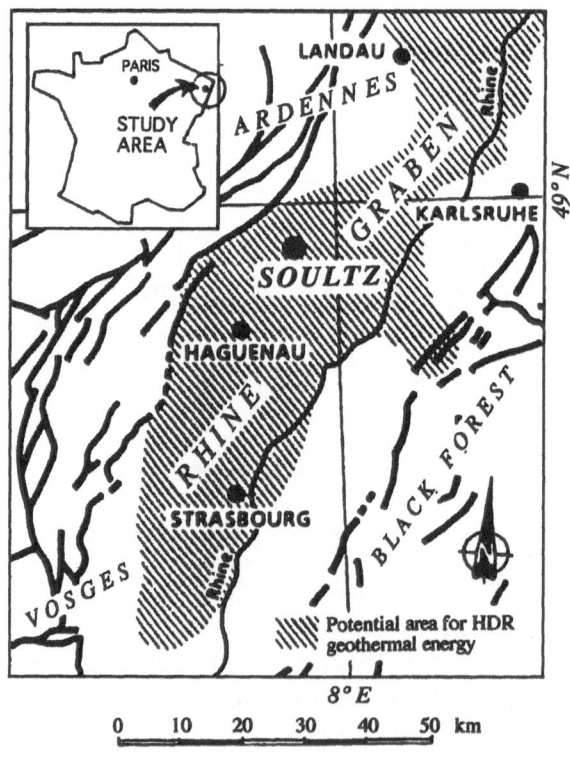

Figure 3
Location of the Soultz HDR field in France.

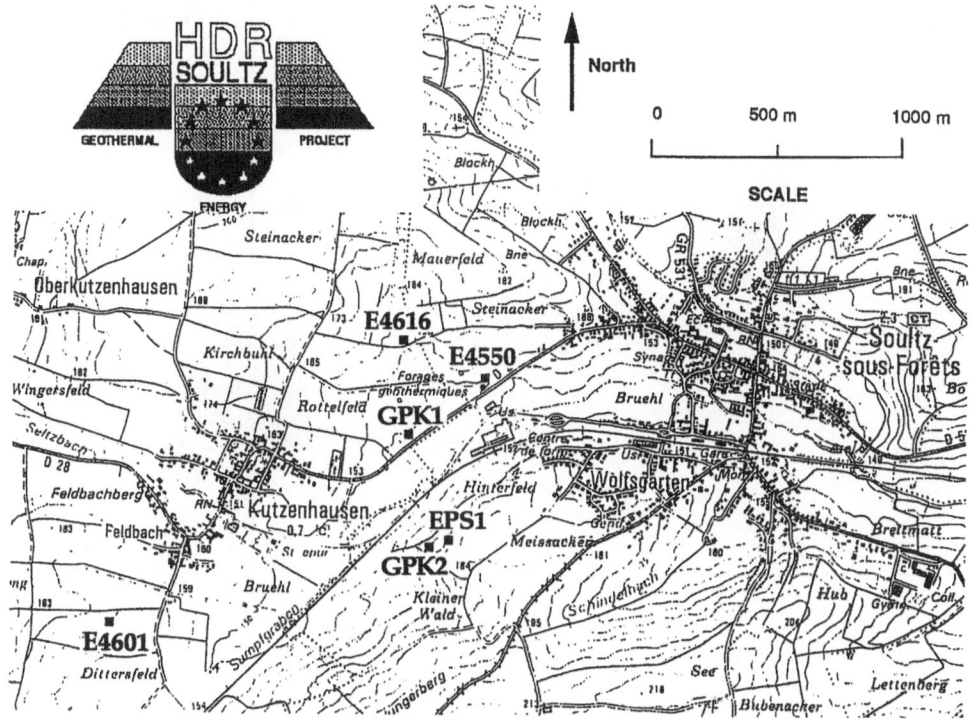

Figure 4
Plan view of the Soultz HDR field.

in Figure 5. From the AE source cloud, we can infer that the artificial reservoir grew in the NNW-SSE direction. The downhole four-component detector was installed inside the granite at the depth of about 1500 m (BARIA *et al.*, 1995). The wave forms are converted to three-component wave forms before analyses.

We use 101 AE wave forms recorded in long data files (about 3.0 s) All of the 101 events were observed only in the E4550 well at the depth of 1483.5 m. Figure 6 shows the AE source location of the 101 wave forms, the observation point E4550, and the orientation of cross sections that are estimated by the AE reflection method in the later section.

We show an example of a typical three-component wave form at the Soultz field in Figure 7. AE events in this field usually have very high signal-to-noise amplitude ratios (S/N) and the dominant frequency is around 200 Hz. We can easily identify the arrival of both *P* wave and *S* wave. The duration of coda of the *S* wave is usually shorter than that in the complex volcanic field, such as the Kakkonda field. This may be caused by the simple geological condition in the Soultz field. But it is still difficult to discriminate reflected waves by a simple observation, as in the case of the Kakkonda field (SOMA and NIITSUMA, 1997).

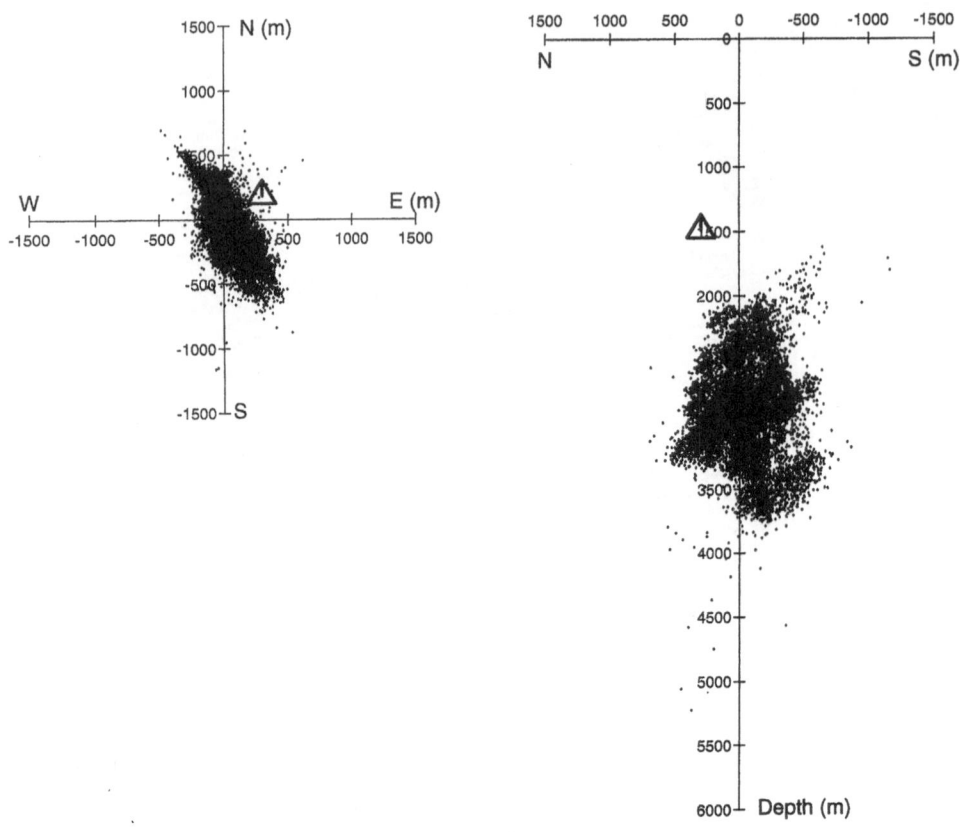

Figure 5
Distribution of all AE events in Soultz in 1993. Δ shows the sensor inside the E4550 well.

3.2 *Detection of Reflected Waves at Soultz*

We analyzed 10l-long AE wave forms in Soultz by investigating the linearity of a three-dimensional hodogram to detect reflected waves inside of the coda. The length of the moving window is 0.1 s (1 wavelength of 10 Hz) since the dominant frequency of reflected waves may be lower than that of direct waves.

An example of the results of this analysis is shown in Figure 8, which was calculated using the wave form presented in Figure 7. In this figure, *P* wave and *S* wave arrivals are clearly indicated. Because of the effect of the length of the moving window, there is a slight difference of arrival time of *P* wave and *S* wave between the raw wave form (Fig. 7) and the wave form of the linearity of the hodogram (Fig. 8). Some peaks of linearity of the three-dimensional hodogram are detected after the direct *S* wave arrival at around 1.8 s, 2.4 s. 2.55 s and 3.0 s (Fig. 8). There is no evidence to exactly identify the origin of these peaks, but there is a possibility of the arrival of other coherent signals such as reflected waves. However,

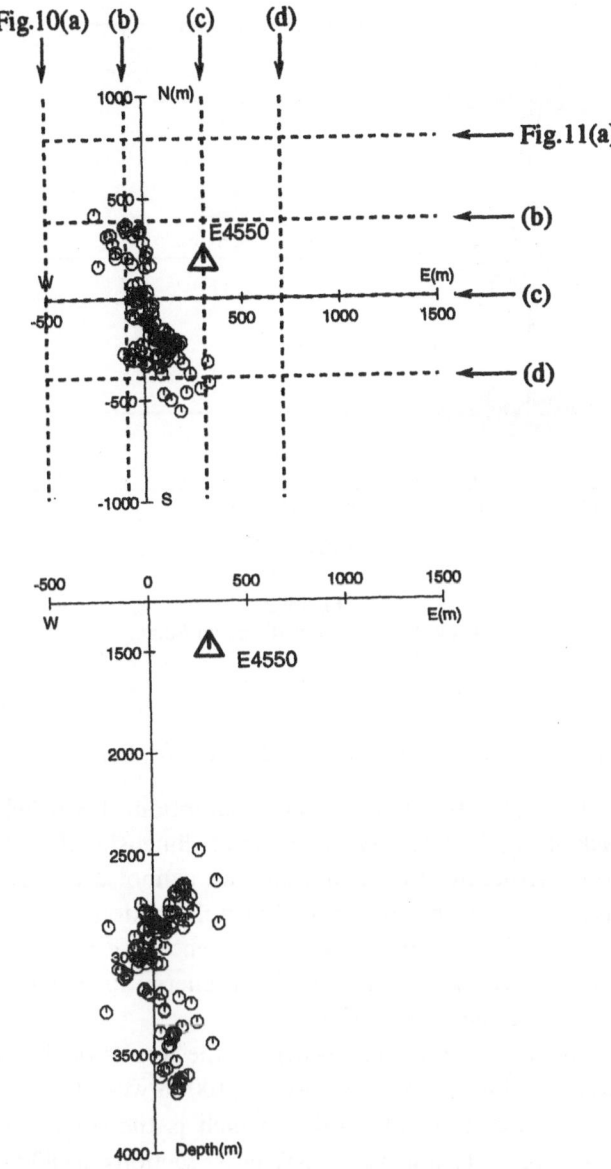

Figure 6

Distribution of 101 AE sources in Soultz in 1993 and the orientation of estimated cross sections in Figures 10 and 11. Δ shows the detector inside E4550 well.

within the coda, the change of the linearity of the hodogram is much simpler than the same results of the Kakkonda field shown in Figure 9. This may reflect the diminished importance of scattering, splitting, small reflection, and so on, due to the simpler geological condition at the Soultz site. We can easily identify the arrival of low energy coherent signals, which are suspected to be reflected waves, from the evaluation of the linearity of three-dimensional hodograms.

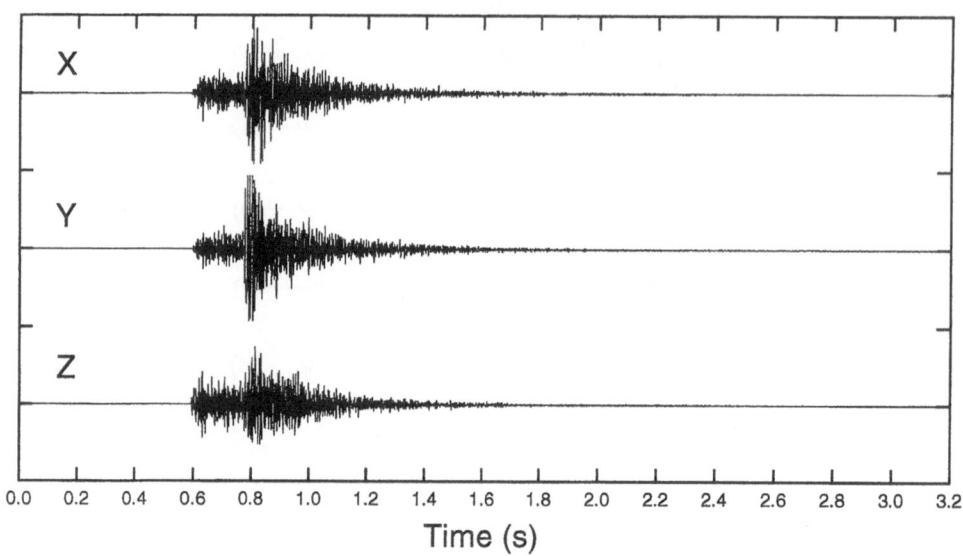

Figure 7
Typical 3C AE wave forms in Soultz.

3.3 Estimation of the Deep Subsurface Structure at the Soultz Site

We performed the 3D inversion with an assumption of S-S reflection using the wave forms which show the linearity of the three-dimensional hodogram to image the deep subsurface structure. In the analysis, we cannot accurately determine the direction of wave polarization when a global polarization coefficient is very low since the shape of the hodogram is spherical. Hence, we set a threshold level of $Cp > 0.1$ for this analysis and assume the signal energy to be 0.0 for values below this threshold to reduce the effects of noise.

Figures 10(a)–(d) and Figures 11(a)–(d) are the results of the inversion. Figure 10 shows N-S cross sections (a) 500 m west, (b) 100 m west, (c) 300 m east, and (d) 700 m east from the well-head of GPK-1, which is the origin of the coordinate system in this field. Figure 11 illustrates E-W cross sections (a) 800 m north, (b) 400 m north, (c) 0 m and (d) 400 m south from the well-head of GPK-1. The reflectors are detected as a little round shape because the spread of the event source locations is not enough for the analysis using only one detector. However, improvement is expected if we use other stations in the future. In these figures, the intensive reflector around AE sources is due to the influence of the continuing linearity of hodograms of direct waves. In Figures 10(d) and 11(d), there are small vertical shapes at the depth of around 2200 m, which are at about 400 m south and about 500 m east, respectively. They may be an image of the side edge of the artificial reservoir or other vertical structures.

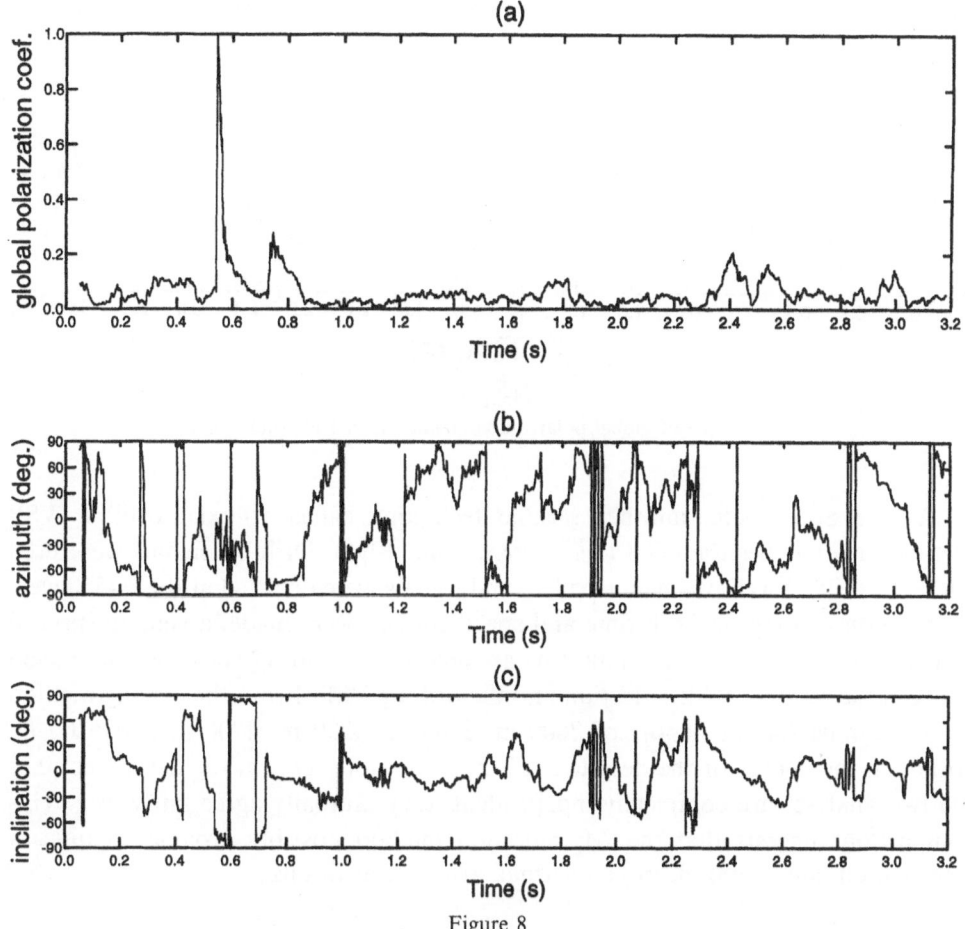

Figure 8

Result of the principal component analysis of the 3C AE wave form in Figure 6. (a) Global polarization coefficient which represents the linearity of 3D hodogram. (b) Azimuth and (c) inclination of the polarization of hodogram.

We can detect some reflectors in those figures at depths of around 3900 m, 4400 m, 4900 m and 5300 m. The reflectors at depths of 3900 m and 4900 m roughly correspond to the arrivals of the linearly polarized signals at 1.8 s and 2.4 s, respectively in Figure 8. There is little geological information regarding the region deeper than the artificial reservoir, i.e. below the depth of about 3900 m. Therefore, the deep estimated reflectors at depths of 4400 m, 4900 m and 5300 m cannot be confirmed with the geological data, but in particular, the reflector at 3900 m may correspond to the bottom of the artificial reservoir shown in Figure 5. The other reflectors possibly show the existence of some structure, such as a natural fracture and some geological boundaries below the artificial reservoir, where the other conventional method could not provide useful information.

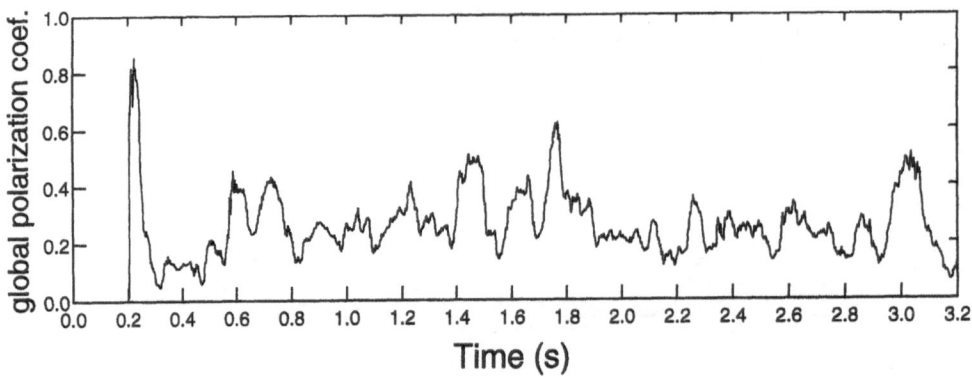

Figure 9
An example of global polarization coefficient in Kakkonda field.

At the Soultz site, another seismic technique called tri-axial drill-bit VSP method also has been applied (ASANUMA *et al.*, 1996). Drill-bit signals were used while the GPK-2 well was being drilled, and signals were detected at the E4550 well. Correlation analyses in both time and space domain were made, assuming spectral whiteness in signal, and some reflectors are detected. Figure 12 shows a comparison between the results of AE reflection and the drilling VSP. From the drilling method there are reflectors at depths of 2600 m, 3700 m, 4200 m, 4700 m and 5200 m. Although the depths of the detected reflectors are not identical to each other since the two analyses are completely independent, they mutually agree fairly well. This comparison suggests that the AE reflection method possibly provides significant information concerning deep geothermal structure at Soultz.

4. Conclusion

We applied the AE reflection method using AE as a wave source to the data from the Soultz HDR site in France, and identified some indicators of deep subsurface structure below the artificial reservoir.

The use of the AE reflection method has many advantages in geothermal fields, such as substantial seismic energy, less wave attenuation, extensive source distribution, and minor cost. In the method, reflected waves obscured by coda can be detected by evaluating change in the linearity of three-dimensional hodograms. 3D inversion is employed to image the deep subsurface structure with a restriction of *S*-wave polarization angles and with a compensation for the heterogeneous source distribution.

In the Soultz HDR site, we could detect reflectors with the assumption of S-S reflection at depths of around 3900 m, 4400 m, 4900 m and 5300 m from the surface. Although there are scant data below the artificial reservoir, the reflector at

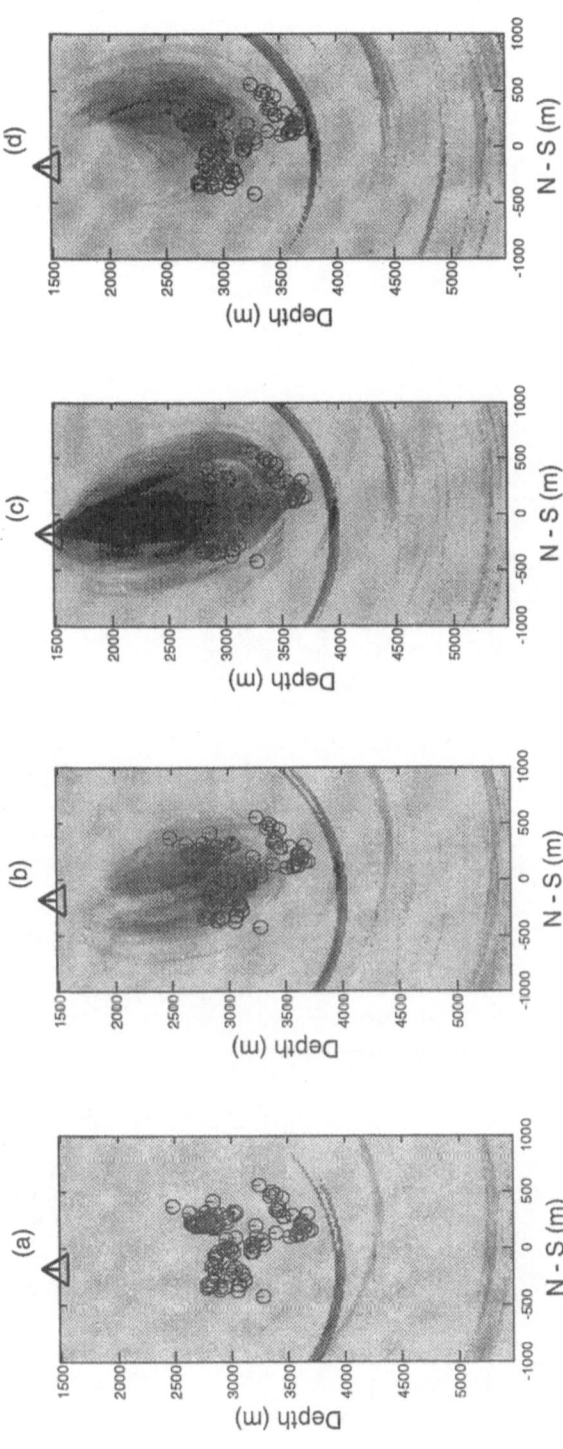

Figure 10

Estimated N-S cross sections by 3D inversion in the Soultz field at the distance of (a) 500 m west, (b) 100 m west, (c) 300 m east and (d) 700 m east from the well-head of GPK-1. Δ and ○ show the projection of the detector and AE sources, respectively.

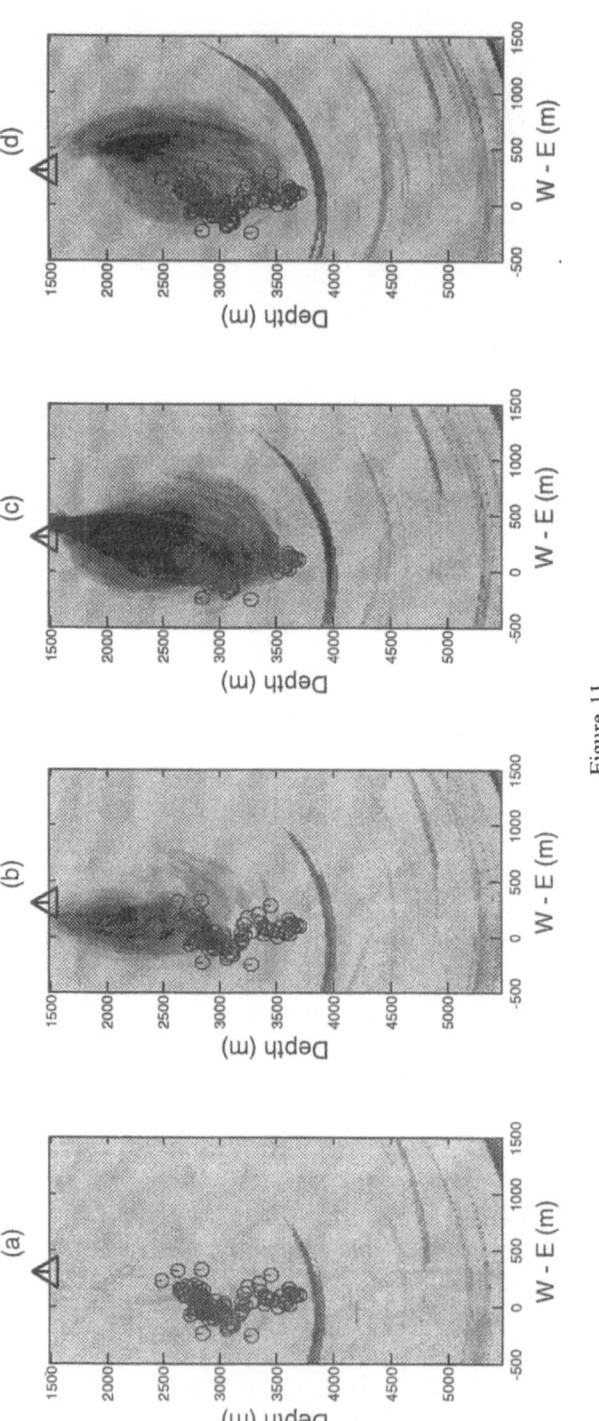

Figure 11

Estimated E-W cross sections by 3D inversion in the Soultz field at the distance of (a) 800 m north, (b) 400 m north, (c) 0 m and (d) 400 m south from the well-head of GPK-1.

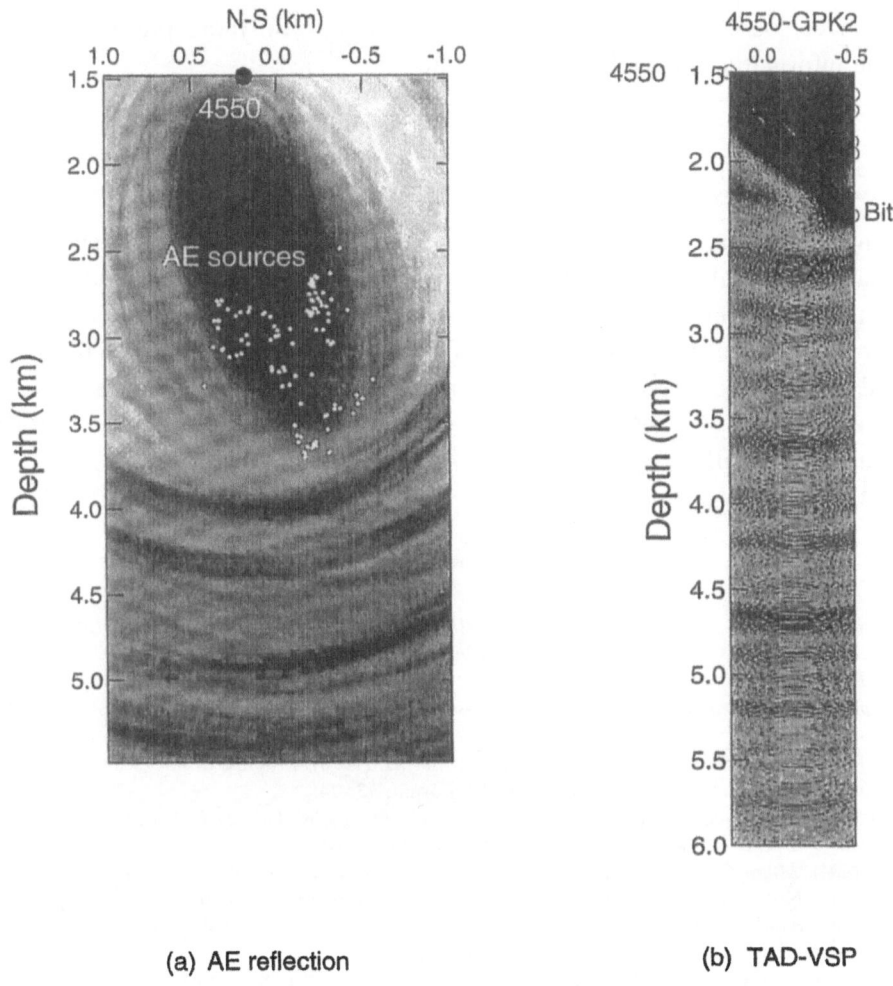

(a) AE reflection   (b) TAD-VSP

Figure 12
Comparison between the AE reflection method and the tri-axial drill-bit VSP method (modified from ASANUMA *et al.*, 1996). (a) AE reflection method, (b) tri-axial drill-bit VSP method.

the depth of 3900 m may be related to the bottom of the artificial reservoir. We could not infer what the other deep reflectors show, although they suggest the existence of deep structure, such as a natural fracture and geological boundaries. Furthermore, small vertical structures were also revealed. It may be possible to detect the vertical fractures both inside and outside the artificial reservoir if we select the optimal AE source distribution. Moreover we compared the present results with those from the tri-axial drill-bit VSP method. Both of the estimated reflectors agreed fairly well with each other.

As a next step, the use of the data observed at the other stations and optimal selection of source distribution would enable us to gain more precise and more reliable information at the Soultz HDR site.

Acknowledgements

This work was carried out as a part of the MTC (More Than Cloud) international collaborative project. We are thankful for support from NEDO (International Joint Research Grant) and the Ministry of Education, Science, Sports and Culture, Japan (Grant-in-aid for General Scientific Research 07455407). We also extend our gratitude to Socomine for providing the data from the European HDR site at Soultz, which is supported mainly by the European Commission, BMBF (Germany) and ADEME (France).

REFERENCES

ASANUMA, H., PARK, J. N., NIITSUMA, H., and BARIA, R. (1996), *Characterization of Subsurface Structure at Soultz-sous-Forêts, France by Triaxial Drill-bit VSP*, SEG Expanded Abstracts, 202–205,

BARIA, R., GARNISH, J., BAUMGÄRTNER, J., GÉRAD, A., and JUNG, R. (1995), *Recent Developments in the European HDR Research Programme at Soultz-sous-Forêts (France)*, Proc. World Geotherm. Cong., 2631–2637.

BARIA, R., GÉRAD, A., BAUMGÄRTNER, J., and GARNISH, J. (1996), *Progress at the European HDR Site at Soultz, France*, Proc. 3rd International HDR Forum, 73–74.

BERKHOUT, A. J., *Seismic Migration—Imaging of Acoustic Energy by Wave Field Extrapolation, vol. 14A, Theoretical Aspects* (Elsevier, Amsterdam and New York 1980).

IIO, Y. (1984), *Micro-Fracture Induced by an Explosion*, J. Seism. Soc. Japan *37*, 109–118 (in Japanese).

NAGANO, K., NIITSUMA, H., and CHUBACHI, N., *A new automatic AE source location algorithm for downhole triaxial AE measurement.* In *Progress in Acoustic Emission III* (eds. Yamaguchi, K. *et al.*). (The Japanese Soc. for NDI 1986), pp. 396–406.

NIITSUMA, H., NAKATSUKA, K., CHUBACHI, N., YOKOYAMA, H., and TAKANOHASHI, M. (1985), *Acoustic Emission Measurement of Geothermal Reservoir Cracks in Takinoue (Kakkonda) Field, Japan*, Geothermics *14*, 525–538.

SAMSON, J. C. (1977), *Matrix and Stokes Velocity Representations of Detectors for Polarized Waveforms: Theory, with Some Applications to Teleseismic Waves*, Geophys. J. R. Astr. Soc. *51*, 583–603.

SOMA, N., and NIITSUMA, H. (1997), *Identification of Structures within the Deep Geothermal Reservoir of the Kakkonda Field (Japan) by a Reflection Method Using Acoustic Emission as a Wave Source*, Geothermics *26*, 43–64.

(Received September 5, 1996, accepted April 17, 1997)

Pure appl. geophys. 150 (1997) 677–691
0033–4553/97/040677–15 $ 1.50 + 0.20/0

┌ Pure and Applied Geophysics

A Miniature Seismic Reflection System for Evaluation of Concrete Linings

F. P. Hassani,[1] A. Sadri[1] and M. Momayez[1]

Abstract—The study presented in this paper demonstrates the application of a miniature seismic reflection (MSR) system as a nondestructive testing tool for evaluation of concrete shaft and tunnel linings. First, the principles of the system are described. Then, results obtained from experimental studies on a concrete shaft lining are presented. Various sections of a shaft lining at different elevations are investigated and their elastic properties at each point are calculated. In another field study, the thickness of the tunnel lining is computed using this system. In both cases, the MSR values are in agreement with results obtained by independent laboratory testing of core samples extracted from the linings.

Key words: Concrete tunnel and shaft lining, seismic reflection, nondestructive testing, MSR system, wave propagation.

Introduction

Shaft and tunnel concrete linings are hollow circular and semi-circular structures that are surrounded by rocks and soil (see Fig. 1). The integrity of linings depends in great part on their original mix, thickness and reinforcement bars. Long-term exposure of concrete linings to hostile physical and chemical variations causes deterioration and defects within the structures. A traditional method of monitoring concrete quality is visual inspection and core extraction. However, visual inspection is not efficient and coring is costly and might create structural damage. In addition, the extracted samples do not reflect the *in situ* behavior of the structure because they represent a small portion of it. The problems such as subsurface deterioration, buckling, fracturing, and the presence of inner voids are not truly identifiable by random coring and visual inspection. To prevent the problems associated with traditional methods, nondestructive testing (NDT) techniques could be used. The assessment of the condition of linings in underground excavations is difficult since most of the deterioration process takes place in the rock side or blind side (in the rock or soil/concrete interface) of the lining. In order to overcome this problem, a data acquisition system based on the seismic reflection principle was developed. Because of the size of the equipment and the range of

---

[1] Department of Mining and Metallurgical Engineering, McGill University, Montreal, Canada.

frequencies used, the new technique is defined as a "Miniature Seismic Reflection" (MSR) system (SADRI, 1996). This system is capable of evaluating the quality of concrete linings and rocks (MOMAYEZ *et al.*, 1995) by providing dynamic elastic properties and thickness. In addition, the system is capable of detecting voids, discontinuities and position of the reinforcement bars within the lining and at the interface with the surrounding rock. At this stage of development the system requires cores extracted from the lining for calibration purposes. In the future, this problem can be resolved by using the MSR system in combination with other NDT techniques such as ground probing radar (GPR).

Background

The propagation of transient stress waves through a heterogeneous bound solid such as concrete member, is a complex phenomenon. However, a basic understanding of the relationship between the physical properties of a material and the velocity of wave propagation can be acquired from the theory of wave propagation in infinite homogeneous isotropic elastic material (TIMOSHENKO and GOODIER, 1970). The use of the MSR system is based on the introduction of a transient stress pulse into a test object by a point impact source. The stress pulse produces compressional (P) and shear (S) waves, which propagate into the test object along hemispherical wavefronts, and Rayleigh (R) waves which propagate on the surface along circular wavefronts.

These waves are partially reflected by internal and external boundaries of the testing object. Seismic waves are reflected by an interface where there is a sudden change in the acoustic properties. In concrete shaft and tunnel linings, reflecting interfaces could be cracks, internal defects, reinforcement bars or contacts between the concrete and rock or soil. In the case of P and S waves, the reflected waves propagate toward the surface and are reflected back into the object. The multiple reflection of these waves produces a periodic transient signal on the surface where the impact was initiated. For the R waves, reflections are caused from surface boundaries.

The amplitude of particle displacements resulting from the impact depends on the direction of propagation. P-wave particle displacements are larger along raypaths with low radiation angles with respect to the angle of incidence and are at their maximum beneath the point of impact. In concrete, when the impact source and the receiver are placed next to each other, P-wave reflections from the top and the bottom of the medium dominate the recorded signal. S-wave particle displacements are larger at an intermediate angle with respect to the angle of incidence. For a concrete with a Poisson's ratio between 0.15 and 0.25, the maximum S-wave energy propagates at an angle of 35° to 45° with respect to the angle of impact (MILLER and MCINTIRE, 1987). Therefore, reflected S waves are stronger farther away from the impact source. In a concrete lining where thicknesses are always smaller than the lateral dimensions, S-wave reflections are best captured at a distance equal to the thickness, away from the impact source. The initial portion of the recorded waveform also includes a dominant R-wave displacement.

In the MSR system a pair of vertical and horizontal displacement transducers are used to detect surface displacements caused by the arrival of reflected waves. Both transducers detect surface displacements generated by P, S, and R waves. However, the vertical displacement transducer is more sensitive to vertical motions caused by the arrival of R waves and P-wave reflections. Similarly, the horizontal displacement transducer is more sensitive to horizontal motions generated by the R and S waves.

The recorded periodic waveforms can be used to compute the time it takes for P and S waves to travel between the surface of concrete and any given reflecting interface. The period of P- and S-wave reflections is related to the path length (or twice the thickness) and their respective velocity in the medium. This period is defined as follows:

$$\Delta t = \frac{2T}{C} \tag{1}$$

where C is the velocity of P or S waves and T is the distance between the surface and the reflecting interface. Therefore, when P- and S-wave velocities in the concrete are known, the thickness can be determined. However, the analysis of the data in the time domain can be complex and time consuming. The analysis of the waveforms can also be carried out in the frequency-domain. Typically, periodic P-

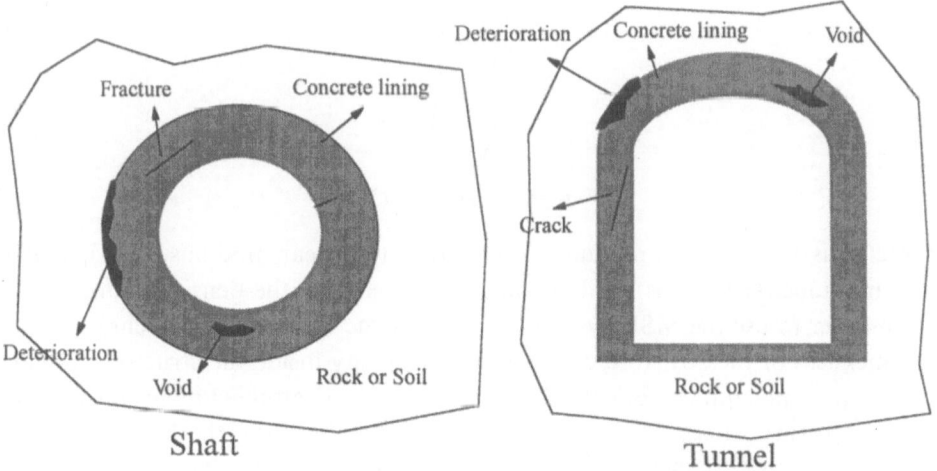

Figure 1
Shaft and tunnel concrete linings.

and S-wave reflection waveforms captured by each transducer are represented by dominant peaks in the spectrum at a frequency equal to $1/\Delta t$. In the frequency spectra calculated from waveforms obtained by either the vertical or horizontal transducer, a high-amplitude low-frequency peak corresponding to the R wave displacement is always present. The dominant P- or S-wave peaks could be from the bottom interface, in the case of sound concrete, and/or from flaws and defects that exist within the concrete. Knowing that period and frequency are inversely proportional, equation (1) can be written as:

$$f = \frac{C}{2T} \tag{2}$$

$$C = 2T \times f, \tag{3}$$

where $2T$ is the two-way path length that each wave travels as a result of reflection from an internal or the external boundary, and f is the frequency of P- or S-wave reflections from the interface. The signals are transformed from time domain to frequency domain, using a Fast Fourier Transform (*FFT*) algorithm.

P- and S-wave velocity measurements are important because P-wave velocity can be used to measure the thickness or to detect the defects within the structure while S-wave velocity can be used to compute the dynamic elastic properties if the density of concrete is known. For a homogeneous isotropic solid, dynamic elastic constants can be calculated from estimates of C_p and C_s using the following equations:

$$E = C_p^2 \times \frac{(1+v)(1-2v)}{1-v} \times \rho \tag{4}$$

$$G = C_s^2 \times \rho$$

$$v = \frac{\left(\dfrac{C_p}{C_s}\right)^2 - 2}{2\left(\dfrac{C_p}{C_s}\right)^2 - 2} \tag{6}$$

$$K = \rho \times (C_p^2 - \tfrac{4}{3}C_s^2) \tag{7}$$

where: E is the Young's modulus (GPa), G is the shear modulus (GPa), K is the bulk modulus (GPa), v is the Poisson's ratio, and ρ is the density (kg/m^3).

In order to use the MSR system as a nondestructive evaluation technique, either the thickness of the structure or the P-wave velocity inside the structure should be known. In many applications the P-wave velocity is determined by testing a portion of the structure where the thickness is known. In this case, the known thickness, and the measured frequency, can be used to calculate the P-wave velocity, by using equation (3). In the case of underground concrete linings it is difficult to obtain accurate information regarding the thickness and quality of the material, since the

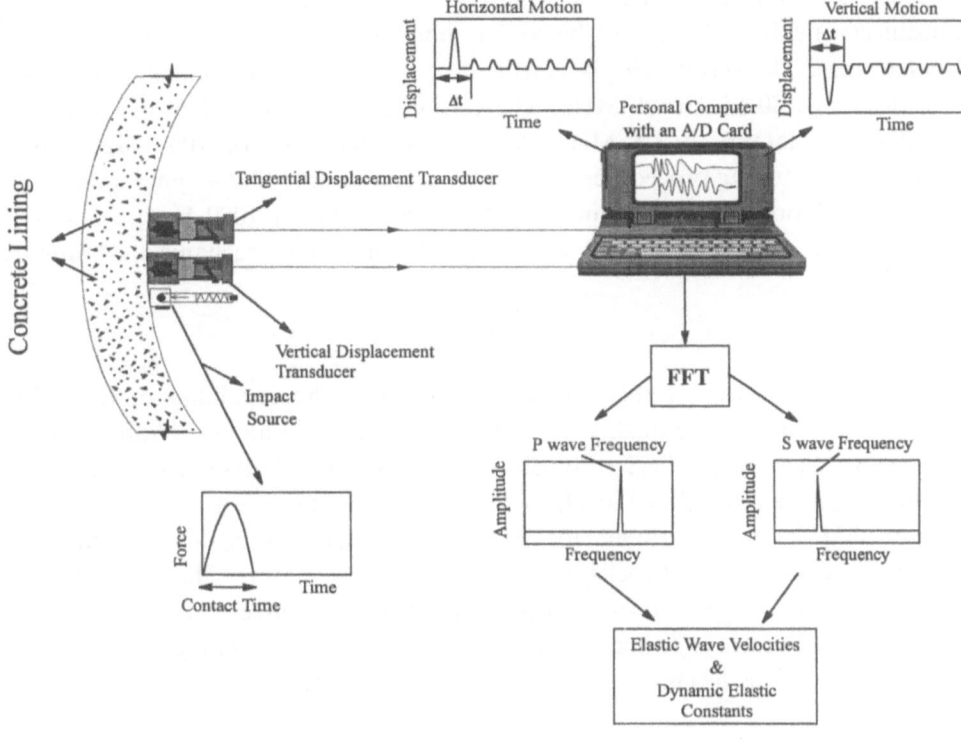

Figure 2
Schematic diagram of the data acquisition system.

condition of the blind side is not known. Therefore, extraction of core samples can provide information about the thickness and the P-wave velocities.

An alternative technique to coring for measuring P-wave velocity would be the measurement of the first P-wave arrival. Here, at least five readings are collected along the surface of the concrete structure and travel times are plotted as a function of station spacing. The slope of the best fit line gives a measure of the P-wave velocity along the surface. P-wave velocities obtained by this method are slightly smaller than most of the reflected P waves due to the heterogeneities within the concrete.

Instrumentation

The MSR system includes a series of spring loaded impact devices ranging between 0.5 mm to 254 mm in diameter, one broadband vertical and horizontal displacement piezoelectric transducer and amplifier, an A/D data acquisition board and a portable computer (Fig. 2). The short duration, low energy transient impacts are responsible for generating low strain rates in the range of 0.5–4.0 $\mu\varepsilon$/s. Both transducers are broadband and behave as point receivers because of their small contact areas. The vertical displacement transducer has a flat frequency response over the range 50 kHz to 1 MHz. The response of the horizontal displacement transducer is flat up to 1.5 MHz. A sampling frequency of 100 kHz is used throughout the testing procedures, and a total signal length of 2048 points was used for the calculations. In the frequency spectra, values below 1000 Hz are removed since they are often related to R waves and electrical and mechanical vibrations.

Case Studies

The main objective of this project was to study the capability of the MSR system for evaluating the quality of underground concrete linings. Shaft No. 2 at the Allan Division of PCS (Potash Corporation of Saskatchewan, Inc.) in Saskatchewan was considered ideal for this study. The site was a production shaft, causing unique time constraints for testing. The external ring of shaft No. 2 is covered by various types of rocks, mainly of sedimentary nature. The inner concrete surface was smooth in various sections and there was no evidence of spalling. Drilling and laboratory testing were conducted to confirm the lining thickness and condition at various levels.

1. Application of the Technique to Shaft Lining

The MSR system was used in this mine to evaluate the *in situ* dynamic elastic properties of the shaft lining at three stations. The shaft, constructed in the 1960s, extends down to 975.4 m located in Prairie Evaporite Formation and is 4.9 m in diameter (Fig. 3). Immediately overlaying the Prairie Evaporite Formation are carbonates, shale and glacial till. The concrete design specification for the shaft lining was sulfate-resistant with 34.5 MPa compressive strength and a density of 2590 kg/m^3. The shaft lining was tested at three depths (640.0 m, 823.0 m and 914.4 m) with variable concrete thicknesses (0.58 m, 0.58 m and 0.57–0.61 m). Thickness values from drilling were used in the calculations. The information about the position and thickness of the reinforcement bars was not available. The lining at the first two test sites at 640.0 m and 823.0 m was surrounded by limestone and the third site at 914.4 m was surrounded by halite.

Experimental Setup

The equipment was placed on top of the personnel career cage and the tests were conducted from the top of the cage. The semi-spherical tipped impact device with a diameter of 2.54 cm was used in all three locations to generate the initial stress pulses. The contact time of impacts was calculated to be between 280 and 350 μs. The readings were taken from two different locations on the accessible side of the concrete lining at each station. For the detection of the *P*-wave displacements, the vertical displacement transducer was placed 5 to 10 cm away from the impact source. To detect the *S*-wave vibrations, the tangential displacement transducer was placed at a distance equal to the thickness of the lining.

Results and Discussion

Figure 4 shows examples of waveforms and frequency spectra collected by the vertical and horizontal displacement transducers at each station. Based on the

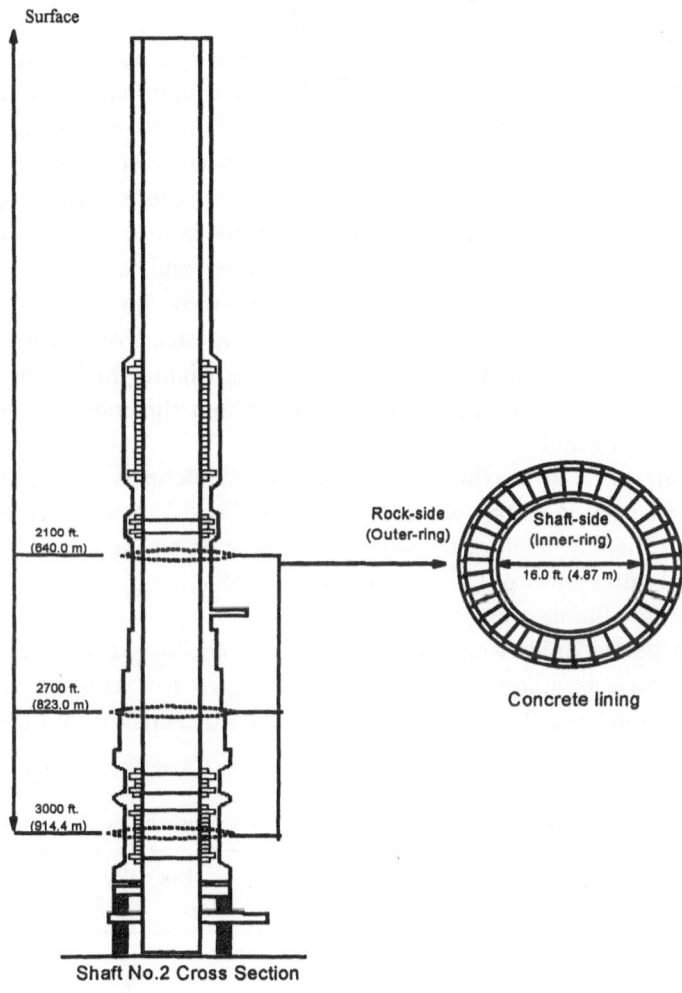

Figure 3
Schematic diagram of Allan Mine and the MSR testing locations (drawing not to scale).

Table 1

A comparison between the P- and S-wave velocity values obtained by the MSR and laboratory measurements

Depth (m)	Thickness Concrete coring (m)	Thickness Construction plans (m)	Laboratory measurements		In situ measurements	
			C_p (m/s)	C_s (m/s)	C_p (m/s)	C_s (m/s)
640.0	0.58	0.61	4620	2750	4418	2662
823.0	0.58	0.61	4640	2680	4248	2209
	0.55	0.74				
914.4	0.61	0.56	4500	2725	4050	2502
	0.57	0.69				
	0.57					

distinct frequency values (largest amplitude frequency peaks) in each spectra, the *P*- and *S*-wave velocities were determined for the three stations and are listed in Table 1. Equation (3) was used to calculate *P*- and *S*-wave velocities. Knowing the density from the extracted cores and the *P*- and *S*-wave velocities, dynamic elastic constants of concrete lining are calculated using equations (4) through (7).

The acoustic impedance, $Z2$ $(C \times \rho)$, of the rocks in contact with the concrete lining for the three testing stations was calculated and is listed in Table 2. The values of acoustic impedance for the concrete were higher than those of the surrounding rock at the 640.0 m and 914.4 m stations. For the testing station at 823.0 m the acoustic impedance of the rock was slightly higher than that of the concrete (Table 2). Because of this it is assumed that the concrete and rock in this area are not well bonded.

Core samples of approximately 3.18 cm in diameter were extracted from the same levels the MSR tests were conducted by Core Laboratories in Calgary for measuring acoustic velocities, densities, and dynamic elastic constants. The system used was an ultrasonic pulse velocity (UPV) device which had automatic pore pressure and confining pressure controllers.

The acoustic velocities were measured at the specific pressures (12.5, 16.5, and 19.5 MPa) corresponding to overburden pressure for samples extracted from 640.0 m, 823.0 m, and 914.4 m, respectively. The densities were measured at each

Table 2

Density, P-wave velocity and acoustic impedance values at the three stations

Depth (m)	Lithology	Density (kg/m^3)	C_p (m/s)	Rock (Z2) (kg/m^2 s) $\times 10^6$	Concrete (Z1) (kg/m^2 s) $\times 10^6$
640.0	Limestone	2370	3850	9.1	12.9
823.0	Limestone	2920	4750	13.9	12.0
914.4	Halite	2110	4200	8.8	11.5

Table 3

Poisson's ratio and dynamic elastic moduli from laboratory and in situ measurements

Depth (m)	Laboratory measurements				In situ measurements			
	v	E (GPa)	G (GPa)	K (GPa)	v	E (GPa)	G (GPa)	K (GPa)
640.0	0.23	48.1	17.9	27.0	0.22	44.4	18.4	26.4
823.0	0.25	47.0	17.4	28.7	0.31	33.8	12.7	29.6
914.4	0.21	46.1	17.5	23.9	0.19	38.3	16.0	20.6

specific pressure and the computed values showed slight variations between 2560 and 2600 kg/m^3. *P*-wave velocity in the cores ranges from 4496 m/s to 4641 m/s, while the *S*-wave velocity ranges from 2688 m/s to 2745 m/s. The C_p/C_s ratio, Poisson's ratio, bulk modulus, and shear modulus are calculated from the velocity and density data. For comparison purposes, the results obtained by the MSR system *in situ* and the UPV technique in the laboratory are listsed in Table 3. It can be seen that the *in situ* results follow the trend of laboratory values, although they are consistently lower except for the Poisson's ratio and bulk modulus values for the station at 823.0 meters and the shear modulus at 640.0 m. Results obtained by MSR demonstrate that *P*-wave velocities obtained from the laboratory measurement are higher by 4.4% at the first station, 8.5% at the second station, and 10.0% higher at the third station. For *S* waves the UPV values are higher by 3.2% at the first station, 17.6% at the second station, and 8.9% at the third station. Both *P*- and *S*-wave velocities obtained in the laboratory and field drop progressively with increasing depth. The calculated dynamic elastic constants obtained *in situ* are generally lower than the values obtained in the laboratory. Similar to the wave velocities, elastic constants decrease as the depth increases. In the frequency spectra obtained by the two transducers, other significant peaks exist (Figs. 4b and 4d). These frequencies are related to the existence of reinforcement bars and potential discontinuities or voids within the concrete structure.

2. Application of the Technique to Concrete Tunnel Linings

In this application, the MSR system is used to examine the thickness of concrete tunnel lining in one of the city of Montreal's underground subway stations. Since the construction of the subway system in 1965, the circulation of saline water behind the lining has caused extensive deterioration to the outer section of the concrete lining. Figure 5 illustrates a schematic diagram of the concrete lining and the deterioration of the outer ring. The deteriorated tunnel section is located about 30.0 meters below the surface. The reduction of the concrete thickness at the rock side is not uniform. Most of the damage has been done at the rock/soil interface of the lining.

In order to estimate the thickness of the lining, the average *P*-wave velocities in the concrete should be known. In this case, the vertical displacement transducer was used. An average *P*-wave velocity of 4330 ± 42 m/s for the concrete was calculated from the areas in which the thickness was known. To examine the results, the calculated thickness values were compared with cores extracted from the concrete lining by the city of Montreal. The results from the laboratory evaluation of the cored samples are listed in Table 4.

Experimental Setup

The access to the site was provided between 2 and 5 a.m., while the subway was closed. The concrete mix was for a 28-day period with a strength of 35.0 MPa. According to the original drawings, there was no indication of reinforcement bars in the roof. This area of the lining was subjected to extensive damage caused by the groundwater infiltration at the contact between the surrounding rock (limestone),

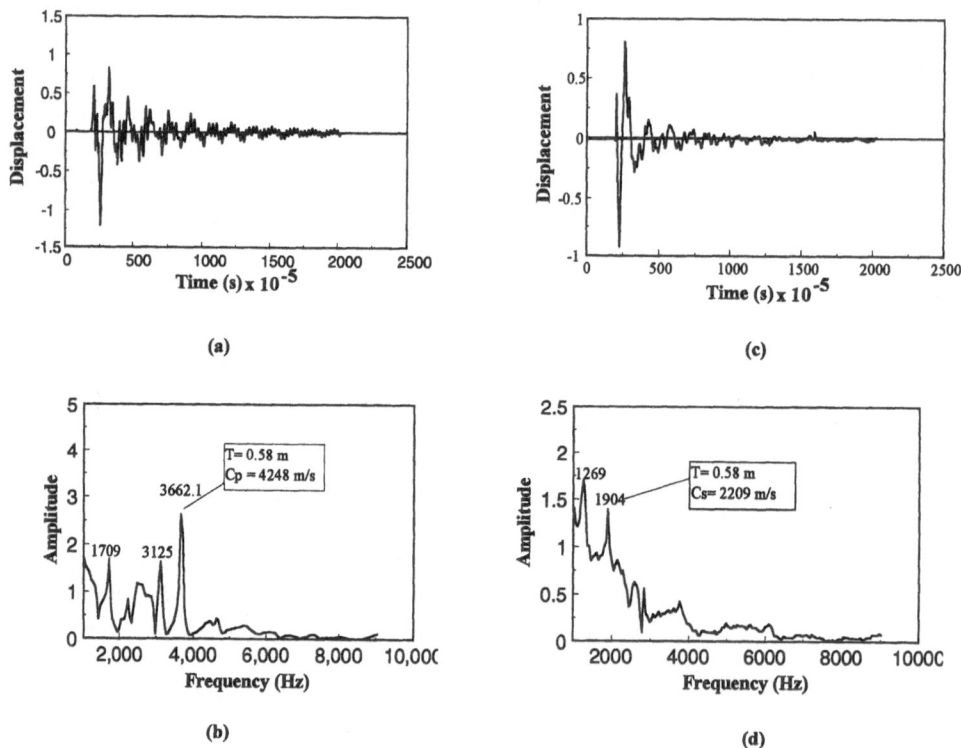

Figure 4
Sample waveforms captured by the vertical *(a)* and horizontal *(c)* displacement transducers and their associated frequency spectra.

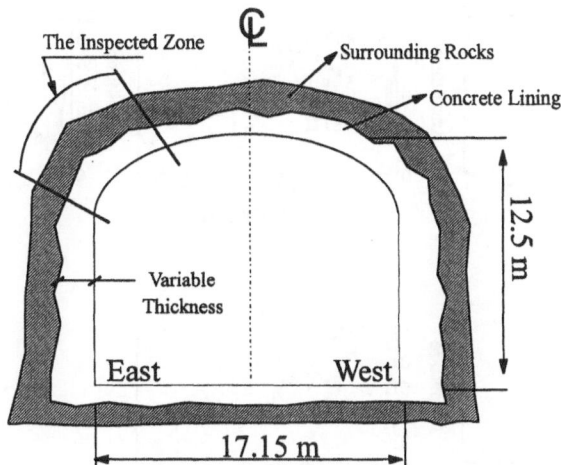

Figure 5
Schematic diagram of Montreal subway station concrete tunnel lining (drawing not to scale).

overburden (till), and the concrete. A hydraulic lift was used to provide easy access to the roof and the sides of the tunnel. There was evidence of spalling and the concrete surface was rough.

For the thickness measurements, only the vertical displacement transducer was used. The 3.0 mm impact device was used to introduce the transient waves into the lining. The contact time of impacts was between 45 to 75 μs. The distance between the impact device and the receiver was kept less than 0.2 times the assumed thickness of the structure (5 to 10 cm) at all times. The thickness of the concrete tunnel lining was evaluated at 20 testing stations. Each station was 0.3 m apart, and in total an area of 2.50 m^2 on the upper east side of the tunnel was surveyed. Four parallel lines (A, B, C, and D) were surveyed. On each line 4 to 6 stations were examined.

Table 4

Results from laboratory evaluation of concrete core samples from the tunnel lining

Sample No.	Thickness (m)	Segregation	Flaw presence	Flaw depth (m)	Rock/concrete separation
1	0.69	no	yes		yes
2	0.51	no	yes		no
3	0.64	no	yes	0.27, 0.39	no
4	0.43	no	yes	0.31	no
5	0.53	no	yes	0.28, 0.48	yes
6	0.56	yes	yes	0.18, 0.31	no
7	0.58	no	yes	0.31, 0.47	yes

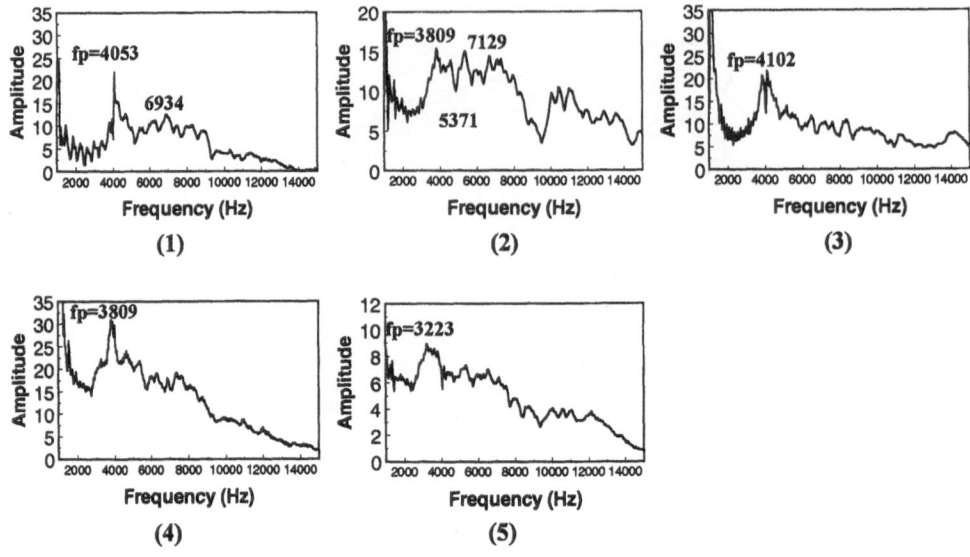

Figure 6
Sample frequency spectra from the concrete tunnel lining.

Results and Discussion

Both surrounding materials (limestone and till) have lower acoustic impedance than the concrete (see Table 5). In the frequency spectra, the single clear high peak corresponded to multiple reflections of *P*-wave signals from the concrete/rock or concrete/soil interface. Figure 6 illustrates examples of the frequency spectra collected from the tunnel lining. A peak around 4000 Hz within the spectra is related to the AC electrical current in the tunnel and is ignored in the calculations. Other peaks in the spectra may be due to flaws, delaminations and discontinuities present between the surface of the concrete and its bottom concrete/rock interface. The *P*-wave velocities were computed experimentally, through a simple calibration procedure in an area of known thickness. Using the average *P*-wave velocity of $4330 + 42$ m/s (density 2310 kg/m^3), thicknesses for the concrete lining in various areas were calculated and are listed in Table 6. These values suggest a thickness variation between 0.45 to 0.67 m.

Table 5

Density, P-wave velocity and acoustic impedance values for concrete tunnel lining

Medium	Dry density (kg/m^3)	C_p (m/s)	Rock (Z2) (kg/m^2 s) $\times 10^6$	Concrete (Z1) (kg/m^2 s) $\times 10^3$
Limestone	2690	2400	6.4	10.0
Till	1601	800	1.2	10.0

Table 6

Thickness and flaw measurements lines A, B, C and D

Station	Thickness frequency (Hz)	Calculated thickness (m)	Flaw depth (m)
A-1	4053	0.53	0.31
A-2	3809	0.57	0.40
A-3	4102	0.53	0.30
A-4	3809	0.57	–
A-5	3223	0.67	–
B-1	4346	0.50	–
B-2	4530	0.53	–
B-3	4199	0.52	–
B-4	4004	0.54	0.33
B-5	4541	0.48	0.28
C-1	4541	0.48	0.28
C-2	4297	0.50	–
C-3	3711	0.58	–
C-4	3662	0.59	0.33
C-5	3662	0.59	–
C-6	3711	0.58	–
D-1	4785	0.45	–
D-2	4590	0.47	–
D-3	3906	0.55	–
D-4	4443	0.49	0.33
–	–	–	0.22
–	–	–	–

A comparison between the calculated (0.45 to 0.67 m) and cored thickness values (0.43 to 0.69 m) indicates a reasonable agreement. The discrepancy between the MSR and cored values is due to the following reasons. First, the cores were not extracted at the precise locations that the MSR tests were conducted. Therefore, the thickness and flaw locations might be slightly different and an exact correlation between the measured and observed values was not possible. Second, the P-wave velocity used for the calculation is an average value and is not consistent throughout the lining. Because of the quality of the concrete, the measured P-wave velocities can be lower or higher than the average value. Therefore, the thickness and flaw locations are detected with a slight discrepancy. The results indicated that the concrete lining has an irregular outer shape at the contact with the surrounding rock. On the frequency spectra a number of distinct frequency peaks suggest the possibility of the presence of voids and fractures in concrete (see Figures 6-1, 6-2 and 6-3) which can be confirmed by cored samples (see Table 4). These points were identified from their high frequency peaks in the spectra which is indicative of potential flaws. Core samples 1 and 2 were extracted from the vicinity of line A. The thickness of 0.69 m from cored sample 1 is close to the value of 0.67 m determined at station A-5 and the thickness value of 0.51 is similar to the value of 0.53 m computed at stations A-1 and A-3. Core samples 5, 6, and 7 were extracted from the vicinity of line B and line C. Thickness values of 0.53, 0.56 and 0.58 from these three cores are similar to the thickness values from stations B-2, B-3, B-4, and C-3 through C-6. Also all three cores show flaws located at 0.28 and 0.31 m, which correlate with the flaws detected in these two lines at 0.28 and 0.33 m. Core sample 4 was extracted from the vicinity of the line D where thickness values between 0.45 and 0.55 m and flaws at 0.22 and 0.33 m were detected. In comparison with the thickness measurement of 0.43 m and the flaw position of 0.31 m obtained from the extracted core, the *in situ* values are reasonable.

Conclusions

The conclusions drawn from the field experimentation are the following: for the shaft lining, the concrete thickness was known, and the dynamic elastic properties of concrete lining were evaluated. The *in situ* results were compared to the laboratory values. The trend of the results is similar for both cases, however, the *in situ* values are generally lower than the laboratory readings. For this tunnel lining, the average *P*-wave velocity of the concrete was calculated in the areas where the thickness was known. On the frequency spectra a number of flaws was identified. Although a direct comparison was not possible, thickness and flaw measurements by the MSR system are in agreement with the results of laboratory measurements carried out on extracted core samples. The results in this field study showed the ability of the technique to evaluate concrete lining thicknesses. As for the concrete shaft lining, a periodical monitoring can foretell potential deteriorations and help prevent structural damage.

The MSR equipment assembly is well suited for testing in an underground environment. The results of the two case studies were encouraging and confirmed that the system can be used as a nondestructive testing technique in order to evaluate the thickness, elastic properties and integrity of the concrete linings in an underground environment. Further hardware and software development are being carried out by the authors to improve the accuracy and ease of use of the system.

Acknowledgment

The authors would like to thank the Saskatchewan Potash Producers Association (S.P.P.A.) for providing access to the Allan Mine. Many thanks to Mr. Moe Molavi, Chief Engineer of Allan Mine and Dr. Arnfinn Prugger of S.P.P.A., for their assistance and interest in this project. The authors acknowledge the financial contributions made by the Natural Science and Engineering Research Council of Canada (NSERC).

REFERENCES

MILLER, R. K., and McINTIRE, P. (1987), *Nondestructive Testing Handbook, Volume 5: Acoustic Emission Testing*, American Society for Nondestructive Testing, pp. 92–122.

MOMAYEZ, M., SADRI, A., and HASSANI, F. P., *Impact-echo: A technique for determining the mechanical properties of rocks*. In *Rock Mechanics*, Proceedings of the 35th U.S. Symposium, 5–7 June, University of Nevada (A. A. Balkema-Rotterdam-Brookfield) pp. 843–848.

SADRI, A. (1996), *Application of the MSR System for Quality Evaluation of Concrete Shaft and Tunnel Linings*, Ph.D. Thesis, McGill University, Montreal, Canada, pp. 8.1–8.26.

TIMOSHENKO, S., and GOODIER, J., *Theory of Elasticity*, 3rd Edition (McGraw-Hill Book Co., New York 1970) 567 pp.

(Received August 30, 1996, accepted July 29, 1997)

Pure appl. geophys. 150 (1997) 693–704
0033–4553/97/040693–12 $ 1.50 + 0.20/0

Pure and Applied Geophysics

Analysis of High Frequency Microseismicity Recorded at an Underground Hardrock Mine

STEPHEN D. BUTT,[1,2] DEREK B. APEL[1,3] and PETER N. CALDER[1,4]

Abstract—This research involved monitoring for high frequency (HF) microseismic activity, in the effective frequency range of 100 to 400 kHz, at an underground hardrock mine. An HF monitor was installed to record activity in and near a pillar at a depth of 2100 m near active mining stopes. Analysis of the recorded events was done in light of production blasting records, the mine-wide microseismic data set for events below a depth of 1800 m and laboratory studies conducted to assist with a general HF data analysis. Data analysis indicated that there were two types of periods of increased activity: the first associated with the passage of the direct seismic waves from the blast through the local area, and the second associated with the arrival of the transient stress change induced by the blast in the local area. Indications of locally increased stress levels towards the end of the monitoring period agreed with the occurrence of two nearby strong seismic events.

Key words: High frequency, induced seismicity, Creighton Mine.

1. Introduction

In recent years, there have been many advances in the application of microseismic monitoring technology in underground mines. Many of these advances have been associated with multichannel, full waveform microseismic monitoring networks. These mine-wide systems typically monitor at frequencies less than 10 kHz. In contrast, high frequency (HF) microseismic systems often monitor in a frequency band greater than 40 kHz and up to several hundred kHz. These monitoring systems have the advantages of spatial filtering (with detection ranges generally less than 10 m, based on observed *P*- and *S*-wave separation) and increased sensitivity as compared to mine-wide systems. Thus, these systems are suited to monitor a local area for the weak HF microseismic activity which is often observed in

---

[1] Department of Mining Engineering, Queen's University, Kingston, Ontario, Canada, K7M 1B6.
[2] Now at Department of Mining and Metallurgical Engineering, Technical University of Nova Scotia, Halifax, Nova Scotia, Canada, B3J 2X4.
[3] Now at Cameco Corp., 2121 11th St. S. W., Saskatoon, Saskatchewan, Canada, S7M 1J3.
[4] Now at Centro de Mineria, Escuela de Ingenieria, Pontificia Universidad Catolica de Chile, Santiago, Chile.

laboratory studies in conjunction with sample stress changes and, particularly, prior to macroscopic sample failure. It is envisaged that these monitors can be used in areas where sensitive monitoring is required or in areas which have been identified by mine-wide systems as being seismically active. This paper will review the microseismic monitoring system used and detail an 11-day field monitoring trial at a deep underground mine.

2. Monitoring System

The monitor used for the monitoring trial was a QSTRESS system. This system was developed at Queen's University and proved to function well in an underground hardrock mining environment (ARCHIBALD *et al.*, 1988; CALDER and SEMADENI, 1990; CALDER *et al.*, 1990). The system is PC based and uses a quartz piezoelectric transducer mounted on the grounded end of a fully bonded steel support element, such as a grouted rebar or splitset. These types of support elements are in good contact with the borehole wall along their full length and function as waveguides to couple the transducer to the intact rock behind the yielded faces of underground openings. Laboratory and field monitoring tests, comparing this mounting procedure to the more traditional but cumbersome practice of attaching the transducer to the bottom of a borehole, indicate that this is a suitable method of transducer attachment. For example, transducers mounted on the ends of steel support elements have a more uniform directional response than borehole mounted transducers and, for the case of fully grouted rebars, microseismic events originating within the rebar or the grout do not become significant until the rebar is under extreme tensile load and begins to yield.

The transducer used was a Physical Acoustics Corporation Micro30 with peak frequency response at 240 kHz and relative frequency responses of -8 dB at 100 and 400 kHz; a low-pass filter and the transducer's natural frequency response significantly attenuated frequencies outside this band. Sampling was conducted at 500 kHz. During operation, recording was triggered on the level of the transducer output and lasted for 4 ms, including 1 ms of signal prior to the trigger point. For data analysis, the recorded waveforms were broken down into several event parameters including the time of the event trigger, the peak rectified time domain amplitude, the event energy and the dominant frequency. Note that all of these event parameters pertain to the signal as received at the transducer and not at the source. The event energy and the dominant frequency were determined from a Fast Fourier Transform (FFT) of a 1 ms long window centered at the trigger point; the event energy was calculated as the area under the FFT and the dominant frequency was assigned as the peak of the FFT. Estimating received event energy by determining the area under the FFT is a standard practice for waveforms in these high frequency bands (SCOTT, 1991). From the peak amplitudes, *b* values for 100 consecutive events were determined. The *b* value is the slope of the recurrence-magnitude relationship from earthquake studies (GUTTENBERG and RICHTER, 1949; SCHOLZ, 1968) and, in practice, varies inversely with the event energies and peak amplitudes.

In a parallel research program, a standalone analog instrument was developed, the Queen's-MIROC Portable High Frequency Monitor (HFM), which functions similarly to the QSTRESS unit (CALDER *et al.*, 1995). At the time of this monitoring trial, however, the HFM was unavailable for extended field monitoring.

3. Data Analysis Guidelines

Laboratory studies were conducted to assist with the analysis of the field data (BUTT, 1996). Triaxial compression tests indicated that flurries of events with relatively higher dominant frequency often preceded macroscopic sample failure and sudden slips on the crosscutting shear fracture (formed at sample failure) during post failure sliding. These higher frequency events were interpreted as the rapid formation of small microcracks prior to sample failure and asperity breakage. Tests involving transmitting constant source microseismic signals across rock fractures at different levels of normal stress indicated that attenuation decreases with increasing fracture normal stress and stiffness. This is consistent with the results of other workers (e.g., PYRAK-NOLTE *et al.*, 1987). Associated with this decreased attenuation is a shift towards higher dominant frequencies for the received events. These results were used as guidelines during field data analysis to indicate changes in the predominant mode of deformation associated with the recorded events or changes in rock mass attenuation.

4. Transducer Location and Mining Operations

Field monitoring was conducted at the Creighton Mine, near Sudbury, Ontario, Canada during December, 1994. Monitoring was done on the 7000 level, at a depth of approximately 2100 m. Note that at the Creighton Mine, development levels are named based on their approximate depth in feet below the shaft collar. The QSTRESS system was installed to monitor a pillar which had been identified by mine personnel as seismically active in response to production blasting operations in that region of the mine (Fig. 1). The transducer was mounted on the grounded end of a steel splitset which was installed in the pillar. Vertical Cratering Retreat mining methods (Lang *et al.*, 1977) with delayed cemented backfill were being used to mine the ore zone between the 7200 and the 7000 levels and in sill pillars between the 6900 to 6800 levels and the 6700 to 6600

levels. The ore zones above the 7000 level and between the two sill pillars were
already mined out and backfilled using mechanized cut and fill methods with
remnant rib pillars left between the stopes. During the monitoring period, a stope
was being mined on the fringe of the mined out region between the 7200 and 7000
levels and one stope each was being mined in the two upper sill pillars (Fig. 1).

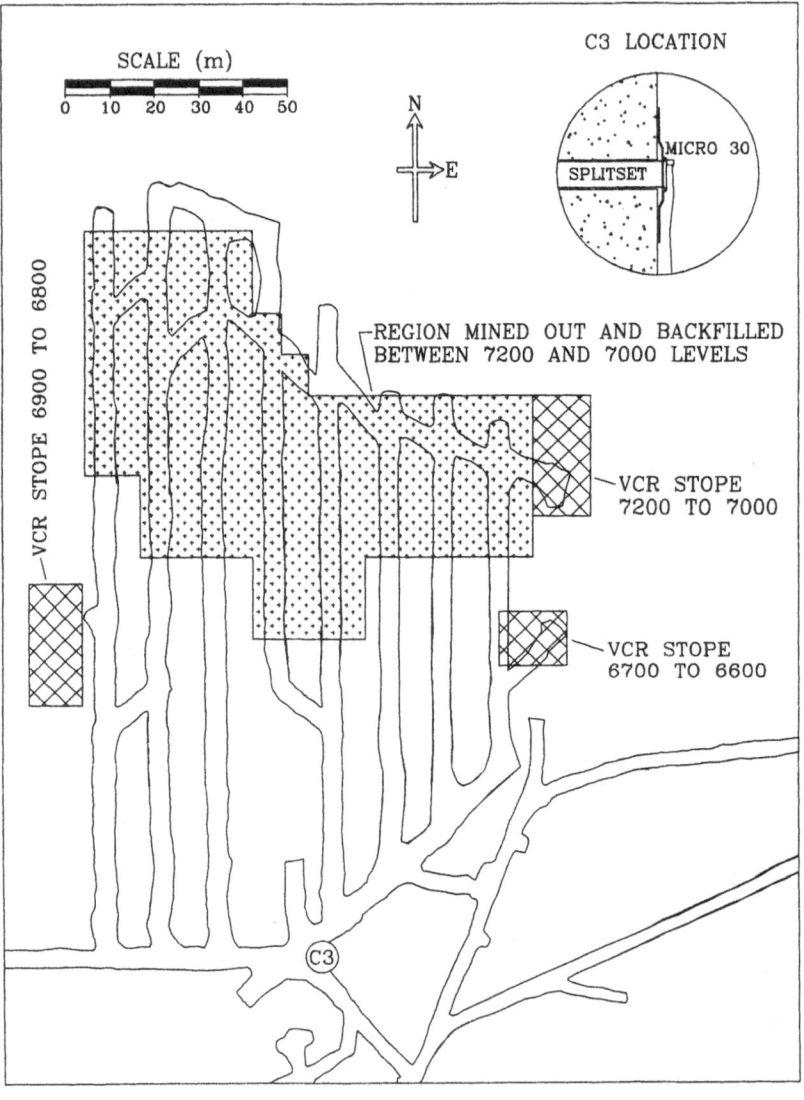

Figure 1
Level plan of the 7000 level of the Creighton Mine. The transducer was installed at the C3 location with
mounting details as shown in the inset, upper right corner.

Table 1

The ranges of intact and rock mass properties of the three rock types in the monitored area (BAWDEN and COULSON, 1993). Intact values are ± one standard deviation

Intact	Granite FW	Sulfide ORE	Norite HW
E (GPa)	57 ± 6	72 ± 9	72 ± 10
Poisson's ratio	0.27 ± 0.03	0.26 ± 0.03	0.28 ± 0.03
density (kg/m^3)	2679 ± 82	3945 ± 484	2878 ± 93
UCS (MPa)	240 ± 16	133 ± 32	210 ± 33
Rock mass			
RMR	62 to 75	58 to 72	57 to 74
E (GPa)	20 to 42	16 to 35	15 to 40
UCS (MPa)	29 to 60	13 to 27	21 to 53

Production blasts, removing 2 to 5 m thick horizontal slices from the bottoms of these retreating stopes, occurred every 1 to 4 days.

During the monitoring period, a 64 channel Electro-Lab MP250 system was in operation covering the 4200 to the 7400 levels of the mine (MORRISON et al., 1989). The complete mine-wide data set for events below the 6000 level was provided with all production blasts clearly identified. Five underground monitoring stations operated by CANMET were used to provide Nuttli magnitudes (NUTTLI, 1973) for the stronger seismic events.

5. Geology and in situ Stresses

The geology in that area of the Creighton Mine consists of granite foot wall, the massive ore zone (nickel and copper sulfides) and norite hanging wall. The ore zone is lens shaped, approximately 150 m thick at its widest point, and dips 80° towards the northwest. The mined out and backfilled region of Figure 1 extends from the foot wall to the hanging wall contacts of the orebody. All the development openings, including the monitored pillar, occur in the foot wall. Table 1 summarizes the intact and rock mass properties of the foot wall, ore and hanging wall rocks. The rock mass stiffness and strength values were derived from the Rock Mass Rating or RMR values (BIENIAWSKI, 1976) and fracture set data following the procedures of HOEK and BROWN (1980). The far-field stresses for the area are based on seven overcore measurements between the 6600 and 7200 levels and are summarized in Table 2. As can be seen, the magnitudes of the principal stresses are at levels comparable to the rock mass unconfined compressive strengths.

Table 2

Principal in situ stresses for the monitored area (BAWDEN and COULSON, 1993). Trends are with respect to mine north in Figure 1

Component	Mag. (MPa)	Trend (deg.)	Plunge (deg.)
Sigma 1	75 ± 14	281	20
Sigma 2	50 ± 15	18	17
Sigma 3	45 ± 10	145	63

6. *Field Monitoring Results*

6.1 *Event Occurrence*

Figure 2 plots the cumulative number of events recorded during the 11-day monitoring period and the times of production blasts. A review of these data indicates that the majority of events occurred as episodes, with event rates of 2 or 3 per second, separated by intervals extending several hours when only single sporadic events were recorded. Further, these events could be grouped into two categories based on their time of occurrence with respect to the production blasts: (1) "immediate" events which occurred immediately after the 7000 level blasts and (2) "delayed" events which occurred several hours after 7000 and 6800 level blasts. These results are summarized in Table 3, which indicates that 94% of the 7087 events recorded during the monitoring period can be grouped into these two categories.

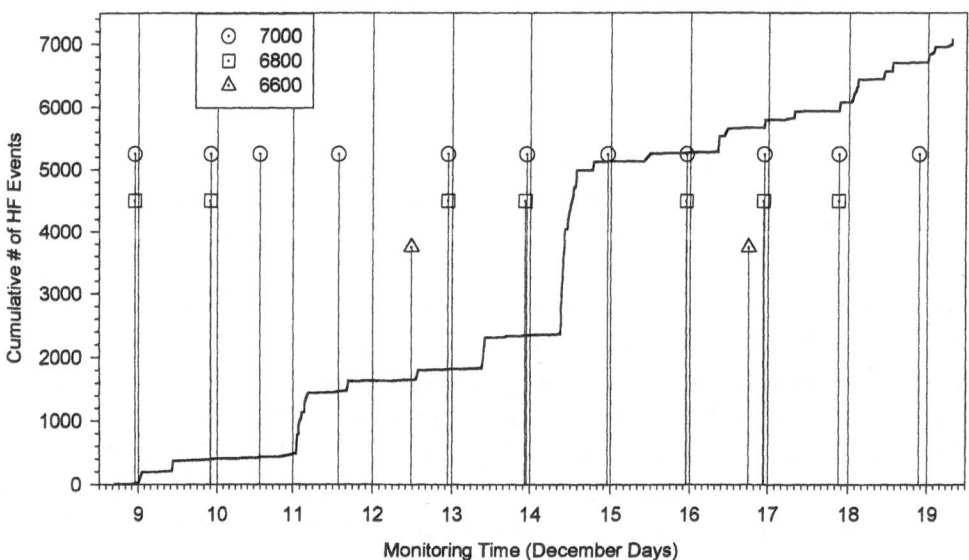

Figure 2

The cumulative number of events recorded during the monitoring period (solid line) and the times of production blasts (symbols).

Table 3

Summary of immediate and delayed episodes of events and times of local production blasts. The time delay for delayed events is relative to the time of the 7000 level blasts

Dec. Day	Time (hours)	Stope (level)	Immediate Events		Delayed Events		
			Delay (min.)	Number of Events	Delay (hours)	Duration (hours)	Number of Events
8	22.89	6800	4.6	5	2.27	0.04	173
8	22.94	7000	1.4		11.66	0.04	160
9	21.92	6800	6.9	9	no identified episode		
9	22.01	7000	1.4				
10	13.71	7000	1.3	7	11.36	3.56	1012
11	13.68	7000	1.3	8	2.89	0.04	159
12	11.51	6600		no correlation			
12	22.76	6800	2.9	9	11.02	0.67	473
12	22.78	7000	1.4				
13	22.51	6800	15.2	6	9.89	4.83	2627
13	22.97	7000	1.4		19.54	0.23	131
14	22.97	7000	1.4	6	12.73	0.04	124
15	22.87	6800	3.4	8	10.01	0.12	248
15	22.90	7000	1.4		12.87	0.04	125
16	18.22	6600		no correlation			
16	22.89	6600	2.7	4 followed by 125 events	8.83	0.08	126
16	22.91	7000	1.4	5.1 minutes later			
17	21.02	7000	1.4	6 followed by 128 events	4.29	1.39	363
17	21.02	6800	1.1	11.3 minutes later	13.61	0.04	123
					15.98	0.04	126
18	21.36	7000	1.4	7	2.67	0.04	122
					4.79	0.04	118
					10.21	0.04	120

With two exceptions, the immediate events occurred as episodes of 5 to 9 events 1.3 to 1.4 minutes after the 7000 level blasts and with no correlation with the other production blasts. The two exceptions to the general pattern occurred on December 16 and 17 when the "regular" number of immediate events was followed a few minutes later by episodes of approximately 130 events. The immediate events are interpreted as the "shake up" of the rock mass near the transducer location resulting from the passage of the direct high amplitude seismic waves from the 7000 level blasts through the local monitored area. The consistent delay of 1.3 to 1.4 minutes was interpreted as the time difference between the QSTRESS and the MP250 clocks since, inadvertently, the two clocks were not set the same. Immediate events were probably not observed for the other production blasts because the direct seismic waves would have passed through highly attenuating backfilled regions before arriving at the monitored pillar.

The delayed events occurred as single episodes of one to several hundred events, or as multiple such episodes separated by several minutes, and they occurred at regular time intervals after the 7000 and 6800 level blasts. Prior to December 16, 7 out of 9 delayed episodes occurred 10 to 13 hours after these production blasts. After December 16, there was a marked decrease in the time interval before the start of the delayed events (3 to 4 hours) and an increase in the number of episodes. The delayed events are interpreted as the arrival of the transient stress change induced by the production blasts in the local monitored area. This interpretation agrees with the patterns of mine-wide events recorded after the production blasts, as is shown in Figure 3 for a blast on December 11; during the first hour after the blast, the mine-wide events are clustered near the production stope but from 11 to 14 hours after the blast, they are spread over a larger volume which includes the transducer position.

6.2 Event Parameters

Analysis indicates that there were two distinct patterns in event parameters for the recorded events: (1) cyclical changes associated with the alternating occurrence of immediate and delayed events, and (2) gradual systematic changes over the duration of the monitoring period. The cyclical event parameter changes, summarized in Table 4, indicate that different modes of deformation were associated with each type of event; the delayed event parameters indicate higher frequency activity, such as microcracking, while the immediate event parameters indicate lower fre-

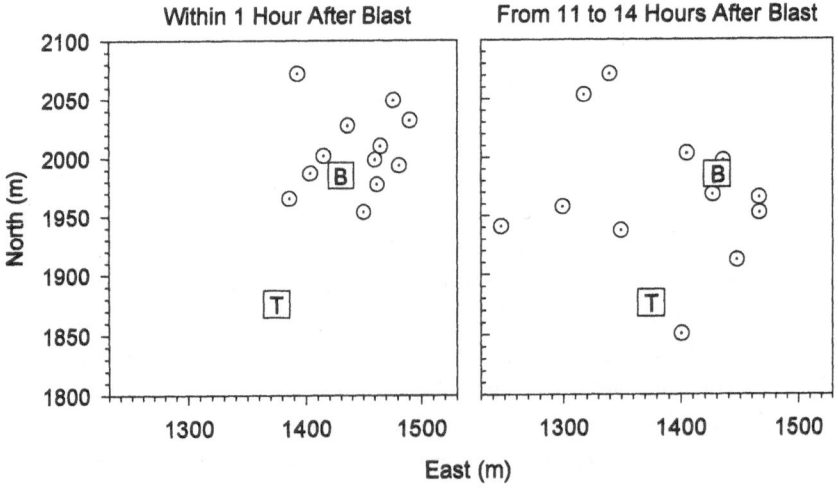

Figure 3
The location of the mine-wide events recorded between the 6800 and 7400 levels at two time periods after the 7000 level production blast on December 11. The blast (B) and transducer (T) positions are as indicated.

Table 4

Summary of recieved event parameters for various types of events

Event category	Energy (Vμs)	b value	Dom. freq. (kHz)
Immediate	0.7 to 2.0	<2.5	210 to 255
Delayed	0.4 to 0.6	>2.5	~350
Immed., Dec. 16 & 17, early events		same as Immediate	
Immed., Dec. 16 & 17, later events		same as Delayed	

quency activity, such as fracture displacement without significant asperity cracking. An interesting observation is that the early events of the immediate episodes on December 16 and 17 had event parameters consistent with other immediate events but the later events (Table 3) had parameters consistent with delayed events. This suggests that the "shake up" associated with the passage of the direct seismic waves from the blasts initiated an extended episode of microcracking.

The gradual changes in event parameters are shown in Figure 4, which indicates that the event energy decreased while the b value and the dominant frequency increased over the monitoring period. These parameter changes are consistent with a decrease in rock mass attenuation, an increase in the predominance of microcracking or a combination of both. This suggests that local stress levels were increasing throughout the monitoring period.

6.3 Indications of Increased Stress Levels

Towards the end of the monitoring period there were several indications of increased levels of stress and/or incipient instability in the local monitored area:
- The number of immediate events increased and, as indicated by event parameter analysis, the later immediate events were associated with increased microcracking activity.
- The delayed events occurred much earlier than during the rest of the monitoring period and the number of delayed episodes increased.
- Gradual changes in received event parameters were consistent with increased local stress levels.

These observations are supported by the occurrence of two strong ($m_N = 3.0$) seismic events near the monitored area on December 18, the first at 14:22 hours (located just above the 7000 level and 142 m west of the transducer location) and the second at 17:51 hours (located 47 m above the 7000 level and 141 m north of the transducer location). Approximately 75% of all the mine-wide events recorded during the monitoring period below the 6600 level occurred shortly after these major seismic events.

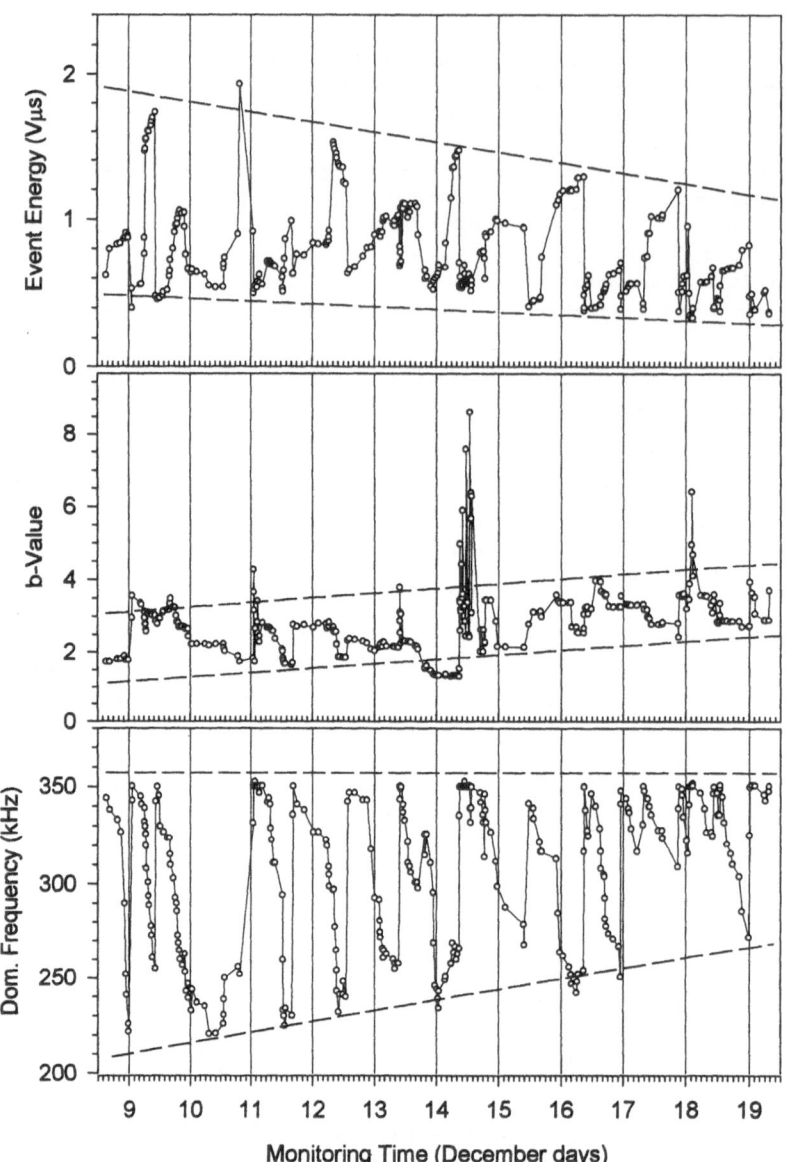

Figure 4
The trends of the event energy, *b* value and dominant frequency over the monitoring period. Note that these event parameters have been averaged over consecutive 5 minute intervals.

7. Conclusions

From the field monitoring outlined in this investigation, several observations and conclusions can be made. These include:

(1) HF monitors can detect local microseismic activity resulting from production

blasts well beyond the limited detection range of the transducers. This local activity comes in two forms:

- Immediate activity due to the passage of the direct seismic waves through the local monitored area;
- Delayed activity resulting from the later arrival of transient stress changes induced by the blasts in the local monitored area.

(2) Indications of increased local stress levels towards the end of the monitoring period, as inferred from changes in the patterns of immediate and delayed events and event parameter analysis, are consistent with the occurrence of two nearby strong seismic events at the same time.

Acknowledgments

The authors wish to thank INCO Ltd. for permission to use the Creighton Mine data, and personnel at INCO for assistance during field monitoring and data analysis. This research was funded by the Canada Centre for Mineral and Energy Technology (CANMET) and the Mining Industry Research Organization of Canada (MIROC). The first author (SDB) received a travel grant from the Queen's University School of Graduate Studies to attend the NARMS Induced Seismicity Workshop. Several anonymous reviews provided suggestions to improve the quality of this publication.

REFERENCES

ARCHIBALD, J. F., CALDER, P. N., MOROZ, B., SEMADENI, T., and YEO, T. K. (1988), *Application of Microseismic Monitoring to Stress and Rockburst Precursor Assessment*, Min. Sci. and Tech. 7, 123–132.

BAWDEN, W., and COULSON, A. L., *Structural and Geomechanical Study of Creighton Mine with Increased Attention Focused on the Levels Below the 6600 Level*, Unpub. report to INCO Ltd., 1993.

BIENIAWSKI, Z. T., *Rock mass classification in rock engineering*. In *Proceedings, Exploration for Rock Engineering* (ed. Bieniawski, Z. T.) (Balkema 1976) pp. 97–106.

BUTT, S. D., *Analysis of Mining Induced High Frequency Microseismicity*, Unpub. Ph.D. Thesis, Dept. of Mining Engineering, Queen's Univ., Kingston, Ontario, 1996.

CALDER, P. N., and SEMADENI, T. (1990), *High Frequency Microseismic Monitoring Trial Near a Rapidly Changing Stress Field in a Hard Rock Mine*, Geophysik 99, 21–34.

CALDER, P. N., ARCHIBALD, J. F., BULLOCK, K., MADSEN, D. (1990), *Case Studies of Microseismic Monitoring for Stress and Rockburst Prediction*, Min. Sci. and Tech. 11, 129–152.

CALDER, P. N., BUTT, S. D., and APEL, D. B., *Development and Evaluation of the Queen's-MIROC Portable High Frequency Microseismic Monitor*, MRL-CANMET and MIROC Contract 23440-2-9170/01-SQ, 1995.

GUTTENBERG, B., and RICHTER, C. F., *Seismicity of the Earth* (Princeton Univ. Press 1949).

HOEK, E., and BROWN, E. T. (1980), *Empirical Strength Criterion for Rock Masses*, J. Geotech. Engng Div., ASCE 106(GT9), 1013–1035.

LANG, L. C., ROACH, R. S., and OSOKO, M. N. (1977), *Vertical Cratering Retreat, An Important New Mining Method*, Can. Min. J.

NUTTLI, O. N. (1973), *Seismic Wave Attenuation and Magnitude Relations for Eastern North America*, J. Geophys. Res. *78*, 876–885.

MORRISON, D., VILLENEUVE, T., PUNKKINEN, A., *Factors influencing seismicity in Creighton Mine*. In *Proceedings 5th Conf. Acoustic Emission/Microseismic Activity in Geologic Structures and Material* (ed. Hardy, H. R.) (Trans. Tech. Publications 1995).

PYRAK-NOLTE, L. J., COOK, N. G. W., and MYER, L. R., *Seismic visibility of fractures*, In *Proceed., 28th U. S. Symp. on Rock Mech.* (Balkema 1987).

SCHOLZ, C. H. (1968), *The Frequency-magnitude Relation of Microfracturing in Rock and its Relation to Earthquakes*, Bull. Seismol. Soc. Am. *58(1)*, 399–415.

SCOTT, I. G., *Basic Acoustic Emission* (Gordon and Breach Science Publishers, New York 1991).

(Received August 27, 1996, accepted May 13, 1997)

Pure appl. geophys. 150 (1997) 705–720
0033–4553/97/040705–16 $ 1.50 + 0.20/0

Pure and Applied Geophysics

Panel Discussion, Workshop on Induced Seismicity, June 18, 1996

SHAHRIAR TALEBI[1]

Introduction

A workshop on induced seismicity was held on June 18, 1996, as part of the second North American Rock Mechanics Symposium (NARMS'96), Montréal, Québec, Canada. The success of the workshop on induced seismicity organized by Art McGarr as part of the 33rd U.S. Symposium on Rock Mechanics in Santa Fe, New Mexico in June 1992 was an indication that NARMS'96 could be an appropriate opportunity for holding a similar workshop. Also, the organizers thought that the timing of such a workshop would be appropriate since the ideas discussed during this workshop could be further developed during the 4th International Symposium on Rockbursts and Seismicity in Mines, held August 11–14, 1997 in Krakow, Poland.

The workshop on induced seismicity was attended by over 50 professionals working in different fields of science and engineering. The four oral sessions and one poster session dealt with seismicity induced by mining, rock failures and source mechanism determination, studies of induced seismicity associated with hard-rock and soft-rock mining, reservoir-induced seismicity and fluid injections, geothermal hot-dry-rock projects and seismicity in oil reservoirs. During a panel discussion, some key issues related to research on induced seismicity and the practical application of the results were discussed.

Panel Discussion

Using the format of the 4th International Symposium on Rockbursts and Seismicity in Mines in Poland, the following issues were selected for panel discussion.

---

[1] CANMET, 1079 Kelly Lake Rd., Sudbury, Ontario, Canada P3E 5P5.

1. Mechanism of Seismic Events and Rockbursts: How satisfactory is the application of a double-couple model in different situations? Should other models be considered or further developed?

2. Monitoring of Seismicity: What is the appropriate compromise between a "black box" system and a software-intensive flexible system? Are there further advances to be made in terms of software and hardware?

3. Geology and Seismicity: How should routine and advanced seismological techniques be used in practical engineering applications?

4. Seismic Hazard Assessment: What are the most useful criteria for predicting the occurrence and level of induced seismicity for various industrial activities? Can maximum magnitudes be forecast reliably?

5. Induced Seismicity: What do different cases of induced seismicity have in common?

Panelists:	Wilson Blake, Consultant, U.S.A.
	Francois-Henri Cornet, IPGP, France
	Art McGarr, U.S. Geological Survey, U.S.A.
	Hiroaki Niitsuma, Tohoku Univ., Japan
	Steve Spottiswoode, CSIR, R. South Africa
Moderator:	Shahriar Talebi, CANMET, Canada
Other participants:	Jean Pierre Deflandre, IFP, France
	Mariana Eneva, Univ. of Toronto, Canada
	Cory Kenter, SIEP, Shell, Netherlands
	Derek Martin, GRC, Canada
	Cezar-Ioan Trifu, ESG, Canada
	Waclaw Zuberek, Univ. of Silesia, Poland

S. Talebi: I would like to welcome you all to our panel discussion. The five major areas that were selected for this discussion by the members of the panel cover a wide range of subjects that are relevant to many of us working on induced seismicity. These five areas have been identified by the organizers of the next rockburst symposium in Poland and we thought that it would be appropriate to discuss them here.

The first area of interest is the mechanism of seismic events and rockbursts. The basic issue raised very often is that some of us are quite satisfied with the application of a double-couple model, but there are indications that this model is not applicable in some situations. Also, some authors promote other possible source mechanisms. The second issue is monitoring of seismicity and I think this is a very important issue. Seismic monitoring systems such as MP250 have been very popular in mines over a number of years. This is basically a "black box" system and is very easy to use for a mine operator. Nowadays we have seismic systems that are a lot more complicated, to the point that some of them actually require a seismologist to make sense of their output. What is a good system to be used and what are the other advances that could be made in terms of hardware and software? The third area is geology and seismicity. Although we saw many examples today, the issue remains as to how some of the routine or advanced seismological techniques can be used from a practical point of view. You saw examples today of how some of these parameters can be useful in terms of hazard assessment. This takes us into the fourth area of discussion. Where are we with regard to predicting the occurrence and level of induced seismicity for various industrial activities. Can maximum magnitude be forecast reliably? The fifth area of discussion is that of induced seismicity in general. What are the common features of what we see in different situations?

I have personally been involved in many different applications of induced seismicity, for example with the mining and oil industries and geothermal Hot-Dry-Rock projects and believe that there are many common observations in different applications where induced seismicity is observed. However, sometimes I get the feeling that there is a communication gap between the rock mechanic community and the seismological community. I believe that we have to get back to the basics sometimes and remember the assumptions that we make before we calculate some of our seismic parameters. This might help communicating the results to others, convincing them of how useful that parameter might be. With that, I invite the members of the panel to start their discussion.

A. McGarr: I am not sure who suggested the first question, maybe it was me. I started studying mining-induced seismicity in the late 1960s and up until about 1991 or so I was a firm believer in the idea that the double-couple model was adequate to explain mining-induced seismicity. This is the same mechanism as seismologists normally assume for an earthquake and, partly because I received my education as a seismologist, I just transferred what I learned in seismology over to mining-in-duced seismicity situation and it was very difficult to persuade me otherwise, until I finally realized that my own data could not be explained by double-couple events, at least in many instances. This was when I started doing full waveform analysis using underground data recorded at multiple sites. What I found was that the events that I studied fell into two categories. In the first category was the minority of events which did entail double-couple mechanisms, in other words, slip across a fault plane with no volume change, and the second category which accounted for perhaps two thirds or a bit more of the total events entailed quite a substantial volume change. The volume change in fact was comparable to the fault-slip component. If you multiply the fault area by the average slip, that product has the same dimension as volume. So the two quantities were comparable for the majority of events. These events were detected in several South African deep gold mines.

Now many mechanisms have been suggested beyond what I have just described here that we have to deal with in the induced seismicity arena.

S. Spottiswoode: I feel that mining events are generated because of the formation of the void and the purpose of a lot of the seismicity in nature is to try to close that void. I think when we interpret our seismograms we have to remember that the voids themselves are a part of the source mechanism. I think that the fundamental mechanism is in fact shear failure. That is the real seismic source mechanism, but if the event occurs very close to the mine workings, you can see a large volume change. A very simple example is the failure of a small pillar that can result in an extremely large volume change. So I think we have to be cautious about assuming that the event is a point source in an infinite homogeneous solid. So I would like to suggest as a point of debate that all events are, in fact, caused by shear failure in mines and that non-shear effects are associated with the voids.

W. Blake: I would agree with what Steve just said. One of the problems we have had in predicting at the Lucky Friday mine is trying to reconcile the first motion studies and the direction of slip; i.e., the focal mechanism with what we see underground, since many times there is a big conflict. We will see the actual movement on the fault and the direction will not agree with the direction that we get out of the seismic solution based on not only instruments in the mine, but also instruments based around the entire mining district. It is a little disconcerting to be spending so much time and money on the seismological data when it does not seem to agree with the actual mechanism observed in the mine. Of course, I have no resolution for this but it is disturbing that probably in 25–30% of the cases the direction of movement that we observe underground is quite different from what we get out of focal mechanism studies.

S. Talebi: Before we pass on to Dr. Niitsuma, I would like to ask you if you take the effect of ray bending due to the complex geology that can be present in a mine into account when you do your focal mechanisms.

W. Blake: We do not do that. We get the fault-plane solution and analyze it in terms of shear movement. But long before that, when you go underground and look and see where certain openings have intersected the fault, you have an absolute clear-cut direction of movement on that fault. When it does not agree with what we get back from the seismologist, it is a little bit disconcerting.

H. Niitsuma: There are two issues that should be considered when dealing with the results of seismic techniques: the origin of seismic activity and the accuracy of source location. For example, in the case of fluid injection shown on this schematic diagram you see a concentration of seismic event locations. What does this area mean? Or this area where there is no seismic activity? Does no activity mean no fluid flow? In hydraulic fracturing, one needs to delineate the newly-fractured layers, areas already fractured and aseismic permeable areas. We have some experience on tensile fracturing tests like the one I am showing you on this next figure. This is a $9 \times 9 \times 10$ meter block of granitic rock on surface, not underground, so there is no confining pressure present. We made hydraulic fracturing in these boreholes and expected that fracturing should occur in pure tensile mode, but we do observe shear events. This next figure shows three examples of focal mechanisms observed. The first example is compatible with a tensile failure; i.e., compressional first motions on all the sensors, but the two other examples show dilatational first motions as well and relate to shear events. We have conducted similar experiments in a mine as well. In this case, the experiment took place at a depth of 500 m and almost all the events were shear events while this experiment also consisted of hydraulic fracturing in the rock mass. Our results indicated large differences in seismic efficiency between shear and tensile failures. Professor Hayashi has calculated on this next figure the seismic efficiency for a 400 m fault buried at a depth of 2 km. I cannot go into details here, but a great difference in the generated wave amplitudes is observed whether this fault moves in shear or tensile mode. In this case, shear cracking creates amplitudes of the order of 10^5 greater than tensile cracking. This means that in hydraulic fracturing, very small amounts of shear motion in a tensile failure process can create shear events of considerable amplitude. My conclusion of this comment is that tensile cracking has a very low seismic efficiency and does create shear events. Therefore, no acoustic emission does not mean no cracking. Also, shear acoustic emission is not always pure shear cracking. So we must be very careful. When we look at the above pattern, then the acoustic emission technology highly depends on the objective of our application. We must be very careful about the distortion of the observed data.

S. Talebi: Thank you Professor Niitsuma. That was a very enlightening contribution you made about this subject. May I ask Francois to make his comment about this point.

F. H. Cornet: I have been involved with microseismic activity mostly associated with fluid injections. But I must say I had to look once in a while at data coming from a mine. I was always surprised that in mines most of the data were never showing a double-couple mechanism. Thinking of the first talk of this morning by Art McGarr as a matter of fact, I understood that in order to explain the large particle velocities, he had to use some kind of bending type mechanism. So, it seems to me that we should not be polarized only on double couples, but maybe on these bending type failures, and they certainly do create volume change. I do not know if that is a mechanism that is observed in the case that I studied but that is something to look at. The second point I would like to make is that in seismology we always "love" double couples, everything that is not double couple we do not look at and I think we are way too biased with this respect. I think Professor Niitsuma's talk was telling us that maybe there is more to be learned. As I was saying in the Los Alamos National Laboratory's hot-dry-rock project, people tried to look at all the events including double couples and they looked at those long-period events. Of course, these events are very tiny, there is very little energy in them. So you do not look at them because they are "dirty signals" and you just want to look at clean signals. But I think there is a lot to be learned by looking at those "dirty signals". In his talk Shahriar described periodic events, but what is causing them? There is some mechanics behind. So I think we have to do some work on non double-couple events and this is a very important topic in my opinion.

S. Talebi: Thank you Francois. Is there any other comment from the members of the panel? Anybody from the floor?

D. Martin: I think the fundamental issue is why do we conclude from microseismic monitoring data that the majority of the source mechanisms are double-couple shear events when all the physical evidence observed around underground openings points to fractures growing normal to minimum principal stress, i.e., extensional events. At AECL's Underground Research Laboratory in Southeastern Manitoba, the Mine-by Experiment was carried out in massive unfractured granite. Even in this environment the majority of microseismic events were classed as double couple yet there was no physical evidence to support it, but there was ample evidence of extensional fracturing.

In Neville Cook's Muler Lecture during the ISRM symposium in Tokyo, he talked extensively about fracture formation and propagation but there was no mention of a sliding or shear crack. In the geological literature there is wide acceptance that even in the formation of a thrust fault, which obviously results from shear movement, the underlying mechanism is extensional cracking at the tip of the advancing fault. Hence, the damage process that allows the fault to form is not double couple. In Canada, our microseismic monitoring in mines typically tracks events with moment magnitude in the range −3 to 0. Visual inspection of these damaged areas also supports the notion that the fractures are extensional. Hence, my question to the panel is why is the seismological interpretation of microseismic events around underground openings, i.e., double-couple shear failure, not compatible with the physical evidence of extensional fracturing?

S. Spottiwoode: I must say I enjoyed the lecture of Neville Cook but I think it was concentrated mostly on the formation of joints. I think in the seismic side we are not looking at the initiation of the fracturing; we are looking at the propagation. In other words it is not the crack tips that are radiating the seismic waves. Perhaps it is not even a shearing process if there is such a thing, not a pure slip: the two surfaces do not just grind smoothly over one another. So it is the final relative motions that we actually see. So it is perhaps true that there is not such a thing as a shear crack.

S. Talebi: Any other comments from the floor?

C. I. Trifu: I liked very much the moment tensor paper that Art McGarr published a couple of years ago in which he studied several events by using full waveform near-field and far-field methods. I am totally favorable to such studies and think that techniques are now available to carry out this type of work. I would strongly encourage, however, the use of these techniques with an appropriate statistical analysis that would evaluate the significance of the results. As such, how confident are we that the inversion with 6 unknowns is more significant than that with 4 unknowns? How confident are we that enough information exist to reliably retrieve the general moment tensor rather than only its double-couple component? For example, Doornbos has spent a great amount of work to formalize the higher order moment tensor inversions, but in the end his analysis showed that there is not enough information in the data that would allow us to proceed along that route. So, the techniques are out there, and I think that we should encourage everybody to carefully use these techniques in their attempt to go forward. On a somewhat related topic, I would like to say that while coming from earthquake seismology, I tried over the last 3–4 years to see if there is any significant change in the behavior of source parameter scaling when comparing small events in mines with relatively large earthquakes. Well, I haven't found too much. Therefore, I would be tempted to say that there is something quite similar in the process of rupture generation, from the scale of large tectonic earthquakes down to that of the rock mass. According to some of my recent results, the average maximum magnitude, which is an important parameter for damage evaluation, scales linearly with the log of the side of cubical volumes taken for the analysis, from magnitude 5, 6 and 7 events in the Cocos-Caribe subduction zone, to events of negative magnitudes in the Canadian mines. A linear scaling has also been found with the log of the time considered for the same data sets. Since it is commonly accepted that large tectonic events are driven by shear, further studies would have to better outline how mechanism similarities and nonsimilarities are responsible for the above trends.

S. Talebi: Thank you Cezar. Is there any other comment about this topic?

J. F. Deflandre: I have just a comment concerning the number of sensors which must be considered to compute focal mechanisms. I am not involved in mining, so generally I have a few sensors, three or more three-components sensors. I do not attempt to study focal mechanisms of my events because I think I do not have enough stations. What about the number which is used in practice? I have the opinion that sometimes people apply a model without the data that should be necessary to do so. Is this just an idea that I have or is it true?

S. Talebi: I would say that it all depends on the situation. You attended the talk of Jim Rutledge who showed us composite mechanisms from only 3 sensors, but if you want to study individual events, you may need quite a few sensors.

F. H. Cornet: That is because you think of double-couple mechanism based on first motion of P waves. If you use the concept of seismic moment tensor, you can use a few three-component sensors and model the source as an inverse problem. You can try to reconstruct the signals you get on four three-component stations and you get a rather good constraint on your seismic moment tensor.

J. P. Deflandre: I have another comment concerning the "dirty" events or periodic events. In their presentation, people usually show the most interesting events, but when you discuss with them after their presentation, they show you lots of poor-quality events that you also have and do not know what to do with. For example I have recorded many periodic events as Shahriar showed today in his presentation. I have tried to understand what the correlation with the other parameters can be. For example I have moved one tool from one depth to another depth in the well. I have rotated the tool and observed, in one case, that the P-wave polarization was related to the tool. So it could be that in this way you can classify them as "tool events". It is a good family of events, but on the other side, these events can occur at a rate of 1500 a day, then no events during a week, then some events coming in, then a few hundreds of events coming, then no events and this seems to be correlated with some physical parameters. So I have the feeling that these events are related to the physical properties of the field.

A. McGarr: I was just going to respond briefly to your first question Jean-Pierre. You were asking how many stations are necessary to define a focal mechanism. I am basically in agreement with what Francois Cornet just said. My own experience was that of just having two stations, but they were installed underground and very close to the events and the systems were very broad, wide dynamic range. So the waveforms were extremely clean and well defined and they provided both the near-field and the far-field information. So there was plenty of constraints provided from the seismograms to define the moment tensor with lots of redundancy, about a factor of 3 redundancy in the equations. You do not need many stations, you really just need good clean data.

S. Talebi: Thank you Art. As I mentioned in my talk, the events that I showed you today have not been dealt with in a research mode. I should say they are very recent and I cannot go into the speculation as to what is their origin. I do not want to speculate about their cause or reason at this time, but I will not rule out the possibility that they can be related to the field somehow. If there is no more comments with the first topic can we move on to the second topic?

The second topic we are talking about is monitoring of seismicity. Mircoseismic monitoring became really a tool used by mining operators, just focusing on the mining application here, with the popularity of the MP250 Electrolab systems. The concept of this "black box" system is to record the signals and then to detect *P*-wave arrivals times automatically, no matter how accurately this is done. Then a number of hypocenters are produced as a result of source location. Thereafter people can put the results on a mine map and make their own interpretation. So this is one concept and then we have the new approach with the development of new systems which are much more sophisticated. They are software-intensive and many more seismological concepts have been incorporated into their development and use. The question now is what is the trade-off? Is the first concept a good concept or how far can we go with the other one. With the software-intensive systems you need very highly educated people to deal with the output. Basically with the first one you do not need a seismologist, but with the second one sometimes you do. The question is whether it is a good idea to have a seismologist or not? Does anyone want to make a comment about this issue?

A. McGarr: Steve and I were just having a little debate as to the significance of this question. I will confess to having suggested this question. The reason I suggested it was because I knew that many scientists and engineers who are developing and promoting seismic recording systems for mine seismicity application are participating here in this workshop and they all have their different philosophies about what the system should be capable of doing and so I was trying to start a little argument between the various users of this type of equipment. I do not know if I will be successful. Another version of this question might be what is a better investment: a good seismologist or some very expensive hardware and software? And this I think clears up the issue. Steve Spottiswoode was suggesting that a black box system is not software intensive. My impression is that black box systems tend to be very software intensive, at least the modern ones. These systems are very sophisticated, very complex and for an ordinary mortal like me who does not do very well when dealing with computers, I sometimes find it pretty hard to imagine what goes on between where the seismometer records the data and what comes out the other end of the computer system. So this is really what is behind my question. I have recorded numerous seismic data personally and the systems used have been fairly rudimentary. As far as I am concerned, all I want is seismograms with a wide dynamic range and wide bandwidth, so that I can get all the information that left the source. But other people who use this data prefer more convenient systems where for example you can calculate a moment tensor for an event just by literally pushing a button on the keyboard. Anyway, that is what is behind my question. I do not know what the answer is and I am sure there is not any single answer to this question. It really depends on the intended use but, nonetheless, I am sure that all of you users of this equipment have opinions. I think Steve is one of these people, so I will just let Steve carry on.

S. Spottiswoode: I think to a large degree the compromise between the software-intensive systems and convenience have to do with the economics of what you are trying to do. Everybody would like to argue that the more data you have the more you can do with it. But it requires an enormous effort to get that right. Also, for all of the work that we are doing one just needs more and more communication effort to get that going. So I think the issue is just how much do you want out before you ask the question how much are you prepared to put into it. I think I am in agreement with your concern Art. Always asking for more is not necessarily going to activate more results at the end. Maybe we should also address the point on further advances. I think that inevitably there is going to be more advances. A particular issue that I have never really seen discussed is whether hardware should not consist of many arrays. For example a few geophones down one hole in which there might be array processing. We did not hear any of that today but I would be surprised if in 5-years time we do not have people discussing array processing.

S. Talebi: Thank you Steve. Maybe we should ask the people who are actually the end-users of this technology to express their point of view. As you said we do our best depending on what they actually want and what they actually need. Is there any comments from the floor about this issue, particularly the people from the mining industry?

W. Blake: Well, to the people of the mines who use these systems it does not matter what type of system it is. It is all a black box. Depending on the capabilities of the system they are interested in knowing where an event occurred, how big it was and whether they can show a plot of it. The end-user at the mine does not really get into checking to see if the arrival time is really accurate, checking to see if the seismic parameters are included. At this point in time we cannot use any of the seismic parameters. On a day-to-day basis we are not interested in focal mechanisms and we are not really interested in any of the seismic parameters. It is just knowing where it was, how big it was, whether a trend is increasing or decreasing. Generally the engineer or technician who is looking at this data is getting, in many mines, one or two thousand events a day. He just does not have the time to go into any detailed analysis. So what they like to see is a print out of event locations. Events may be located by stope, this number of events from such and such a stope, so you can compare over time how the mine is behaving, e.g., is it a normal number of events after a blast or an abnormal number of events. What the mining industry is actually utilizing is really very simple. We would like to have all these bells and whistles so that you push a button and have this data filtered to take out one type of focal mechanism or do a PCA analysis or look at dynamic stress drop and all these things that may give a little edge on whether something is going to happen or not. The person working in the mine just does not have the time to get really involved in data analysis and the mine does not want him to. His job is to keep track of the data and basically in the small mines there is not a group, it is just one person and maybe one person and a technician. So, this person has to go down to keep the geophones running, he has got to keep his wiring going and he also has to go down and communicate with the miners to find out what their opinions are about where the activity is located. Is this a significant event? Is this an unusual occurrence? So the actual amount of time spent on data analysis at all the mines that I go to is basically a very small percentage of time. We are just glossing over the data and unless something really dramatic happens we are not looking at the data.

F. H. Cornet: I must say once again that I am not working for the mining industry, but I have been impressed by the quality of the questions that are being asked in this discussion and the one we are discussing now. You want to know where the events are. I am afraid you are not going to know where the events are with a "black box" system. The first exercise a seismologist has to do is to play with the velocity model. Depending on the velocity model you have different possible locations. In the case of the mine data mentioned earlier, I have seen three different students working on those data and as a consequence the events have been located in different areas of the mine. In the case of the Los Alamos National Laboratory's Hot-Dry-Rock project, when the project started they had one down-hole three-component tool and you had the "potato-chip" theory of fracturing, until they found out that there was resonance between one component and the other and that the locations were wrong and a second tool was needed. When I think of a mine, from what I understand about a mine, there are changes in the stress field and therefore changes in velocity. If you change your velocity field you change your source location whether you work with a black box system or not. So I am afraid that the answer to the question of black box versus seismologist is that it is way cheaper to have a good seismologist and a few tools than to have a very good black box that does not integrate the velocity model.

J. P. Deflandre: I completely agree with Francois Cornet concerning the black box. I have a very beautiful black box system. I can do everything from the implementation at the site to complete data processing using my black box system. I just have to adjust parameters. I can compute source parameters, analyze wave polarization, etc. But my main problem with locating the events is using a good velocity model. I had to use, in one case, 8 layers with different velocities. I take into account the polarization computation at the sensor and reflection effects at each change of acoustic impedance and I have to converge these results with the computation of travel time between P and S waves. If I do not do all that, I will have large errors in my source locations. Francois said that depending on the velocity model you can get different results. This is typically the case in another site I know about in France where at first the events were located above a reservoir. They tried again with a new software and triaxial seismic data and now they changed their interpretation. So the most important thing is to well locate the events. If you are not sure about the results, do no try to interpret what you see.

S. Talebi: Thank you Jean-Pierre. Let's move on to the next topic which is related to this one. Here we are talking about using different seismological techniques of which source location is the most prominent one. We are talking here about two types of techniques: what we consider to be routine in our seismic analysis and what we call advanced. I just want to make a comment here going back to the subject of communication gap between the rock mechanics community and the seismological community. Sometimes a seismologist talks about a seismic parameter, for example stress drop or apparent stress. I am not sure if at the other end of the conversation the rock mechanics worker really realizes how many assumptions were made before these parameters were calculated and what they really mean. Sometimes it could be quite misleading. I am not going to give any examples. We have one particular case here with the use of seismic doublets that we can talk about in the context of engineering applications. How should these techniques be used?

W. Blake: I think seismic monitoring of particular mining activities is vital. If you are in a rockburst-prone mine and it is popping and cracking all the time it is the only way you get a handle on where the trouble spots are, where they are shifting, what happens when you blast. We pull people out of the mines or out of the stope very frequently, if it is too active you have to pull them out. Normally the miners may have pulled themselves out already. With respect to source location again, black box systems do a great job for locating. We are not interested in a point. We can tell what stope it occurred in, normally we locate blasts within ten meters. So it is not very difficult to locate whether you have an automatic picker or you have a voltage threshold. The key is to have a good array and to have enough geophones around so that you have a lot of redundant information. Source location is not a problem. I think you really find out how stress is transferring, where it is going and what areas of the mine can contain it and what areas of the mine cannot contain it.

J. P. Deflandre: I agree with you. What I said before was not concerning mining because you have lots of sensors. But in the case of deep downhole sensors we have the problem of a layered velocity model.

S. Talebi: Thank you. How about focusing on the particular case of doublets? Basically, what you have here consists of two events identified from their very close locations and similar signatures. Using a method like the one explained by professor Niitsuma's colleague, you can get a higher degree of accuracy in your source location results. Is there any application of this type of advanced technique in other cases of induced seismicity?

H. Niitsuma: I would like to show you a diagram about seismic source location error. Seismic source location has some errors because of three factors: detection of the wave onset, velocity structure, and our arrangement of the sensors. By using the doublet approach, we can detect very precisely the relative time delay from cross-spectral analysis. So we can make precise relative mapping. Sometimes, absolute mapping itself is not so important, but precise relative source location is useful. In this case we can largely reduce the error associated with time picking. I mean that the onset of P wave is sometimes difficult to detect precisely, but the relative time delay is easier to pick, for instance just by comparing the two waves, not the onsets, but the peaks. This simple technique can considerably increase picking accuracy. Another method is joint hypocenter determination, which can adjust the velocity model or station correction and everybody knows this technique. The other way is to use robust mapping, which is insensitive to the velocity structure. Zonal location techniques and some other kinds of technique are insensitive to the velocity structure. The "hypocenter tomography" has been developed at Los Alamos by Mike Fehler and his colleagues. By using hypocenter tomography we can also make an inversion and evaluate the velocity structure. Some new techniques such as AE reflection method, coda analysis and drill bit VSP are also noted on this figure. We are now mainly concerned with AE activity and source location, but acoustic emission data contains a lot of information, for instance shear-wave splitting, shape of coda and so on. So, future technologies will provide substantially more information, I believe.

S. Talebi: Thank you Dr Niitsuma. Let's move on to the next topic, the issue of hazard assessment. As you know, the huge amount of investment in trying to predict earthquakes has had very limited success. The question is basically what is the most useful criteria for predicting the occurrence and level of induced seismicity for various industrial activities and more specifically can the maximum magnitude be forecast reliably.

A. McGarr: This is a fairly complicated question but I will try to make a start on it. I think the two primary criteria for predicting the occurrence of induced seismicity due to industrial activities have involved stress change as a consequence of those activities. The stress changes in turn cause earthquakes. The general finding has been that fairly modest stress change can cause a remarkable amount of induced seismicity. Now my experience in the South African mines was somewhat the opposite. There the rock is extremely strong and stable. There we are quite certain that the stress changes were large and there was not too much of a mystery as to why there is induced seismicity. In many other situations fairly small stress changes lead to, or can have at least a large influence on, induced seismicity. The most recent development has involved interaction between natural earthquakes in California for example. One example is the Loma Prieta earthquake in 1989 south of San Francisco influenced the seismicity on numerous faults within about a hundred kilometers. Using dislocation theory the stress changes on those faults where seismicity was either increased or decreased were of the order of one bar or less which is very small compared with the stress changes that we think occur during an earthquake. That is a phenomenon that we still do not understand. Reservoir-induced seismicity is another example of very small induced stress leading to substantial seismicity at least in many cases. So that is I think, a major area of ignorance for earth sciences. How do such small changes in stress lead to such large changes in seismicity? It is so far just an observation and I think we can make use of it in predictions but it would be much better to understand it. I think I will leave it at that.

S. Spottiswoode: Actually I have some comments unrelated to Art's comments on the subject. I think the title "seismic hazard assessment" could be misleading to a degree in the mining environment where it is not the seismic event that poses the hazard itself but the effect on the underground excavations. This has been a problem I think in many mining districts, certainly all over South Africa where you can get an event of magnitude 3 that causes little damage to the mine and then a magnitude 1 which causes quite a lot of damage. This is controlled by a number of factors such as the proximity of the source location to the mining excavations, but an important one is just how the excavation responds to seismic waves. We have been measuring the ground velocities right on the skin of the excavations and find there is extremely large variations of ground velocities even over scales of about 1 meter. For example, when we place one instrument next to a support element and another one a meter away we find large variations. In order to understand the hazard due to or resulting from seismicity, we also need to consider the kind of mechanisms that Art McGarr was talking about in his presentation; i.e., what happens at a very local scale. So I think the issues are not only related to seismology, but that rock mechanics people have not studied the situation very closely. We have to consider this when we are talking about hazard.

M. Eneva: Art McGarr gave the example of the Loma Prieta earthquake. I would like to mention that this example is only related to the static stress changes so you have, by necessity, involved very small distances and this can hardly be qualified as close to any of the remote triggering "fever" which is going on in the earthquake seismology community. For those who need other mechanisms, not static stress changes, the June 1992 Landers earthquake is the one that I think raised much more debate and made many people who did not want to hear about earthquake migration just a few years before suddenly interested in the subject. Now you can see a number of papers on that subject.

A. McGarr: Thanks Mariana. I agree with you about that earthquake and the apparent dynamic triggering. We now know what happens, but we do not understand why. Mechanisms have been proposed but more data are going to be necessary before we resolve the specific mechanism.

C. Kenter: I have a very practical question. We are considering for one of our reservoirs which is at a depth of 4000 m a gravity stimulation. We are looking into the possibility of letting the reservoir collapse into a 500 m-wide area around a 20–50 m deep cavity. Now are you in the panel or anybody else in the audience aware of a method that can reliably predict what magnitude of earthquake that will induce? So it is very well defined: it is a reservoir 4000 m deep and you have underneath the reservoir a cavity of 500 m in diameter and about say something like 20–50 m deep and all of a sudden at the push of a button you let the whole thing collapse. I want to know what is going to happen with the reservoir, what is going to happen with the overburden and what type of earthquake do I find at what distance to that particular location?

A. McGarr: I think if you can be confident about the total volume reduction and also about the response of the rock mass to that volume reduction whether it deforms seismically or aseismically or some combination, then I think you could, without too much uncertainty, predict the amount of seismic deformation, the total seismic moment of the seismicity associated with this engineering operation. I will just mention a minor success I had a few years ago. It is really minor, but there was this deep well in Germany that was recently stopped at a depth of about 9000 m and the last operation they did on this was a big hydrofracture at the bottom, similar to what Francois was showing earlier along the French German border, but much deeper. One of the seismologists involved in that operation ran into me at an AGU meeting about a month before they were going to do that and asked if I could estimate how large the largest earthquake was going to be, so that they know how to set their instruments. I went back to my office and did some calculations and I told him your magnitude will probably not be more than 1.5 and in fact that was about the largest event they recorded. So these things can be predicted but you have to be lucky and make the right assumptions. My calculation was based on the amount of fluid that he told me was going to be injected in this massive hydrofrac operation.

W. Zuberek: I would like to mention that we have found, in our area of monitoring in Poland where we recorded about 1000 tremors per year with events of magnitude above 1.7, that the change in seismicity in one place can affect the seismicity in a completely different place. So we think that this system can have a behavior like critically-organized chaotic behavior. In this case it could be very difficult to predict the individual event.

S. Talebi: Thank you. How about the last subject. We try to make it more general here and put everything under the umbrella of induced seismicity, then try to see what is the common factor of these events and the way they all occur. Francois.

F. H. Cornet: The one thing I can think there is in common is that it takes rock mechanics to understand. In other words we should not be concerned only with seismology. I think the problem we are concerned with is extracting from the seismic data what is relevant to rock mechanics problems. Now what you are providing in terms of density of events in a location is already very significant. Maybe we can do more than that. I must say I am very much in favor of Professor Niitsuma's doublet approach because if you have a more accurate location, you start identifying your structures. Now how relevant those structure are to our problem? I do not know the answer, but it is already very important to know that they exist. There are some other techniques for identifying structures. I am not going to go into a description of these techniques, but think that we can learn more about structures. Another aspect we can learn about by working on velocity structures is the issue of how the velocity is changing with time and the relationship between changing velocity and stresses which is definitely very important. For example, shear-wave splitting tells us about the opening of fractures, those parts which are dilated and those parts which are not dilated. By the way that is a tool which we have been using for ourselves to convince ourselves that induced seismicity is not telling us where we put a lot of fluid because we did not see very significant changes in velocities. So I think the common point between mining and fluid flow is really rock mechanics.

S. Talebi: Thank you Francois. Any other comments? It seems that we have exhausted the list. I just want to give the chance to anybody interested in raising any other issue of importance at this time. Is there any other subject or issue that we can briefly discuss?

All right. I would like to thank the members of the panel, all the session chairmen and all the participants who have contributed to this meeting. Thank you all very much.